INFORMATION in ACTION

A Guide to Technical Communication

Second Edition

M. Jimmie Killingsworth

Texas A&M University

Jacqueline S. Palmer

Texas A&M University

Allyn and Bacon

Boston   London   Toronto   Sydney   Tokyo   Singapore

Vice President: Eben W. Ludlow
Series Editorial Assisstant: Linda M. D'Angelo
Senior Marketing Manager: Lisa Kimball
Editorial–Production Administrator: Donna Simons
Editorial–Production Service: Omegatype Typography, Inc.
Composition and Prepress Buyer: Linda Cox
Manufacturing Buyer: Megan Cochran
Cover Administrator: Linda Knowles
Cover Design: Susan Paradise
Interior Design: Glenna Collett
Electronic Composition: Omegatype Typography, Inc.

Copyright © 1999, 1996 by Allyn & Bacon
A Viacom Company
160 Gould Street
Needham Heights, MA 02494

Internet: www.abacon.com

Between the time Website information is gathered and then published, it is not unusual for some
sites to have closed. Also, the transcription of URLs can result in unintended typographical errors.
The publisher would appreciate notification where these occur so that they may be corrected in subsequent editions.

Library of Congress Cataloging-in-Publication Data

Killingsworth, M. Jimmie.
    Information in action : a guide to technical communication /
  M. Jimmie Killingsworth, Jacqueline S. Palmer. — 2nd ed.
        p.   cm.
    Includes bibliographical references and index.
    ISBN 0-205-28587-2
    1. Communication of technical information.   I. Palmer, Jacqueline S.
  II. Title.
  T10.5.K47   1999
  808'.0666—dc21   98-22375
                                                              CIP

Printed in the United States of America

10 9 8 7 6 5 4 3 2 1   VHP   03 02 01 00 99 98

Brief Contents

Contents

Guidelines at a Glance

List of Samples

Preface

Information in Action: A Guide to Technical Communication offers students in every technical field practical guidelines for communicating with audiences ranging from scientific experts to the general public. This book treats communication as a form of social action—the making of knowledge and the building of communities. Its underlying focus is on how we translate information into meaningful action to solve problems and make the world a better place.

This second edition of *Information in Action* builds on the social and rhetorical emphasis of the first edition. As part of a general updating of the research base for the current edition, topics such as international communication, intercultural exchange, workplace diversity, gender, and ethnicity appear not only in the expanded Part One, which treats the contexts of technical communication, but also in the chapters on style, graphics, document design, and editing. New discussions of research using the Internet and the World Wide Web round out the new globalist perspective of the book.

Although it uses research-based approaches and introduces new theoretical perspectives on technical communication, *Information in Action* is organized around familiar practices in writing instruction—assignments based on common genres of technical writing, with an emphasis on form and style. The book's innovative material—dealing with research methods, collaborative learning, ethics, a graphics-oriented approach to document design and development, and more—is woven into a recognizable structure.

An Action-Oriented Approach

The approach to technical communication embodied in *Information in Action* reflects theoretical work in the field and an adherence to a belief in philosophical pragmatism. The key idea is that we understand things better through experience and practice than through contemplation or observation. We also get to know things best by experiencing their functions and behaviors rather than by observing their physical attributes or static qualities. Accordingly, *Information in Action* engages students quickly in the activity of writing, advocating a hands-on approach. In addition, it demonstrates how information and different forms of communication operate within a sociocultural context. It describes a report, for example, as a document that aids decision making, not as a piece of writing with three main parts and twelve subordinate parts. Throughout, the text conveys the conviction that information is an active part of our experience and not merely a bundle of independent words and numbers. Likewise, the classroom is a place to immerse students in the experience of technical communication, not a place to impart static principles, formulas for success, and rigid formats for writing.

The CORE Method for Document Development

The most significant innovation introduced in the first edition of *Information in Action* was the CORE method of document development. The CORE method capitalizes on the advantages of the process approach to writing instruction while sidestepping its drawbacks. The process approach can bog down students by involving them in an endless round of invention, drafting, and revising. The movement toward a completed product is slow and uncertain. But the CORE method streamlines the process approach. Retaining an emphasis on critical thinking, it shows students how to work through clearly defined, yet flexible and open-ended, steps—with definite objectives for the end product. Students begin by doing an *audience-action analysis,* then draft a fairly straightforward and unadorned *core document*—a position paper, a concept paper, or a task analysis. They can define their analysis of audience needs and adjust the content at this early stage in the process—before they generate a long, complex document that is difficult to revise. From this stage, students can proceed to further research, graphical development, oral presentation, document design, and editing, seeking feedback from potential users or model readers at each stage. The end product is an action-oriented document that is highly responsive to users' needs. This new edition features a chapter organization that stresses the work of prewriting, with new explanations of the analytical processes of the CORE method, and provides a separate chapter on writing core documents to show more clearly how this step fits into the process of developing information products. By keeping the theoretical underpinnings in the background, the new edition strives to increase the usability and practicality of the CORE method.

Organization

Information in Action is divided into three parts. Part One provides a detailed approach to the various contexts of technical communication. Chapter 1 introduces the key rhetorical concept of the book—the separation of the *context of production* from the *context of use*—to talk about the kinds of problems that arise when producers of information products fail to envision, analyze, and satisfy the needs of the information users. In urging students to do audience-action analyses and write audience-action profiles, the CORE method provides a means of bridging the gap between the two contexts. Chapter 2 further develops technical authors' sensitivity to differences among users and fellow workers by covering ethical, social, and cultural issues that frequently arise in technical communication. Chapter 3 shows more specifically how to use the CORE method to build context sensitivity and user awareness into information products, beginning with a management plan that unites the members of a communication team in the context of production around the goal of meeting needs and expectations in the context of use.

Part Two shows how traditional topics in document development fit within the CORE method. Chapter 4 discusses the process of researching and developing information, introducing students to the conventional tiered-access system of organizing information. Chapter 5 introduces the idea of writing short core documents on the way to producing complex information products, providing a more efficient, industry-based process of engaging trial audiences and writing incrementally. Chapter 6 examines ways to use and produce graphics, emphasizing the rhetorical functions that each type of graphic can perform. Chapter 7 covers document design, showing how graphics and written text can work together to achieve the best effect. Chapter 8 simplifies the task of preparing an oral presentation by focusing on graphics as an organizing principle and providing guidance in rehearsal and delivery for different audiences. Chapter 9 presents a compendium of useful guidelines on writing style to be used in drafting and revising, and Chapter 10 discusses editing as the last phase of document development.

Part Three applies the CORE method to the development of special types, or *genres,* of technical communication. Chapter 11 discusses technical correspondence in several contexts and media—memos, letters, e-mail, and computer conferencing. Chapter 12 deals with technical reports as decision-making and problem-solving documents. Chapter 13 covers special techniques needed for reporting experimental research in engineering and the sciences. Chapter 14 introduces techniques for writing proposals, and Chapter 15 deals with instructional manuals.

Features

Information in Action offers features that promote learning of the CORE method and developing a sense of purpose that will inform students' practice in technical communication.

- Learning objectives at the beginning of each chapter help students to set performance goals.
- "Guidelines at a Glance" features in almost every chapter provide clear directions for planning and writing at each stage of document development.

These specially highlighted lists also serve as criteria for review and analysis of draft documents.

- Full-length annotated student and professional samples provide realistic and accessible models for students to emulate, as well as guidance in developing critical and analytical skills.
- Interesting cases based on actual events and real scenarios examine ways in which ethical and rhetorical factors operate in the world of business, industry, government, and academia.
- Exercises at the end of each chapter develop students' critical thinking skills and reinforce learning of key concepts in the technical writing process. The exercises focus on both individual and collaborative efforts. Collaborative exercises cultivate analytical skills and provide practice in all phases of the writing process.
- Photographs and figures throughout the text illustrate principles under discussion and provide a human dimension to the technical writing process.
- A two-level bibliography at the end of each chapter directs students to additional reading and research materials.

Instructor's Resource Manual

In addition to all of the features above, the text package includes an *Instructor's Resource Manual.* This comprehensive supplement features alternative syllabi, plus teaching materials for each chapter: overviews that elaborate on key concerns; creative teaching suggestions arranged in a highly accessible format; reading tests and answer keys; criteria for writing and scoring each of the preliminary and final information products; and a set of transparency masters.

Acknowledgments

I am joined in this second edition of *Information in Action* by my frequent coauthor Jacqueline S. Palmer, who, as an experienced technical writer, education specialist, and technical writing teacher, has added immensely to my understanding of the work. Jackie and I have benefited greatly from the continued support of Eben Ludlow, Doug Day, and the staff at Allyn and Bacon, as well as the encouragement of our students and colleagues at Texas A&M, which remains little short of amazing. Special thanks go to our department head Larry Mitchell, my research assistant Yuen Wen, and my staff assistant Shannon McDonald. Our daughter Myrth also helped not only with her patience and good humor, but also with putting the pieces together.

We remain grateful to the reviewers who helped us with the first edition: Carol Barnum, Southern Polytechnic State University; Doug Catron, Iowa State University; William O. Coggin, Bowling Green State University; Dean Hall,

Kansas State University; Dan Jones, University of Central Florida; Carolyn Miller, North Carolina State University; Barbara Olds, Colorado School of Mines; Carolyn Plumb, University of Washington; Scott Sanders, University of New Mexico; Brenda Sims, University of North Texas; and Katherine Staples, Austin Community College. We also got help and continue to receive regular counsel and encouragement from many others, including John Battalio and James Frost of Boise State University; Hillary Hart of the University of Texas, Jim Kalmbach of Illinois State University, Stuart Selber of Penn State, and above all, our colleagues at Texas A&M University: Kathryn Alexander, Diana Ashe, Valerie Balester, Gary Floden, Joanna Gibson, Cecelia Hawkins, Chris Holcomb, Marty Jacobsen, Dave Pruett, Elizabeth Tebeaux, and Jeff Todd, among many others.

Along with all of those good people who helped us to develop the first edition, we want to thank particularly the reviewers of this edition: Barry Batorsky, DeVry Institute; David Fleming, New Mexico State University; Dan Jones, University of Central Florida; Jamie Larsen, North Carolina State University; Ann Marie Mann Simpkins, Purdue University; and W. J. Williamson, Michigan Technological University.

The Collaboration Continues

Whether you are a student or an instructor, we invite you to join the team of people who helped us develop and improve this text. If you have comments, questions, or suggestions for additional topics or activities, please share them with us. In future editions, we'll try to make *Information in Action* even more responsive to our users' needs. You can write to us at the Department of English, Texas A&M University, College Station, Texas 77843. Or send e-mail to killingsworth@tamu.edu.

MJK
JSP

Contexts for Technical Communication

LINE

▼▼▼▼▼▼▼▼▼▼▼▼▼▼▼▼▼▼▼▼▼▼▼▼▼▼▼▼▼▼▼▼▼▼▼▼▼▼▼.▼▼▼▼
Information—shared, articulated interpretations of experience—frees us from the ruts of instinctual behavior, or emotional behavior, or stimulus-response behavior, or crowd-governed behavior, giving us entry to the free-form world of invention. Technology is the product of this freedom, and language is its medium.

—D. B. Smith, "Axioms for English in a Technical Age"

▼▼
Lonely as is the craft of writing, it is the most social of vocations. No matter what the writer may say, the work is always written to someone, for someone, against someone.

—Walker Percy

▼▼
Technical writing accommodates technology to the user.

—David Dobrin

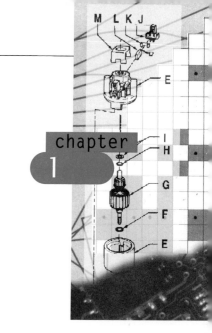

The Rhetorical Situation: Context of Production and Context of Use

CHAPTER OBJECTIVES
▼▼▼▼▼▼▼▼▼▼▼▼▼▼▼▼▼▼▼▼▼▼▼▼▼▼▼

After you have worked through this chapter, you should be able to do the following:

- Recognize the general rhetorical situation of technical communication as the production and use of information
- Understand the differences in the needs and concerns of producers and users of information
- Put your understanding to work by writing a draft audience-action profile based on an analysis of the concerns of producers and users in a typical situation involving technical communication

Technical communication is the process by which researchers and technical experts develop information products for users. Whenever you write a report, a proposal, a manual—even a letter—you have to discover and develop the information a user needs and then design and deliver a product. The information product has value if it helps the user perform an action efficiently and with sound results. Transforming information into action, then, is the goal of technical communication and the overall theme of this book.

To do the work of technical communication, you will need to use all of your skills as a researcher, speaker, illustrator, and writer. But before you write your first sentence, a great deal of planning is needed. The chances of succeeding in your communication will improve greatly if you spend some time thinking about the information you wish to convey, the people who will use it, and the actions they will perform. Part One of this textbook will help you find ways to structure and improve the efficiency of your prewriting habits.

In Chapter 1, we introduce technical communication as a kind of writing or speaking that responds to the needs of a particular moment—the scene of a potential action, sometimes called the *rhetorical situation.* To understand the rhetorical situation in the prewriting phase, we recommend the use of an *audience-action analysis,* a question–answer session about the producers and users of information that results in a written *profile* of the situation at hand. By the end of Chapter 1, you will be able to write a draft profile that presents the basic characteristics of the rhetorical situation for a typical technical information product.

Chapter 2 shows you how to broaden your understanding of the situation, adding information about the social, political, and cultural contexts of your communication. Adding ethical and cultural "filters" to your audience-action analysis will enrich the profiles you write about rhetorical situations. In Chapter 3, you learn how to incorporate your profile into a management plan that takes account of the specific forms and media used in your communication. With the management plan completed, you're ready to begin writing and developing your work, the process for which is detailed in Part Two.

The Rhetorical Situation

Any writer or speaker, "technical" or not, works within a rhetorical situation: You need to say something to someone on some occasion. Rhetoric is the art of finding the right way to deliver the message that will produce the best results for everyone involved.

As a technical writer or speaker, you face a special kind of rhetorical situation: Someone wants to solve a problem or satisfy a need and doesn't have the technical knowledge to do so. It is up to you to provide that knowledge in a usable form. You may need to:

- provide instructions to help a group of users perform special tasks on a computer,
- present the surprising results of a scientific experiment to researchers from other labs working on a project similar to the one you represent,
- give a recommendation on city planning in a report that makes the results of an engineering test understandable to people on a citizens' council,
- or propose a more equitable way of distributing goods donated to a local charity for needy people in your town.

A communication that responds effectively to rhetorical situations such as these must first of all be *context-sensitive*. Literally speaking, a *context* is what surrounds a single bit of communication, a word or phrase, for example. When people complain about being quoted "out of context," they mean that users of their language have distorted their intended meaning by taking words away from the surrounding words that help to explain them. A college president might say, for example, "Education these days is no good unless we provide a solid foundation in ethical understanding to guide our technological actions." If a newspaper reporter quotes only the first part of the sentence—"Education these days is no good"—and tries to claim that the speaker was cynical about the university's mission, the reporter has given the exact words but, by discarding their original context, has distorted their intended meaning to the point of misquoting the president and risking charges of slander and libel.

Any single statement is surrounded not only by other words but also by the broader context of a social situation—the immediate occasion for producing the language. This social situation is in turn embedded in a still broader context— the society or culture that permits such occasions to arise and have meaning for the participants. See Figure 1.1.

Consider the words of the college president in our previous example: "Education these days is no good unless we provide a solid foundation in ethical understanding to guide our technological actions."

Context 1 is the speech in which the words first appeared. Within this context, the president's comments take on the purpose and tone of the overall speech, which is an appeal for funding, a proposal that projects past successes into a possible future.

Context 2 is the occasion for the speech. For example, it may have been a meeting of contributors to the university. Perhaps they have responded well in the past to pleas for contributions that support computer labs—hence his reference to technological education. Now the president is hoping to convince them to support a new project that promotes ethical deliberation. He uses his past successes to set up success in the future.

The context of the occasion is embedded within Context 3, the larger cultural context he shares with his contributors. If the school is sponsored by a religious group—a Catholic church or Jewish synagogue, for instance—the idea of ethical education already forms a strong background, which he may choose

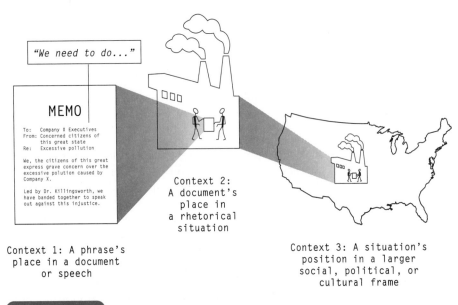

FIGURE 1.1

Three broadening views of context.

to bring into the foreground by appealing to the school's founding traditions. Likewise, if the school is a state institution, he may quote Jefferson, Lincoln, or some other secular "saint" on the ethical aims of state education. How far to go in bringing this contextual background into the foreground of the present occasion is one of the big decisions this communicator has to make in developing his communication.

The problem becomes more difficult when the situational and cultural contexts of the communicator and audience differ more radically. Imagine a Christian engineer from the United States as a Peace Corps volunteer promoting a new method of irrigation to Buddhist farmers in Nepal who have followed the same methods of production for centuries, or an environmental consultant explaining a program to prevent rain forest damage to a culturally varied audience—representatives from a number of countries from the United Nations. With each gradation of difference between the communicator and the audience, the decisions about how to communicate, how to put information into action, become more complex.

Our first two chapters help you confront and sort out these complexities and discover a flexible but reliable method for creating context-sensitive documents. In this chapter, we present a model of the situational context of technical communication, a general picture of the most typical instances of information production and use in technological societies. Chapter 2 broadens our model to take account of ethical, political, and cultural contexts.

Technical Communication Skills

As a producer of information products, you'll need to develop a set of skills that relate to three general areas of human endeavor, or arts:

1. **Rhetoric:** the art of using words effectively in speaking and writing. Rhetorical skills provide ways to build relationships through the exchange of information. They fall into two general categories: *analytical skills,* which enhance your ability to review situations and audiences with an eye to designing and improving both your own documents and those of others; and *productive skills,* which enhance your ability to implement designs through writing, speech, graphics, and other forms of communication.
2. **Management:** the art of controlling or directing work. Management skills provide ways to build better relationships with fellow workers and handle communication processes. Management skills can help you make the best use of your time and energy in dealing with assignments. Effective management begins with a set of positive attitudes: a willingness to plan your work and to review and revise it; an openness and flexibility in the face of new demands; and a willingness to cooperate with others.
3. **Ethics:** the art of setting standards of conduct and making moral judgments. Ethical sensitivity points you toward decisions that reflect a sense of responsibility for others. Ethical communication requires that we consider the value of the information we offer as well as the timing and accessibility of our presentation. Ethical skills should work as a filter in all of our decision making, both in reviewing and in implementing communication designs.

An action-oriented approach to communication should be rhetorically effective, well managed, and ethically sound. This book is designed to help you develop your skills in these areas. Later chapters will build on the introductory discussions in this chapter.

Context of Production, Context of Use

Technical communication typically takes place in situations in which knowledgeable producers provide users with information they need to perform actions. Producers and users work in distinct, but overlapping, fields of action—the *context of production* and the *context of use.* The context of production is the environment in which writers and researchers work. The context of use is the environment in which decision makers and users read the writers' and researchers' products—reports, proposals, and manuals—call for changes or more information, and then convert the information into action.

To understand what it means for a producer and a user to exchange information intended for action, consider the example in Case 1.1.

case 1.1   A TYPICAL EXCHANGE OF INFORMATION

When you deposit money in a bank, the transaction is registered orally (when you speak to the teller), in written form (on a deposit slip), and in electronic form (in the bank's computers). You can access this information in any of these forms. A record of the face-to-face transaction appears on the bank's security camera; an audio tape may record the conversation; the written transaction shows up on the deposit slip you save for your records and on your monthly bank statement; and the electronic transaction is recorded in the bank's database, where it can be accessed at any time. These records allow both you and your bank's employees to reconstruct the experience and then use the information to verify past actions and influence future actions.

A complete contextual account of this interchange of action and information would be extraordinarily complex. It would have to include all of your accumulated knowledge and experience, all of the bank teller's, and all of the bank's databases. It would also have to consider much information that is omitted in the bank's record-keeping procedures—the state of mind, intentions, and emotions of both you and the teller, for example.

However, if we focus on the single transaction—the moment when the teller passes the written deposit slip to you—we can define a situation that is common to every information exchange. There is a producer of information, in this case the teller who writes out or types the deposit slip. There is a form of communication, the deposit slip itself. And there is a user, or consumer, of the information—you, the bank customer.

Figure 1.2 is a linear model of this transaction that shows the producer as the active party and the user as a passive receiver of output.

FIGURE 1.2

A linear model of an input–output transaction. The producer is active and the consumer is passive.

An interactive transaction model. Communication requires an active producer and an active user.

In real life, receivers play a more active role in the transaction than this model suggests. The bank had to go through an extensive refining process, collecting feedback from customers and making adjustments in the information before it arrived at this deposit slip. If someone were to hand you a whole computer printout when you needed only a single sentence or number, you would complain and ask for only what you need.

Information has value only when it is relevant to the particular needs of a user. Any communication that provides *less* information than the user needs has no value because it cannot help him or her make a decision and take action. Any communication that provides *more* information than the user needs is ineffective because the user will have a hard time accessing the valuable information, sorting the wheat from the chaff.

Figure 1.3 presents an improved model that depicts the user as an active party in the communication. In some technical communication situations, however, the process is actually circular. The user often generates information for the producer, and the two parties exchange roles back and forth until they arrive at a mutually satisfactory information set.

Once you understand that your reader is an active user you will have an easier time selecting, designing, and presenting information in a form that makes use easy. Consider computer documentation, for example. Knowing that readers are likely to use manuals as they work at a computer station, writers and designers take a number of steps to enhance usability:

- Instead of dense paragraphs full of information, they set up their sentences as numbered lists of brief instructions that are easy to follow and return to after the user's eye leaves the page.
- They design the page with plenty of white space and illustrations that also help the user to make transitions from the page to the screen.
- They publish the book with a binding that allows it to lie open on the user's desk.
- They build instructions into the computer software itself, transferring instructional information from the print manual to the screen in pull-down menus, dialog boxes, and help files.

Likewise, the author of an engineering report will craft different sections of the document to accommodate the needs of different active users:

- The executive summary is pitched at nontechnical managers. It emphasizes the most important conclusions and recommendations of the report.
- The results, or findings, section provides technical details so that engineers and technicians can reproduce or improve upon the actions reported.
- The cost section is set up in tables and columns so that accountants and auditors can review the budget easily.

Understanding how your reader will use the information you provide is key to producing an effective document. Computer manual writers, for example, incorporate features for ease of use at a workstation.

Users and producers bring different needs, desires, and expectations to a transaction. Their different needs must somehow merge if the communication is to be effective. From the producer's standpoint, the communication should be based on accessible information and should be relatively easy to produce. From the user's perspective, the information should be legible, relevant, and fast and easy to interpret. Producer and user share an interest in the accuracy of the information and in its value in the world of technological action.

If we consider these different needs and desires, we must make our model more complex. Figure 1.4 shows that each party operates within a context—the context of production or the context of use.

Both the context of production and the context of use contain databases, or stores of processed information. In the context of production, the producer draws on the database to supply the information needed by the user. In the context of use, the user records the information in a different database.

Both contexts also contain actions—experiences that may be recorded as information. The information given in the communication will not only be saved for the record but also guide important actions in both contexts.

In the example of the bank transaction (Case 1.1), you would decide whether to deposit or withdraw more money by comparing the information on

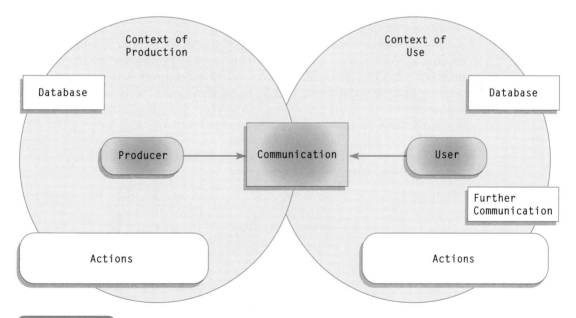

FIGURE 1.4

A contextual transaction model. Communication occurs within a context of production and a context of use.

the deposit slip with the information in your database (probably a checkbook in this case). Likewise, in the context of production, the bank would add the information to its database to decide whether you can withdraw more money or perhaps whether you are qualified to take out a loan.

In the context of use, the communication between the producer and the user often leads to further communication with a third party. For example, after seeing the balance in your account, you might call another bank to have funds transferred.

Information is the base from which communications and actions proceed. If the information is incorrect, poorly conveyed, or misunderstood, an epidemic of errors will infect the information flow. If the communication is slow, frustrating, or unpleasant for either the producer or the user, these negative emotions are also likely to affect the flow from communication to action. That's why accuracy and goodwill are crucial elements of effective technical communication.

case 1.2   AUDIENCE AWARENESS IN THE SOFTWARE DOCUMENTATION INDUSTRY

To illustrate the distinctions between the context of production and the context of use, let's examine a model that Henrietta Shirk, a technical communication researcher, developed to show the typical action of any computer program. This "software paradigm" is shown in Figure 1.5.[1]

For the documentation writer, the most important thing to remember, according to Shirk, is that software producers and software users are interested in different parts of the paradigm. The software producer is mainly concerned with processing; the software user focuses almost exclusively on input and output, conceptualizing the processing as a kind of "black box." For example, users of a spell check program don't really need to know whether the program uses PASCAL or machine language. They need to know only which word is misspelled (output) in the text that they have entered (input). By contrast, the programmers back at the company—the people responsible for writing, maintaining,

Input ⎯⎯⎯⎯⎯⎯→ Processing ⎯⎯⎯⎯⎯⎯→ Output

FIGURE 1.5

A simple model shows the typical action of a computer program.

---

1. Henrietta S. Shirk, "Prologue for Teaching Software Documentation," *Perspectives on Software Documentation: Inquiries and Innovations,* Thomas T. Barker, ed. (Amityville, NY: Baywood, 1991) 25–44.

and updating the program—have a definite interest in the program's code and internal structure.

Because of this difference, communications that take place within the company (internal communications) should be very different from those between the company and its customers (external communications). Internal communications include problem statements, engineering specifications, project reports, and memos; external communications include reference guides, user manuals, training manuals, and online help.

Too often, Shirk says, external communications do not adequately transform the information contained in internal communications for use outside of the company. But by recognizing the need to shift the focus from processing to input and output, software producers are taking a big step in the right direction. ●

Conducting the Audience-Action Analysis and Writing the Profile

In later chapters, you will learn more about how to conduct research and analyze an audience of users, design information products for that audience, write and produce documents according to your designs, and review and revise documents with your audience in mind. For now, our contextual model provides a foundation for conducting an initial analysis on which to base your later work.

Any *analysis* involves identifying the components or parts of an object or a task to gain a better understanding of the whole. In this sense, a dissection in a biology lab is an analysis, but so is the set of drawings you make in an engineering design lab or even the paper you write in a literature class that discusses a short story in terms of its setting, action, characters, and theme—the parts of the whole story.

case 1.3 ANALYZING AND PROFILING AUDIENCES AND ACTIONS FOR A SOFTWARE DEVELOPMENT PROJECT

The diagram in Figure 1.4 on page 11 represents an analysis of a generic rhetorical situation. It identifies various components of the situation: a context of production, a context of use, a producer, a user, a communication, and various actions, databases, and so on. One approach to developing an audience-action analysis of your own is to draw such a diagram with the components made more concrete, using the following list, for example:

- producer = technical manual writer
- user = customer, an office worker

- context of production (cp) = computer company specializing in insurance software
- context of use (cu) = insurance company offices
- cp databases = engineering specs, technical documentation files, user test files
- cp actions = software development and technical writing
- cu databases = company files, old documentation, etc.
- cu actions = word processing for insurance claims, policies, etc.

Figure 1.6 shows how you might diagram such a rhetorical situation.

Another approach is to create a list of such information and write up an *audience-action profile* to use in communication planning. To get started on your profile, answer the questions in Guidelines at a Glance 1, keeping notes as you go, listing the components of the rhetorical situation you are entering, and thinking through as many of the problems as you can. On the basis of these notes, compose a short document that profiles the situation, again following the directions in the Guidelines.

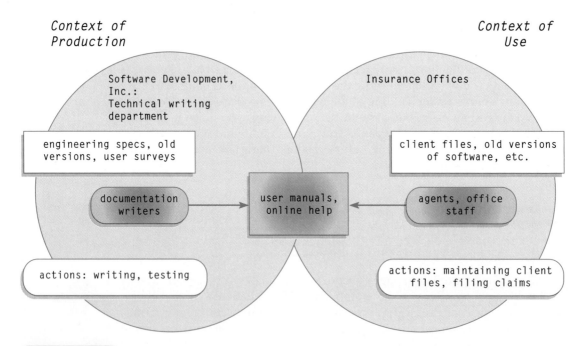

FIGURE 1.6

Diagram of the contexts of production and use for a software documentation project.

GUIDELINES AT A GLANCE 1
Doing the audience-action analysis and writing the profile

Sample questions for context of production:
- Who is responsible for producing the information product?
- What information is needed for production?
- What actions are needed for production?
- Why is the information product being produced?
- How will the information product be produced?
- What form will it take?
- To what extent will technology be used in production?

Sample questions for analyzing the context of use:
- Who will use the information?
- What is the reading level of the users?
- Are there technology considerations?
- What actions is the user expected to perform on the basis of the information?
- What information does the user need to perform the actions effectively and efficiently?
- What kinds of information product(s) (letter, report, manual, proposal, web site, etc.) is the user likely to need to accomplish the communication tasks?
- What features will the information product(s) need to help the user act in the desired manner?

Writing the profile:
- Paragraph 1: Discuss the most important issues in the context of production (based on answers to questions in the analysis).
- Paragraph 2: Identify and describe your chosen audience. Discuss the important issues in the context of use.
- Paragraph 3: Discuss how you as a producer plan to take account of user needs.

Sample 1.1 on page 16 shows a completed draft of an audience-action profile for the team project described in Case 1.3. Samples 1.2 and 1.3 (pages 17 and 18) are draft profiles for individually authored student projects. Sample 1.3 shows the kind of expanded profile you will learn to write in the next two chapters.

As you learn more about the processes and products of technical communication in Parts Two and Three of this textbook, you will be able to write more specific audience-action profiles. Chapter 2 will give you additional general direction, and Chapter 3 will show you how to incorporate your profile into a management plan. But by using only Guidelines at a Glance 1 and a little common sense, you should be able to compose a good draft profile that gets you started on the projects or exercises you have before you.

<u>An Audience-Action Analysis and Profile
for a Software Development Project</u>

The technical writing department of Software Development, Inc., has the assignment of producing the user manuals and online help for version 2.0 of ClientBase, a databasing program used exclusively by insurance offices. The software allows agents to collect, arrange, and access client information and file claims for their clients. In our company (the context of production), we use a team approach, uniting interface designers, illustrators, writers, and editors in the documentation effort. The documentation team works with the engineering specifications for the new version (2.0), copies of the old version (1.2) and its documentation, and surveys returned by users of Version 1.2. We have also researched the documentation practices of our two leading competitors. We create a plan and a draft copy of the new manuals and prototypes of the new screens and online help. Then we test the prototypes with trial users and revise the draft as needed.

The context of use for this documentation is a typical insurance office. Agents and clerical workers with no computer experience beyond word processing and often little more than a high school education maintain client files and enter claim information via computers and modems. They are networked with their home offices, which receive the claims, process them, assign adjusters or make payments as required.

The problem we face is to make the computer screens and the support documents as simple as possible to use while allowing the users to take full advantage of all the software features. The interface of Version 1.2 walks the user through each step of each function ("create client file," "file claim," etc.). The user selects functions by pull-down menus, each of which includes a help session. When the user activates the selected function, the program asks the user to enter each item of information in a pop-up box. The surveys returned by users of Version 1.2 found the program so easy to get started on that they never used the step-by-step tutorial provided in the manual for beginning users. They said they only opened the reference section when they had problems, and they complained that the manual didn't provide enough help with how to handle problems. So we have decided on a three-part approach to revision:

- build as many instructions as we can into the interface,
- omit the tutorial in the manual, and
- include more information and instructions on troubleshooting.

SAMPLE 1.1   Draft for an audience-action profile

**Audience-Action Profile for a Report on the Use
of Shark Cartilage as a Method of Treating Cancer**

by Jerad Hayes

The Context of Production

There are five main activities necessary for producing my document: researching, analyzing data for usefulness, organizing useful data, writing the document, and proofreading. Each of these takes time.

Since the subject has been around for a while, there should be some books on it in the library; however, some web-based research would be appropriate because shark cartilage is being sold over the web. Any data I collect from the web will have to be sifted carefully to ensure it is not biased by commercial interests. The books that I obtain in the library should be less biased. I do not know any experts in the field of shark cartilage usage for cancer treatment, so it is not likely that I will be able to do an interview as part of my research.

One of the major limitations of the production of this document is the time constraints. I could research a large amount of data in one field in hopes of spurring the interests of many scientists in that field, or I could research a small amount of data in varying fields to spur research across fields.

The Context of Use

The target audience of this document is a group of scientists I am trying to convince to do additional research in the field of shark cartilage. The only information that the users need to know is that which sparks interest and motivates them to perform further research.

After analyzing all of the data, I must carefully think about organizational tactics to maximize the effectiveness of the document. The data should flow logically so that the reader will have an easy time understanding the document and will be left with a good impression. Since the document is directed towards other scientists, I can assume that some technical jargon will be understood by the reader.

Taking User Needs into Account during Production

The target audience may be too broad, since it includes chemists, biochemists, doctors, etc. For this broad audience, I will need to address an array of topics so that individuals from different fields might find information to motivate them. However, I might make more of an impact if I write several documents on shark cartilage research, each targeting a different scientific field.

SAMPLE 1.2   Draft for an audience-action profile of a student project (individual author)

A Proposal for Recruiting College Students on the Internet

by Jeremy Bennett

Context of Production

I am an account executive for a college publication called *Campus Recess Magazine.* Besides selling advertising, it is my job to help the publisher run certain aspects of the business. Because *Campus Recess* is such a lean organization, the account executives often must bring certain business issues to the publisher's attention.

I have the responsibility of preparing a proposal for the publisher, who is also the owner. The proposal will be designed to convince the publisher to recruit employees through the Internet. Almost 100% of the employees working for *Campus Recess* are college students, a very successful strategy for the company. However, there has been a significant problem making the college student population aware of opportunities offered by the company in any given interviewing period. There are several production objectives important to the completion of the proposal.

The first task is to set the objectives of the proposal. This step is instrumental in guiding my research. Next, I would conduct secondary research. The secondary research will include learning about websites and their different functions, learning about recruiting, and exploring the success rate of the two together. This information can be found online, in databases, in periodicals, and also in books. Online material will include both good and bad examples of recruiting college students over the Internet. It will also be important to gather quantitative material and produce charts and graphs.

I will then conduct primary research consisting mostly of interviews. The interviews will be with managers of human resource offices and college students and will focus on recruiting on the Internet and their personal experiences. I will also interview veteran employees of *Campus Recess* and attempt to establish a recruiting model for the company. Finally, I will make inquiries regarding previous reactions the publisher has had to various suggestions. This will greatly aid in understanding my user.

I will then organize the research data into a usable outline, draft the document, and add appropriate graphs and tables. My peer support team will offer extensive constructive criticism, after which, I will begin editing the document. After editing the document, I will attempt to make it appear as professional as possible. This will be accomplished by color laser printing on the highest quality bond paper, with graphics and a glossy bound cover. The final step will be to submit the document to the publisher. My peer support team will review drafts weekly.

There are a number of management issues important to the production of this document. For instance, the publisher/owner of the magazine operates in five college markets. He is based in Austin. Therefore, most business is conducted over

SAMPLE 1.3   Student audience-action analysis and profile for a proposal to a specific individual (draft copy)

phone, computer, and fax machine, which makes him a very busy man. The document cannot include wasteful or immaterial information because he will lose interest.

The next important issue is money. The company is still experiencing growing pains and has not fully penetrated all five markets. This makes all suggestions requiring a large amount of money a waste of time. Finally, the user is a veteran of print media and knows little else. The document must be inviting, simple, colorful, and written with a strong sense of style for the publisher to take it seriously. The hierarchy of the organization is surprisingly flat. The owner/publisher is at the top, with the editorial director, artists, copywriters, account executives, distributors, and administrative staff reporting directly to him. There is one large group of contributing writers that reports to the editorial director, but that is the only real depth in the organization.

Context of Use

The publisher will use this document to gather information and make an informed decision about whether or not to use the Internet as a recruiting tool in the future. The user of the document is an entrepreneurial person with a short attention span. He enjoys a fast-paced business environment. He is also easily excitable and easy to persuade in the proper setting. The publisher is deeply concerned with media and media-oriented businesses. Taking into account his background, I must remember not to load the document down with wasteful information. It is also important for the proposal to contain a great deal of colorful graphics in a sleek, professional package.

The user already uses the Internet. He is paying one of his artists a small fee to maintain the site, which costs him next to nothing since he trades advertising for cyberspace time with Cybercom in Austin. The problem is that no one visits the site because no one knows it exists. The object of the proposal will be to convince the publisher to expand his site with a comprehensive recruiting section. The publisher graduated from the University of Texas ten years ago and started *Campus Recess* roughly six months after graduation. He has surrounded himself with college students for this entire time period. Company culture reflects the owner's fondness for his college years, and it is widely held that he started this company so that he would never have to leave. Despite the fact that the publisher is 33, he stills regards himself as a student. He has a moderate understanding of business terminology, below average understanding of computer terminology, and an expert level of media jargon.

There are certain skills the publisher will need to process this information. The most important are basic conceptual business skills, which he does possess from running a successful company for ten years. Competitive products my employer could use in recruiting students are the following: *Campus Recess,* the corresponding school's newspaper, local newspapers, school placement centers, and strong networking. The easiest is *Campus Recess,* because of its cheap cost

SAMPLE 1.3 continued

to the publisher. However, this option has proved to be largely ineffective because students do not use the publication for this purpose and therefore do not process the information correctly.

The intended user actions are: adding a recruiting section to the current website, promoting the site more often through current radio contracts, and posting messages on electronic bulletin boards frequented by many college students. There are several unintended actions: the publisher could reject the idea, he could add the new section without corresponding promotion, or he could post messages on newsgroups having virtually no college student patrons.

It may be necessary to include books and manuals dealing with the most effective ways to recruit college students on the Internet. If these materials are not available, it may be necessary for the owner to use a consultant who may better aid him in recruiting over the Internet. To efficiently perform the desired task, the owner must be told the most effective and efficient method to add to and maintain his site, promote his site, and post messages on electronic bulletin boards. The document will be well written, to the point, in memo format, with color pictures, graphs, charts, and a sleek cover binding.

Conflicts/Ethical

The user has very little cash to spend as most of it is tied up in his growing business. However, after producing the extended website, he must continue to pay radio advertising costs. The advertisements must generate capital themselves or they are a waste of money. As an account executive, I have several ethical obligations in producing this document. If I am dishonest and do not disclose the entire truth in an attempt to make myself look better, I am destroying the trust my publisher has in my decision-making abilities. Therefore, if I encounter conflicting information, it is my responsibility to make it known. It is also my obligation to present the least costly aspect to each of my suggestions. Due to my employer's financial position, it would be irresponsible to convince him to waste money.

Socially, I must understand that the publisher enjoys college and college students. I must also understand that he thinks in terms of media and media business. I should be very careful in producing the document and avoid annoying him with "business school lingo." I have a responsibility to cite all sources from which I take more than three words.

There are two implications of violating *Campus Recess'* ethical codes and policies. The first would be my termination. The second is akin to the Golden Rule. My publisher has been completely honest and has trusted me in all affairs. I would hate to disrupt that relationship by violating his trust.

SAMPLE 1.3 continued

Select one of the following problems (or divide into teams of four or five members, each team taking a problem). Using Guidelines at a Glance 1 (page 15), perform an audience-action analysis and write a profile like the one in Sample 1.1. Put yourself into the position of the producer (in number 1, for example, you might take the role of the designer at the bicycle company). Imagine that the reader of your profile (not the user of the information product you are developing) is either an upper-level supervisor of your activities (one not familiar with your usual practices) or a new member of your organization.

EXERCISE
1.1

1. A bicycle company recently developed a new kind of brake for hard mountain riding. The new system is much more powerful and reliable than earlier designs but requires users to be able to coordinate hand and foot movements in a way that may be foreign to them. Engineers are consulting with physical therapists, marketing specialists, artists, and manual writers to build consumer interest in the product. What communication strategies are they likely to use?
2. The Spanish department at Central High School found a software system that has a documented success rate of 40 percent in improving student performance on tests of conversational Spanish. The school enthusiastically authorized the purchase of the system. The problem is that students have been lethargic in using the system in the computer lab during their free study periods, and classroom use alone is not enough to achieve the desired results. Several students have complained that they don't understand the directions in the manuals and the instructional tapes that come with the product. What communications can the teachers use to motivate students and make it easier for them to use the product more frequently?
3. A construction company is trying to convince the city council to increase funding for a roads project. After winning the bid for the project, the company discovered a geological feature on the site that makes the use of the plans they originally proposed prohibitively expensive. They've developed an alternative plan but still need more money to make it work. The company has hired consulting geologists and legal specialists to help them make their case. What communication tasks will they need to perform?
4. A group of chemists has discovered a neutralizing chemical that will prevent infusion of toxins into local water supplies if cotton farmers spray it on their crops two weeks after spraying the insecticide that is used most frequently in the region. Tests show no decrease in the effectiveness of the insecticide with use of the chemical. The chemists are working with agricultural extension agents, the local farm and ranch board, and the Environmental Protection Agency in an effort to promote voluntary use of the neutralizing chemical, which is provided at no charge to farmers, using government funds. How do they sell the idea?
5. A group of safety engineers working for an airline company has found that a certain model of aircraft cannot effectively deploy its de-icing mechanisms under very cold conditions. The planes have been flying in such conditions on a regular basis, and luckily, no accidents have occurred. But multiple tests

confirm the risk. The safety group has decided to recommend replacing the aircraft with another model. They are worried that management will be slow to act on this expensive recommendation. After all, the planes have experienced no difficulties in actual flying conditions. What do the engineers need to do?

6. In hard economic times, a paper company decides to close two of its older plants because they cannot match the level of production at other plants. The managers of Plant A, one of the plants targeted for shutdown, are working with their environmental engineers and public relations officers to argue that the plant deserves to stay open because of its crucial position in the small town in which it operates and because it has a better environmental record than any of the company's other plants. In addition to appealing to management to stay open, the advocates of Plant A are considering asking the company to sell the facility to another company. How can they make their case?

Recommendations for Further Reading

Bateson, Gregory. *Steps to an Ecology of Mind.* New York: Ballantine, 1972.
Jones, Dan, ed. *Defining Technical Communication.* Arlington, VA: STC Press, 1996.
Schriver, Karen. *Dynamics in Document Design: Creating Text for Readers.* New York: Wiley, 1997.

Additional Reading for Advanced Research

Anderson, Paul V., R. John Brockmann, and Carolyn R. Miller, eds. *New Essays in Technical and Scientific Communication: Research, Theory, Practice.* Farmingdale, NY: Baywood, 1983.
Blyler, Nancy Roundy, and Charlotte Thralls, eds. *Professional Communication: The Social Perspective.* Newbury Park, CA: Sage, 1993.
Dobrin, David. *Writing and Technique.* Urbana, IL: NCTE, 1989.
Killingsworth, M. Jimmie, and Michael Gilbertson. *Signs, Genres, and Communities in Technical Communication.* Amityville, NY: Baywood, 1992.
Smith, D. B. "Axioms for English in a Technical Age." *College English* 48 (1986): 567–579.

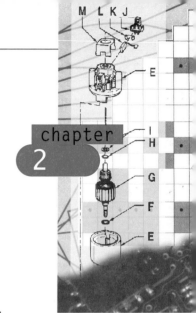

Ethical, Political, and Cultural Issues in Technical Communication

CHAPTER OUTLINE
▼▼▼▼▼▼▼▼▼▼▼▼▼▼▼▼▼▼▼▼▼▼▼▼▼▼

Collaborating with Others in the Context of Production

Issues of Access and Integrity in the Context of Use

The Wider Context: Dealing with Cultural Differences in International Communication

Expanding the Audience-Action Analysis and Profile

CHAPTER OBJECTIVES
▼▼▼▼▼▼▼▼▼▼▼▼▼▼▼▼▼▼▼▼▼▼▼▼▼▼

After you have worked through this chapter, you should be able to do the following:

- Recognize the need to work with others in producing effective technical communications and develop ethical as well as efficient procedures for managing collaboration
- Recognize the value of free access to reliable information, abiding by ethically rational procedures for using information in decision-making processes
- React sensitively to cultural differences in communication practice, making allowances whenever possible for social and cultural demands in the use of information
- Add ethical, political, and cultural "filters" to your audience-action analysis

Chapter 1 showed how any act of communication is embedded in a series of social contexts. You begin your work in a context of production, collaborating with fellow workers in developing an information product to be used in an entirely different context. Using an audience-action analysis in the pre-writing phase of your work helps to align the demands of the context of production with human needs in the context of use. Developing an awareness of the user's needs in a model of your rhetorical situation *before you begin to write* is a first step toward strengthening the analytical and productive skills required in good technical communication.

This chapter provides an opportunity for an extended look at the social contexts of technical communication. Relating to others in acts of communication always creates practical problems and also raises ethical issues that affect you both as a colleague of other people in your work environment—the context of production—and as a supplier of information to clients in the context of use. This chapter challenges you to become both a good team member and a reliable citizen in the world of technical communication. To achieve these goals, we invite you to add ethical and cultural "filters" to your audience-action analysis, developing good pre-writing habits that hone your skills as an interpersonal communicator.

The ultimate goal of ethical analysis in technical communication is to develop good *policies* for the development and use of information. Because it leads toward policymaking, ethical and cultural analysis overlaps with the *politics* of information, which deals with the power that different individuals and groups are able to exert on each other in rhetorical situations. We will see that ethical and political issues change somewhat when considered from the different vantage points of the context of production and the context of use.

Collaborating with Others in the Context of Production

Almost every survey of professional writing done in the last two decades suggests that technical writers typically produce documents collaboratively. A 1982 survey done by Lester Faigley and Thomas P. Miller, for example, showed that 73.5 percent of all respondents "sometimes collaborate with at least one other person in writing." Lisa Ede and Andrea Lunsford, in a survey of 1400 professionals in industry, business, education, and government, found a similar frequency of collaboration.[1]

Technical communication researcher Deborah Bosley defines collaborative authorship as "two or more people working together to produce one written

1. Lester Faigley and Thomas P. Miller, "What We Learn from Writing on the Job," *College English* 44 (1982): 557–69; Lisa Ede and Andrea Lunsford, *Singular Texts/Plural Authors: Perspectives on Collaborative Writing* (Carbondale: Southern Illinois University Press, 1990), especially Chapter 2.

document in a situation in which a group takes responsibility for having produced the document."[2] *Responsibility* is the key word; everyone connected with the project has a share in it. Collaborative authors must do the work of writing together and receive whatever credit or blame accrues to the project after it is produced.

But the locus of responsibility is not the only thing affected by collaboration. The very process by which documents are produced can also change. Individual authors may still spend a great deal of time alone, bent intently over computer keyboards or yellow legal pads. Increasingly, however, collaborating authors are blending the activities of writing and speaking to share their roles as researcher, author, editor, and document designer with others. Some collaborators even sit at keyboards and compose together with the help of software designed for computer conferencing and collaborative writing. The question, then, is not *if* technical communicators collaborate to produce their documents, but rather *how* they do it.

Two Models of Collaborative Authorship

When most Americans—reinforced in their individualism by years of education— are asked to work in teams, their first impulse is to resist the idea. Their second impulse is to divide the big task into little tasks that can be handled individually. Classical economics (as conceived by Adam Smith), workplace tradition, and the century-old practices of "scientific management" support this *division of labor* model for group writing. Division of labor is the hallmark of "Fordism," the use of assembly lines with minute tasks assigned to highly specialized workers. The model is known for its efficiency—its productivity and speed.

However, division of labor has begun to lose much of its appeal in industry as the workforce becomes more educated and demands more meaningful work. Today, a new model is moving to take center stage in industrial management practices—an *integrated teams* model. All team members work together at every stage of a developing project and maintain a view of the whole project, from beginning to end, in all its complexity and all its component parts.

For collaborative writing projects, the division of labor and the integrated teams models offer distinct advantages and disadvantages. Tables 2.1 and 2.2 summarize the pros and cons of each model.[3]

In most situations, the integrated teams approach works best, because it results in a product with fewer "seams" and more consistent quality. But there are degrees of integration. Rarely are writing teams fully integrated. Although

---

2. Quoted in Ede and Lunsford, *Singular Texts/Plural Authors: Perspectives on Collaborative Writing* 15.

3. The tables are derived from M. Jimmie Killingsworth and Betsy Jones, "Division of Labor or Integrated Teams: A Crux in the Management of Technical Communication?" *Technical Communication* 36 (1989): 210–221.

TABLE 2.1

Advantages and disadvantages of the division of labor model of collaborative authorship.

Advantages	Disadvantages
1. Few meetings are required, allowing individuals to proceed at their own pace and to work on other projects as time permits.	1. Strong project management is required to keep all individuals on task and to pull everything together in the end. In student projects, a leader must emerge. Others may come to resent the leader. And the leader may grow resentful at having to take up the burden of keeping the project on track.
2. All individuals have clear tasks to accomplish, with little overlap or confusion among roles.	
3. Theoretically, at least, the people best able to carry out each task are assigned to that role (the best researcher does the research; the best writer, the drafting; the best artist or software specialist, the graphics; and so on).	2. Resentment may also arise if some members of the team think others have easier assignments, so they come to care less about the project and do not give their best effort.
4. Evaluation of individuals is easier, since it is obvious who has done what job in producing the report (if the art is bad, the artist is to blame, for example).	3. A weak link in the team can cause the entire project to appear weak in the end. Even if everything is strong in the final product but the graphics, the product as a whole suffers.
5. Production is usually faster than in models that require fuller integration of team members.	4. Because each member cannot see the project as a whole developing, the product may lack coherence or evenness, so in the end the project manager must put in extra effort to "smooth out the seams."

Source: Used with permission from *Technical Communication,* published by the Society for Technical Communication, Arlington, Virginia.

TABLE 2.2

Advantages and disadvantages of the integrated teams model of collaborative authorship.

Advantages	Disadvantages
1. Everyone in the project has a clear idea of how the project as a whole is going, so the parts of the product generally fit together better. An integrated team equals a better integrated product.	1. More time is required for meetings.
	2. Individuals must be willing to devote additional time to developing good personal relations within the group.
2. All individuals tend to feel "ownership" in the project and thus feel better about making a fuller commitment to it.	3. Individuals may be unclear about their roles in the project or may feel that they are being manipulated by stronger team members.
3. Team members can compensate for "weak links," so final quality tends to improve.	4. Hard-working team members may resent slackers who don't show up for meetings, and so forth.
4. In the process of sharing roles and specialized knowledge, team members can arrive at a better understanding of matters outside their own areas of specialization.	5. The product must be evaluated as a whole because it is a true team effort, so it is difficult to award individual praise or blame.

Source: Used with permission from *Technical Communication,* published by the Society for Technical Communication, Arlington, Virginia.

some writers work best with their coauthors when they actually sit down together and write sentence after sentence, many find such a prospect daunting. You may prefer to have an initial meeting, do some document planning and outlining, then divvy up assignments, agreeing to get back together before you write very much or proceed very far on the project. Then, with bits of writing and research notes in hand, you can sit down with your team members and go over each idea in greater detail.

In other words, combine the two models: The integrated team model provides the foundation for the process, but at various points the team divides the labor at hand. Notice that this hybrid approach avoids the rigidity of the old division of labor model, with its assignment of hard and fast roles based on specialized abilities, but it compensates somewhat for the slowness and inefficiency of "writing by committee."

Table 2.3, based on responses to the Ede and Lunsford survey, shows that in practice, many collaborators combine the two models for planning, drafting, and revising a document. All six approaches listed in the table borrow elements from both the division of labor and the teams model.

The Ethics of Collaboration in the Process of Developing Information Products

Like any other social activity, collaborative authorship involves ethical considerations. *Ethics* has to do with an individual's sense of what is right and good in any social situation. Even communicators who perform effectively and efficiently—who write well, speak clearly, and have a strong sense of the rhetorical needs in a given situation—still face some difficult challenges when ethical problems arise. Questions of team members' responsibilities, rights, and

TABLE 2.3
Professional approaches to organizing writing teams.

1. The team plans and outlines together. Each member drafts a part. The team compiles the parts and revises together.
2. The team plans and outlines. One member writes the entire draft. The team or group revises.
3. One member plans and writes a draft. The group or team revises.
4. One person plans and writes the draft and submits it to a team of revisers who work without consulting the original author.
5. The team or group plans and writes the draft. A single team member revises it without consulting the other team members.
6. One member assigns writing tasks. Each member carries out individual tasks. One member compiles the parts and revises the whole.

Source: From Ede and Lunsford, *Singular Texts/Plural Authors*, 63–64.

Team members may work on pieces of a project individually, then integrate the results of their work at regular team meetings.

privileges can cause a project to grind to a halt if they are not answered to everyone's satisfaction.

Within the context of production, the main concern is for individuals to recognize and carry out their responsibilities to the organization for which they work and to the colleagues who rely on them. Since collaboration is the norm for developing information products in technological societies, the ethics of teamwork come into play with every project.

From the start of a project, team members should reach an agreement about their processes of interaction. Who is responsible for doing what? This agreement should be part of either a code of conduct set forth in company policy or the terms of a contract or management plan. Is there a clearly defined team leader? What are the processes by which the team comes to a final decision? It may seem better to answer these questions as they arise in the course of work, but in fact, you can save a great deal of time and worry by laying out some ground rules early on. Questions such as these should be foremost in your mind as you do the exercise in ethical analysis at the end of this chapter and the exercise on writing a management plan at the end of Chapter 3.

Consider as binding all agreements you make at the beginning of a project. For example, if you have agreed to complete your part of a writing assignment by Tuesday, and on Monday you suddenly feel the demands of a last-minute assignment from another project, don't forget that your team is counting on you. If you need help, ask for it, but do not simply set aside your responsibilities and arrive at a team meeting empty-handed, with nothing to offer but excuses. Make plans to complete work with time to spare so you can avoid this kind of problem.

On the other hand, try to be flexible with your demands of others. Offer help when the need arises. Be firm with deadlines, but be understanding when someone makes a reasonable request for additional time. Never react too quickly when someone reports a personal shortcoming. Listen to the reasons involved and try to sympathize, adjusting the schedule where possible.

If a team member becomes entirely unproductive or destructive to your efforts, however, at some point you will have to take corrective action. After trying to solve the problem by adjusting the schedule or conceding to some of the person's demands, consult with other team members about how they have reacted to the troublesome partner. If you are in agreement with them over the nature of the problem—late work, for example, or work that does not meet the quality standards of the group—approach the person with suggestions for more reliable performance. If the problem continues, seek arbitration. In student projects, go to the course instructor and request a team conference. In the workplace, talk to your supervisor about the need for additional authority to deal with the problem. If you are the team leader, you have an added responsibility to ensure the effective functioning of the group.

Problems may arise from factors other than weak performance on the part of a single member. Personal differences between two team members or *office politics* may cause conflicts. For example, the members may be competing for a new position that is about to come open in the department, so they feel that they might get ahead in the race by making the competition look bad, degrading their opponent's contributions to the team project and taking every opportunity to criticize the other person harshly. Obviously, such behavior has no place in the rational performance of a job. But people are not always rational. They may not even be aware of their own motives for behaving as they do toward others. Their fellow team members would do well to recognize conflicts early on, urge cooperative action, and if necessary seek arbitration quickly without waiting for the conflict to totally destroy the team's ability to function.

Conflicts may also go deeper than simple differences between conflicting personalities. They may arise from an inability to deal with *workplace diversity.* As education extends its reach to new social classes and groups—as more women, minorities, and recent immigrants enter the world of high technology, for example—and as technical communication seeks to reach a global market, you are much more likely to have team members with backgrounds, habits, and values quite different from your own. Team efficiency demands that you listen carefully and react sensitively to contributions that may seem odd at first, perhaps because they are delivered in language that is unfamiliar to you or with emotions that seem unreasonable to you.

For example, you may be surprised to hear a team member describe a project for building a waste incinerator on a rural site as "unconscionable." Such language may appear to be out of place on a team whose main considerations are economic and technical. But what if you learn that the speaker grew up near the proposed site, had a family that had lived there for generations and formed a deep attachment to the site in question? When such emotions are involved, it is very difficult to look at decisions with the coldly rational eye of economic and technical considerations. And what if the situation is intensified further by the ethnic background of your colleague? Let's say he is a Native American whose tribal traditions teach that the site in question has sacred value. He may be struggling with problems of identity, trying to reconcile his loyalty to his home community with his determination to be an effective scientist or technologist. If you and his other colleagues treat the matter of site location insensitively, you may become a focal point for his conflicted feelings, the very symbol of the things he has been taught all his life to avoid. One day you may find that he rejects everything you suggest for no clear reason. The project grinds to a halt.

To take another example, imagine that one day during a coffee break, a male member of a five-person team with only one female member tells a joke about the length of women's skirts. When work resumes, he notices that the woman team member has fewer contributions to make. "Is there a problem?" he asks. "No problem," she says. But in fact she is feeling that her "otherness" in the group, obvious to everyone from the start, is now painfully apparent. Her gender has been brought to the forefront. She might have said, "I don't appreciate sexist jokes," but that would have brought even more attention to her. In a field like engineering, where women make up a small minority of the work force, her situation would be all the worse. She's supposed to be "one of the boys." Now she feels alienated. Her effectiveness is diminished and her self-confidence shaken, all because of a "harmless" little joke. It wasn't even dirty! Yet in some contexts, it could be considered a serious ethical violation, an instance of sexual harassment. Policies on harassment often prohibit actions that cause people to feel singled out because of their gender. Even if such action doesn't violate policy, however, it does hurt. And it violates most codes of ethics, even one as seemingly cold and rational as utilitarianism, which urges that we judge the consequences of our actions by how well they further the interests of all involved, always aiming toward the greatest good for the greatest number. It is never in the best psychological interest of a person to be treated as a social outcast or to be made to feel uneasy.

Ethics demands that we treat fellow workers with respect, even if we have trouble sympathizing with their perspectives. On communication teams, respect translates into clear behaviors. During the process of developing information products:

- Listen carefully to what others say, and try to put yourself in their place.
- If you can't sympathize with other team members, ask them to help explain their views on the conflicts you are experiencing.
- If the conflict still remains, seek outside arbitration, and abide faithfully by the results and agreements that come from this arbitration.

Rather than trying always to win, to get what you initially think is best, try achieving *consensus* among your partners, finding ways to meet as many people's goals as possible. You do not always have to agree with other people, but you should always treat their position with respect, consider it fairly, and reject it only after careful consideration and negotiation.

Even when the development of your information product is complete, you still have some ethical obligations to consider. You need to make sure that the right people get credit for their contributions to the team. When you write the preface or acknowledgments in a technical document, for example, or when you report to supervisors on your activities, always credit the people who helped put the document together—and not only your team members, but also those who served as sources of information, and support people such as editors, librarians, artists, and printers. Showing respect in these ways is not only the right thing to do; it is also quite practical. It paves the way for effective cooperation on the next project.

Issues of Access and Integrity in the Context of Use

As you begin to think beyond the context of production, about what users will do with your information products, new ethical and political concerns come to light. Again we are dealing with people's obligations and rights, but the focus shifts somewhat. In the context of use, we focus on issues of access and integrity, our client's right to know and to act on information, and our own obligation to provide reliable information in an appropriate form.

In a modern democracy, reliable information and the freedom to act on that information are directly connected to political power. Whoever has access to the best information has the power to act effectively or to control action. Technical communication—the purpose of which is to provide access to up-to-date information for action and decision making—therefore carries a large ethical and political burden. The stakes are high for all participants. If the aim of democracy is to distribute opportunities for exercising power widely among the people, then it stands to reason that open access to information is the best policy.

But life in a democracy is not so simple. Since citizens have the right to pursue their own interests, they may need to limit access to information at times to advance these interests. If, for example, I have a design for a new microchip that will make millions of dollars for my company, don't I have a right to keep the information about the design secret until the product is released? Most people would say yes; the market system would fail if producers were not granted this kind of competitive edge.

Under certain conditions, however, there may be limits on my freedom to guard product information, designs, or research. In times of war or other emergencies, for instance, the government may decide that all information about high technology is pertinent to national security, so I would have to grant access, at least to qualified government agents.

Utilitarian Ethics

In their attempts to resolve conflicts between public and private interests, technical decision makers generally rely on utilitarian ethics. The British philosopher Peter Singer defines utilitarianism as the philosophy that requires a person to "choose the course of action that has the best consequences, on balance, for all affected."[4]

Utilitarianism is certainly not the only ethical perspective operating in modern democracies. Alternatives include consensual ethics, which defines the good as the solution that everyone in a group can agree on at a given moment in time, and the ethics of inherent value, which argues that goodness is universal and that only one choice can ever be viewed as right, once the blinders of individual interest and political bias are removed.

But utilitarianism dominates the technical scene, partly because technical information lends itself to making choices about costs and benefits. The best choice, according to utilitarianism, is an action that furthers not only the actor's own interests, but everyone else's as well—the classic "win–win" situation. Unfortunately, win–win situations are rare in the competitive marketplace. Instead, most ethical players seek to minimize suffering or avoid erecting barriers for others as they act to achieve their own desires.

Ethical Rationality

Even with these seemingly simple goals in mind, making ethical decisions is not easy. There are no simple formulas for calculating ways to minimize the hurt or maximize benefits for people who are interested in our information and our actions. Instead, we must reason through our decisions about proper action and very often debate the advantages and disadvantages of different courses of action. In these activities, we call on rationality.

Rationality is more than simple logic. As every reader of science fiction knows, "logical" behavior does not necessarily lead to ethical actions. It may seem entirely logical, for example, to slaughter a third of the world's population to relieve the pressure on strained resources. But even if we set aside the premise that all human life is inherently valuable, the policy is not ethical. It is obviously not in the best interests of the one third that gets slaughtered, nor even in the best interests of the slaughterers, who would probably suffer guilt

---

4. Peter Singer, *Practical Ethics*, 2nd edition. (Cambridge, England: Cambridge University Press, 1993) 13.

or some other form of psychological pain from the action. By almost any code of behavior, wanton mass murder is wrong.

To be rational, a decision must consider certain underlying principles for human action—that life is inherently valuable, for example, or that all citizens have a right to pursue happiness. Rationality also requires a method for arriving at decisions about the fitness of principles and policies so that people can adjust their form of personal and political governance to accommodate historical change. The Constitution of the United States is a rational ethical system, as are the by laws of an organization or the policies and procedures of a company. These systems not only tell us how to act but also provide means for changing the standards of action.

Figure 2.1 shows a model for making rational, ethical decisions about technical communication issues. Each element in the model represents a phase in the process and a problem to be solved:[5]

- **Data:** In technical decision making, the first question is always the same: Where can we get the most reliable data? In every instance, we rely on relevant facts that we or others have gathered through experimentation or by observing related actions in the past.
- **Information:** Next comes the analytical problem. How do we select among the facts and arrange them into relevant categories?
- **Knowledge:** This presents a synthesizing problem. What are the possible solutions (options for action) based on the available information? We can state each option for action as a position or a claim, a reasoned argument that, given certain information, we should act in a certain way.
- **Backing and Warrants:** This is really a constellation of problems and questions. Responses to these questions constitute a rational, ethical filter for selecting final options from among the many possibilities generated in the knowledge-making part of the process. To what authorities does the decision have to answer? What is the backing—principles, laws, precedents, and policies—of those authorities? What warrants—values, rules, and assumptions—have they used to make past decisions? Which of those decisions are relevant to this one?
- **Decision:** This is a judgment problem. Ultimately, we must decide which position is best, technically and ethically, drawing on all of the solutions we have recognized up to this point.

---

5. This decision-making model is based on philosopher Stephen Toulmin's description of rationality and on the concept of information development summarized nicely in *Knowledge-Based Systems: A Manager's Perspective* by G. Steven Tuthill and Susan T. Levy (Blue Ridge Summit, PA: Tab Books, 1989), 32. For a clear account of Toulminian rationality, see Stephen Toulmin, Richard Rieke, and Allan Janik, *An Introduction to Reasoning,* 2nd edition (New York: Macmillan, 1984). We developed the use of the model to describe decision-making processes in technical communication in collaboration with Professor Patricia Carlson of Rose Hulman Institute of Technology and Dr. Susan Dressel, formerly of Los Alamos National Laboratory.

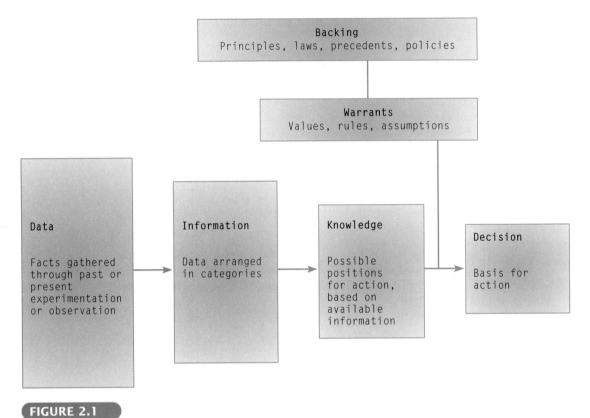

An ethical decision-making model.

The most rational and ethical way to use the model is to solve each problem in order: Keep your mind open until all the information is in; analyze and weigh the information; consider many possible options for action as they relate to the precedents, policies, rules, and principles of relevant authorities; and finally decide on a formal position from which to act.

In real life, however, decision makers often skip the first set of problems, make hasty decisions based on scant information, and then scramble to substantiate them by digging up appropriate warrants and supporting information. In his book *Who Will Tell the People?*, political journalist William Greider suggests that this kind of rationality is not really legitimate; it is pseudo-rationality. He argues that it is unethical to make judgments before gathering data and information. If you get the steps out of order, he says, you are more likely to overlook, ignore, or distort evidence that goes against your position. In Washington, D.C., information service organizations working for lobbyists, and even senators and representatives, often hunt up facts that support their clients' positions, ignoring or subverting any information they find that undermines their position.

This sort of *rationalization* is very different from true *rationality*. Rationalization mimics rationality but only makes excuses for action; true rationality gives good reasons for action.

Ethical technical communication depends on our preserving—and observing—this difference. In creating information products, we ought to work from the most complete data set available, developing information carefully and fully in the research phase. When we present information in documents such as recommendation reports and proposals, we should develop positions and claims with the idea that many options for action have good claims on the attention of our information users. Although we may begin with a favorite position, we should strive for objectivity. Even objective writers cannot entirely eliminate bias when human judgment is involved, but we can present our audience with clear reasoning about the many choices involved. The decisions we recommend should therefore consider as many options as possible and make a strong case that the final recommendation has the greatest benefits for the most people and creates the fewest disadvantages.

Values in Technical Communication

Every different context of production and context of use has a set of values that provides the backing and warrants for making good decisions. In addition, some ethicists argue that technical communication has its own set of values that distinguish professional communicators from people who fill other roles—from managers, for example, or politicians. One possible set of values for technical communicators appears in Table 2.4. As you review the cases in the following sections, and as you work through the exercises at the end of the chapter, ask yourself how well this set of values conforms to your own and how well it covers the behavior you would expect from a technical communicator. Where do our values of openness and integrity fit into this scheme, for example?

Two Famous Cases

It is a sad truth that, despite the ethical integrity of individual authors and social groups—and despite a strong effort to bring the best technical information into consideration when important decisions are being made—the world of technological action continues to be troubled by miscommunication, misunderstanding, and bad decisions. Far too often, negligence and carelessness lead to actions with dire consequences.

No doubt, technology is a risky business. Accidents cannot be helped; they are beyond everyone's control, and everyone is a victim in an accident. But when something goes wrong because of poor judgment, bad information, a

TABLE 2.4

Values for technical communicators.

Value	Behavior
Honesty	Tell the truth.
Legality	Obey the law (in matters of copyright, for example).
Privacy	Reveal confidential information only with permission.
Quality	Create products that best serve the user.
Teamwork	Work together to achieve win–win goals with fellow workers.
Avoiding Conflict of Interest	Remain loyal and observe fair play.
Cultural Sensitivity	Recognize the values of diversity not only in markets but also in the workplace by showing respect for fellow workers and products users.
Social Responsibility	Preserve and protect the public good.
Professional Growth	Develop and maintain a high level of personal skill.
Advancing the Profession	Do whatever possible to represent the whole profession and contribute to its growth and improvement.

Source: Based on Lori Allen and Dan Voss, *Ethics in Technical Communication: Shades of Gray,* New York: Wiley, 1997, 38.

poor decision, or faulty communication, we enter the realm of cause and blame, liability and responsibility—the realm of ethics.

Two prominent events in recent history—the partial core meltdown at the Three Mile Island nuclear facility in 1979, and the 1986 explosion of the space shuttle *Challenger*—have been described as "accidents" by well-meaning analysts and news reporters. But recent analyses suggest that, instead of accidents, these cases were disasters in the field of communication ethics. The Nuclear Regulatory Commission's report blamed a "breakdown of communications" and "crucial misunderstanding" for the event at Three Mile Island, and the Presidential Commission's report on the space shuttle disaster pointed to "a serious flaw in the decision-making process."[6]

6. See especially Carl G. Herndl, Barbara A. Fennell, and Carolyn R. Miller, "Understanding Failures in Organizational Discourse: The Accident at Three Mile Island and the Shuttle *Challenger* Disaster," *Textual Dynamics of the Professions: Historical and Contemporary Studies of Writing in Professional Communities,* ed. Charles Bazerman and James Paradis (Madison: University of Wisconsin Press, 1991), 279–306. The quotations that appear in this chapter and the sample texts are taken from this article. Herndl, Fennell, and Miller in turn build upon the work of J. C. Mathes, especially his analytical report *Three Mile Island: The Management Communication Failure* (Ann Arbor: College of Engineering, University of Michigan, 1986). A summary of Mathes's findings appears in *Technical Communication and Ethics,* ed. R. John Brockmann and Fern Rook (Washington, DC: Society for Technical Communication, 1989). Whereas Mathes blames the effects of the disaster on miscommunication in an analysis that emphasizes forms and media of expression, Herndl, Fennell, and Miller emphasize social context and argumentative structure in their analysis. We borrow from both of

case 2.1 THREE MILE ISLAND

In testimony to the Presidential Commission, it became clear that well before the incident at Three Mile Island, the engineering department at Babcock & Wilcox (the builder of the reactor) had recommended changes in the instructions on how to operate the reactor. The problem centered on the high-pressure injection (HPI) system, which injects water into the core when pressure falls below an acceptable limit. If operators disable the automatic HPI too quickly or bypass it, the core is "uncovered" and liable to meltdown. Several months before the Three Mile Island accident, an accident had occurred at another plant, in Toledo. Operators at that plant, following their training, disabled the HPI system before the system leak had been isolated. They avoided disaster only because the Toledo facility was running at 10 percent power. Realizing the implications of this event, J. J. Kelly of the engineering division of Babcock & Wilcox wrote a memo calling for a change in customer training so that operators would learn not to shut down the HPI too early. The memo appears in Sample 2.1.

The tentativeness of the original memo suggests that Mr. Kelly knew he was on shaky political ground. Notice especially the sentence "Since there are accidents which require the continuous operation of the high pressure injection system, I wonder what guidance, if any, we should be giving to our customers on when they can safely shut the system down following an accident?" But then he goes on to make very definite recommendations in strong imperative sentences (a and b).

His original concerns about company politics were apparently well founded. A manager in the Nuclear Services division, who was in charge of customer training, cast doubt on the procedural recommendations made by Mr. Kelly and even belittled his technical advice. As a result, the recommendations were ignored.

When it became clear that no action was being taken on the problem, another (apparently higher-ranking) engineer, Mr. Bert Dunn, wrote a memo. His memo, which appears in Sample 2.2, states the problem and the recommendations in no uncertain terms. He sent yet another memo later. But the people in Nuclear Services appeared to have held their ground in spite of these repeated warnings, and no changes in operating procedures were passed on to the plant operators. Unable to reconcile the advice coming from the engineers with that coming from his own division, a manager in the Nuclear Services branch recommended that the matter be studied further before the changes in training were made.

---

these approaches in the brief analysis given here. We have also been influenced here and throughout this chapter by Jack Griffin's essay, "When Do Rhetorical Choices Become Ethical Choices?" which is reprinted in the Brockmann and Rook collection, *Technical Communication and Ethics*. For other good discussions on the *Challenger* disaster, see the recommended reading list at the end of the chapter.

The Babcock & Wilcox Company
Power Generation Group

To: Distribution
From: J. J. Kelly, Plant Integration
Cust.   Generic
Subj: Customer Guidance on High Pressure Injection Operation
Date: November 1, 1977

DISTRIBUTION

B. A. Karrasch   D. W. LaBelle
E. W. Swanson   N. S. Elliott
R. J. Finnin   D. F. Hallman
B. M. Dunn

Two recent events at the Toledo site have pointed out that perhaps we are not giving our customers enough guidance on the operation of the high pressure injection system. On September 24, 1977, after depressurizing due to a stuck open electromagnetic relief valve, high pressure injection was automatically initiated. The operator stopped HPI when pressurizer level began to recover, without regard to primary pressure. As a result, the transient continued on with boiling in the RCS, etc. In a similar occurrence on October 23, 1977, the operator bypassed high pressure injection to prevent initiation, even though reactor coolant system pressure went below the actuation point.

Since there are accidents which require the continuous operation of the high pressure injection system, I wonder what guidance, if any, we should be giving to our customers on when they can safely shut the system down following an accident? I recommend the following guidelines be sent:

a) Do not bypass or otherwise prevent the actuation of high/low pressure injection under <u>any</u> conditions except a normal, controlled plant shutdown.

b) Once high/low pressure injection is initiated, do not stop it unless: T_{ave} is stable or decreasing <u>and</u> pressurizer level is increasing <u>and</u> primary pressure is at least 1600 PSIG and increasing.

I would appreciate your thoughts on this subject.

SAMPLE 2.1   Kelly's original memo

The company eventually adopted the changes, but not until after the incident at Three Mile Island on March 28, 1979. It fulfilled the worst fears of engineers Kelly and Dunn. A partial meltdown occurred when the operators did not use the HPI system correctly. Before the Presidential Commission, Mr. Dunn testified, "Had my instructions been followed at [Three Mile Island], we would not have had core damage; we would have had a minor incident."

The Babcock & Wilcox Company
Power Generation Group

To: Jim Taylor, Manager, Licensing
From: Bert M. Dunn, Manager, ECGS Analysis (2138)
Cust.:
Subj: Operator Interruption of High Pressure Injection
Date: February 9, 1978

This memo addresses a serious concern within ECCS Analysis about the potential for operator action to terminate high pressure injections following the initial stage of a LOCA. Successful ECCS operation during small breaks depends on the accumulated reactor coolant system inventory as well as the ECCS injection rate. As such, it is mandatory that full injection flow be maintained from the point of emergency safety features actuation system (ESFAS) actuation until the high pressure injection rate can fully compensate for the reactor heat load. As the injection rate depends on the reactor coolant system pressure, the time at which a compensating match-up occurs is variable and cannot be specified as a fixed number. It is quite possible, for example, that the high pressure injections may successfully match up with all heat sources at time t and that due to system pressurization be inadequate at some later time t 2.

The direct concern here rose out of the recent incident at Toledo. During the accident the operator terminated high pressure injection due to an apparent system recovery indicated by high level within the pressurizer. This action would have been acceptable only after the primary system had been in a subcooled state. Analysis of the data from the transient currently indicates that the system was in a two-phase state and as such did not contain sufficient capacity to allow high pressure injection termination. This became evident at some 20 to 30 minutes following termination of the injection when the pressurizer level again collapsed and injection had to be reinitiated. During the 20 to 30 minutes of noninjection flow they were continuously losing important fluid inventory even though the pressurizer was at an extremely low power and extremely low burnup. Had this event occurred in a reactor at full power with other than insignificant burnup it is quite possible, perhaps probable, that core uncovery and possible fuel damage would have resulted.

The incident points out that we have not supplied sufficient information to reactor operators in the area of recovery from LOCA. The following rule is based on an attempt to allow termination of high pressure injection only at a time when the reactor coolant system is in a subcooled state and the pressurizer is indicating at least a normal level for small breaks. Such conditions guarantee full system capacity and thus assure that during any follow on transient would be no worse

SAMPLE 2.2   Dunn's first memo

continued

than the initial accident. I, therefore, recommend that operating procedures be written to allow for termination of high pressure injection under the following two conditions only:

1. Low pressure injection has been actuated and is flowing at a rate in excess of the high pressure injection capability and that situation has been stable for a period of time (10 minutes).

2. System pressure has recovered to normal operating pressure (2200 or 2250 psig) and system temperature within the hot leg is less than or equal to the normal operating condition (605°F or 630°F).

I believe this is a very serious matter and deserves our prompt attention and correction.

SAMPLE 2.2 continued

case 2.2 THE SPACE SHUTTLE *CHALLENGER*

The similarities between the Three Mile Island case and the *Challenger* case are remarkable. In both disasters, differences in judgment and a political contest between internal groups—not any absence of rationality or information—caused a failure to make the right decision. In the Three Mile Island case, the Nuclear Services division at Babcock & Wilcox failed to make changes in its training regimen despite recommendations from colleagues in another department. In the *Challenger* case, a similar lack of coordination between technical staff and management had even more disastrous results.

The fatal problem with the *Challenger* was caused by a failure of O-rings used as seals in the spacecraft. NASA knew about the faulty parts well before the explosion on January 28, 1986, as the memo in Sample 2.3 (written by a junior NASA analyst) indicates, but NASA's technical experts, such as the senior engineer who authored the memo in Sample 2.4, were not sure of the extent of the problem. They therefore relied on the recommendation of the manufacturer of the rocket motors, Morton Thiokol, Inc., of Utah.

On the day before the ill-fated launch, the disagreement between engineers and managers in the parent company became clear in a teleconference involving officials at NASA and Morton Thiokol. Throughout the conference, the Thiokol engineers argued that the cold weather forecast for the launch could make the problem with the O-rings even worse than it already was. But management, citing launches in similar conditions in the past, apparently believed that the engineers were being overly cautious.

Their differences arose from their different kinds of experiences, which translated into different kinds of warrants. The engineers argued their position

NASA
National Aeronautics and Space Administration

7/23/85

```
TO:      BRC/M. Mann
FROM:    BRC/R. Cook
SUBJECT: Problem with SRB Seals
```

Earlier this week you asked me to investigate reported problems with the charring of seals between SRB motor segments during flight operations. Discussions with program engineers show this to be a potentially major problem affecting both flight safety and program costs.

Presently three seals between SRB segments use double O-rings sealed with putty. In recent Shuttle flights, charring of these rings has occurred. The O-rings are designed so that if one fails, the other will hold against the pressure of firing. However, at least in the joint between the nozzle and the aft segment, not only has the first O-ring been destroyed, but the second has been partially eaten away.

Engineers have not yet determined the cause of the problem. Candidates include the use of a new type of putty (the putty formerly in use was removed from the market by EPA because it contained asbestos), failure of the second ring to slip into the groove which must engage it for it to work properly, or new, and as yet unidentified, assembly procedures at Thiokol. MSC is trying to identify the cause of the problem, including on-site investigation at Thiokol, and OSF hopes to have some results from their analysis within 30 days. There is little question, however, that flight safety has been and is still being compromised by potential failure of the seals, and it is acknowledged that failure during launch would certainly be catastrophic. There is also indication that staff personnel knew of this problem some time in advance of management's becoming apprised of what was going on.

The potential impact of the problem depends on the as yet undiscovered cause. If the cause is minor, there would be little or no impact on budget or flight rate. A worst case scenario, however, would lead to the suspension of Shuttle flights, redesign of the SRB, and scrapping of existing stockpiled hardware. The impact on the FY 1987-8 budget could be immense.

It should be pointed out that Code M management is viewing the situation with the utmost seriousness. From a budgetary standpoint, I would think that any NASA budget submitted this year for FY 1987 and beyond should certainly be based on a reliable judgment as to the cause of the SRB seal problem and a corresponding decision as to budgetary action needed to provide for its solution.

> Richard C. Cook
> Program Analyst
>
> Michael B. Mann
> Chief, STS Resources Analysis Branch
>
> Gary B. Allison
> Director, Resources Analysis Division
>
> Tom Newman
> Comptroller

SAMPLE 2.3   Memo on the O-ring problem by a junior analyst at NASA

NASA
National Aeronautics and Space Administration

Jul 17 1985

MPS.

TO: M/Associate Administrator for Space Flight

FROM: MPS/Irv Davids

SUBJECT: Case to Case and Nozzle to Case "O" Ring Seal Erosion

As a result of the problems being incurred during flight on both case to case and nozzle to case "O" ring erosion, Mr. Hanby and I visited MSFC on July 11, 1985, to discuss this issue with both project and S&E personnel. Following are some important factors concerning these problems:

A. Nozzle to Case "O" Ring Erosion

There have been twelve (12) instances during flight where there has been some primary "O" ring erosion. In one specific case there was also erosion of the secondary "O" ring seal. There were two (2) primary "O" ring seals that were heat affected (no erosion) and two (2) cases in which soot blew by the primary seals.

The prime suspect as the cause for the erosion on the primary "O" ring seals is the type of putty used. It is Thiokol's position that during assembly, leak check, or ignition, a hole can be formed through the putty which initiates "O" ring erosion due to a jetting effect. It is important to note that after STS-10, the manufacturer of the putty went out of business and a new putty manufacturer was contracted. The new putty is believed to be more susceptible to environmental effects such as moisture, which makes the putty more tacky.

There are various options being considered such as removal of putty, varying the putty configuration to prevent the jetting effect, use of a putty made by a Canadian manufacturer which includes asbestos, and various combinations of putty and grease. Thermal analysis and/or tests are underway to assess these options.

Thiokol is seriously considering the deletion of putty on the QM-S nozzle/case joint since they believe the putty is the prime cause of the erosion. A decision on this change is planned to be made this week. I have reservations about doing it, considering the significance of the QM-S firing in qualifying the FWC for flight.

It is important to note that the cause and effect of the putty varies. There are some MSFC personnel who are convinced that the holes in the putty are the source of the problem but feel that it may be a reverse effect in that the hot gases may be leaking through the seal and causing the hole track in the putty.

Considering the fact that there doesn't appear to be a validated resolution as to the effect of the putty, I would certainly question the wisdom of removing it on QM-S.

B. Case to Case "O" Ring Erosion

There have been five (5) occurrences during flight where there was primary field joint "O" ring erosion. There was one case where the secondary "O" ring was heat affected with no erosion. The erosion with the field joint primary "O" rings is considered by some to be more critical than the nozzle joint due to the fact that during the pressure build up on the primary "O" ring the unpressurized field joint secondary seal unseats due to joint rotation.

SAMPLE 2.4   Memo on the O-ring problem by a senior engineer at NASA

The problem with the unseating of the secondary "O" ring during the joint rotation has been known for quite some time. In order to eliminate this problem on the FWC field joints a capture feature was designed which prevents the secondary seal from lifting off. During our discussions on this issue with MSFC, an action was assigned for them to identify the timing associated with the unseating of the secondary "O" ring and the seating of the primary "O" ring during rotation. How long it takes the secondary "O" ring to lift off during rotation and when in the pressure cycle it lifts are key factors in the determination of its criticality.

The present consensus is that if the primary "O" ring seats during ignition, and subsequently fails, the unseated "O" ring will not serve its intended purpose as a redundant seal. However, redundancy does not exist during the ignition cycle, which is the most critical time.

It is recommended that we arrange for MSFC to provide an overall briefing to you on the SRM "O" rings, including failure history, current status, and options for correcting the problems.

Irving Davids

cc:
M/Mr. Weeks
M/Mr. Hamby
ML/Mr. Harrington
MP/Mr. Winterhalter

SAMPLE 2.4 continued

based on their experience with handling the damaged parts; the managers argued their position based on their experience with flight decisions and program needs. Ultimately, the decision fell to the "decision makers," the managers, who made the wrong recommendation.

The Moral of the Two Cases

We have given only the briefest outline of what happened in each case, but even this sketch is enough to illustrate the complexity of decisions and communication in a modern technological and administrative culture. As events unfold, it is not always easy to assume responsibility, even when you know you should. You can tiptoe into a tense political situation (like Mr. Kelly in Sample 2.1) or speak boldly and plainly (like the authors in Samples 2.2 and 2.3) and still find yourself unable to gain the political edge needed to do the right thing.

Once a disaster occurs, it's notoriously difficult to lay blame on any individual for any particular action because so many people must cooperate to make anything happen, and actions are so complex. Again, everyone has an explanation, complete with information and warrants. Ultimately, everything comes down to judgment. And we must recognize that, whereas cases of bad judgment make the news, cases involving good judgment occur every day.

Even in these cases, we can find much to admire in the actions of people like Kelly and Dunn of Babcock & Wilcox, Cook of NASA, and Roger Boisjoly, the leading expert on seals at Morton Thiokol. Cook and Boisjoly argued vigorously against the recommendation that the launch of the *Challenger* proceed, only to be overruled in the end by Thiokol management.

Though it is tempting to grow cynical when we learn that heroic efforts of this sort fail to influence the process of decision making, we should instead see these men's efforts to communicate their knowledge as strongly and as effectively as possible as a model for communication that is ethically motivated as well as technically sound.

The Wider Context: Dealing with Cultural Differences in International Communication

You might think that rhetorical situations like those faced by the engineers involved in the Three Mile Island and the *Challenger* episodes could hardly have been more difficult and complex. But what if the projects had been located in other countries? What if the workers spoke a different language? What if Morton Thiokol had been a German or Japanese contractor with NASA?

Such variations are possible, and even likely, in the world of high technology with its global reach. With every expansion of technology into the international marketplace, the need for cultural sensitivity becomes a greater concern for technical communicators.

In recent years, researchers have added greatly to our understanding of international communication, but the topic still remains controversial. In fact, the very possibility of international communication has been disputed. Even if you know the language of another country and have a sense of its history, you may completely miss the subtleties of culture. The cultural differences between the Eastern and Western, or the Northern and Southern, hemispheres present technical communicators with their greatest challenge. Edward T. Hall, one of the world's most distinguished cultural anthropologists, goes so far as to say, "Any westerner who was raised outside the Far East and claims he really understands and can communicate with either the Chinese or the Japanese is deluding himself."[7]

With over 5000 languages actively used in the world, several hundred of which may be used in science and technology, you can't expect American English to do all your work for you. Nor can you expect much help from general guidebooks such as Roger E. Axtell's well-known *Do's and Taboos around the World*, and the companion volumes *The Do's and Taboos of Body Language around*

7. Edward T. Hall, *Beyond Culture* (New York: Anchor, 1976), 2.

the World and *Do's and Taboos of Using English around the World.* From books like these we learn, for example, that

- the meaning of simple signals varies widely among cultures: hand gestures that seem innocent or positive, even something as innocuous as an open-palmed wave of the hand, may be interpreted as obscene in certain cultures;
- colors which indicate one thing in one culture may mean something else in another: red may mean caution or danger in the West, but not in the Far East, where it suggests joy, rebirth, or festivity;
- the same words may mean something entirely different as you move from one English-speaking country to another: "napkin" means "diaper" in England.

Even guidebooks of several hundred pages barely scratch the surface of the many cultural differences an international communicator faces. At best, these interesting books can show you what you're up against in attempting to develop worldwide information products.

Still, if you want to open the widest possible access to your information, you will inevitably seek an international audience. Just putting up a home-page on the World Wide Web gives you a global reach, which might lead to opportunities for business and study that you had not anticipated. So, lacking an anthropologist's or experienced translator's cultural understanding of the various readers you might attract, what can you do to improve your chances of effective communication?

You can solve the problem partly by effectively testing an early version of your information product. The graphics expert and consultant in international communication William Horton says, "The only insurance against cultural miscommunication is testing with expected viewers."[8] However, testing can guarantee success only for a very localized audience, only the people best represented by the tested subjects. Remember that within cultures, individual people can vary greatly.

So what can you do in the early stages of product development, as you plan and write the first versions of your document? We suggest that, before you write one word or plan one graphic, you adjust your audience-action analysis to determine how far you need to go in fitting your information product to an international market. In Table 2.5, we give three possible approaches—localization, internationalization, and globalization—as defined by Nancy Hoft in her book *International Technical Communication.* Each approach has its own requirements that influence how you plan and develop your communication.

Each approach has advantages and disadvantages. Localized products improve sales, overcome cultural differences and product resistance, and help the producer quickly gain a foothold in new markets, but they are expensive and slow to develop. Internationalized products are cheaper and faster to produce and distribute but are not as well suited to particular niches in the market.

8. William Horton, *Illustrating Computer Documentation* (New York: Wiley, 1991), 210.

TABLE 2.5

Approaches to international communication.

Approach	Definition	Requirements
localization	creating or adapting an information product for use in a specific country or specific market	For each different context, a new version of the product accommodates translation into a new language and changes in currency, date, and time formats; in addition, it takes account of deeper cultural characteristics such as learning styles, gender and class sensibility, and communication taboos.
internationalization	designing an information product to be easily localized for export anywhere in the world	The product consists of two kinds of information: core information and international variables. The core information must be translated but otherwise remains relatively unchanged in each different version, while the variables are localized.
globalization	creating an information product to be used in many cultural contexts without modification	The product, usually a short communication such as an airline safety card, uses globally understood signs and images to relay simple messages.

Source: Based on Nancy Hoft, *International Technical Communication* (New York: Wiley, 1995), 11–31.

Globalized products are ready for immediate distribution anywhere but are quite limited in what they can communicate.

The trend among technical communicators these days clearly favors internationalization, especially since the process of internationalizing has some benefits beyond marketing considerations. Identifying the core information of a document has rhetorical benefits. It helps the producer focus on the most essential points and highlight these points with easily translatable writing and the clearest possible, most globalized graphics available. For example, in a computer manual, the core information would be the steps for actually performing tasks—setting the computer up, creating a file, saving the file, etc.—while the variables would involve motivational material such as photographs showing satisfied users and prefaces that promote the product's "user-friendliness." By attending to the core information first, the producer is more likely to create an effective, action-oriented document that helps the user get the job done. This approach fits nicely with the more general CORE method of producing documents that we develop in subsequent chapters.

At the first stage of product development—the audience-action analysis—your main task in internationalizing your information product is to locate areas of cultural sensitivity and possible bias. Whenever possible, plan to make the core information free of detectable bias.

GUIDELINES AT A GLANCE 2

Adding ethical, political, and cultural filters to the audience-action analysis and profile

Sample questions for analyzing the context of production:
- What are the responsibilities of each team member in the production process?
- How is each team member held accountable?
- Do any potential conflicts threaten to emerge, and how will these be managed?
- What contracts, codes, or other governing devices are in effect?

Sample questions for analyzing the context of use:
- What will users do with your information product? What decisions will they make; what actions will they take? How will you ensure that the information you provide gives a sufficient basis for these decisions and actions?
- What ethical obligations do you as a producer have to the user of the information product, and how will these obligations affect the presentation of information?
- Will you be required to make recommendations or give guidelines for action? How will you account for alternatives to the recommended actions?
- Could anyone's reputation be damaged or safety threatened as a result of the information you reveal, the recommendations you make, or the instructions you give? Can you see ways to craft your recommendations so that you minimize the potential for suffering and maximize the benefits for all concerned?
- What are the cultural characteristics of your chosen audience? Will the market include international users or users whose ethnicity is different from your own? What steps will you take to limit the possibility of cultural bias causing ineffective or offensive communication?

Writing the profile:
- Paragraph 1: Discuss the most important ethical and political issues in the context of production (based on answers to questions in the analysis), and state how you propose to deal with these issues.
- Paragraph 2: Discuss the important ethical, political, and cultural issues for the context of use, and state how you propose to deal with these issues.

Expanding the Audience-Action Analysis and Profile

To create information products sensitive to the widest possible context of use, you should perform ethical, political, and cultural analyses in the pre-writing phase of product development. To analyze your audience and the actions they will take using the information you provide, begin with the questions in Guidelines at a Glance 2. Do not feel limited to these questions. Make your analysis as complete as possible.

Then follow the directions for writing an extension of the audience-action profile we first described in Chapter 1 (pages 13–15). An example of such an extension, which builds on Sample 1.1 from Chapter 1 (page 16), appears in Sample 2.5.

The team that created Version 1.2 of ClientBase has been reassembled for the development of Version 2.0. In that project, the engineering representative (Jones) was at first hesitant to make changes suggested by the technical writers (Smith and Vallejo) in the user interface of the program itself. For efficiency's sake, Jones wanted to keep programming changes to a minimum and consign rhetorical improvements to the print document and online help sessions. But user confusion in preliminary testing convinced Jones of the need to keep the interface consistent in approach with the documentation. We are glad to have Jones back on the team so that we do not have to fight this battle again. The team is fully integrated and ready to work together. The only possible problem is Vallejo's commitment to continue intensive work on the QuestTime project during the beginning of the ClientBase revision. She has agreed to devote Tuesdays and Wednesdays to the ClientBase project except during July when all her time must be given to QuestTime. We will request additional staff during that month to compensate for Vallejo's absence. Smith will take training responsibilities during that time.

In the new version, we will try to solve some ethical problems we encountered in Version 1.2. We were convinced by our marketing representative (a visiting member of the team) to "go easy" on the number of warnings and cautions we included in the instructions. The argument was that too many "negative messages" would hurt sales. But our surveys revealed a high level of user frustration over not being warned of the sensitivity of the program during certain data entry routines. Unwary users who had lost data felt betrayed by the documentation. Our new approach, oriented toward minimal documentation with maximum troubleshooting instructions, should solve the problem for users. But we fear a hard sell will be needed for marketing. So we are requesting that a marketing representative be added to our team early on to gain an insider's view of our approach and to discover a marketing strategy that makes the most of the approach.

To ensure effective communication among British users, we will take the same approach we took in Version 1.2, creating a "B" version for use in Great Britain. After developing the American version, the team will create Version 2.0B by screening the core information to be sure no Americanisms exist in the instructions and by changing date, time, and currency formats in both the interface and the documentation. We are also aware that, if Version 2.0 succeeds in our

SAMPLE 2.5 Extension of a draft for an audience-action profile (building on Sample 1.1 in Chapter 1, page 16)

American and British markets, we may explore German markets as early as Version 2.1. Initial cultural analysis suggests that mere translation may not be sufficient to satisfy German users. They tend to prefer generous documentation with great technical detail (see Hoft, *International Technical Communication,* New York: Wiley, 1995, pages 336–37). Very likely, our minimalist approach will not work for this market. We may be able to write an expanded and updated version of the documentation for Version 1.2 and then have it translated into German for testing. We recommend that this effort begin as soon as possible after testing the American and British versions. To avoid high costs, we can find German users to test a prototype segment of the German version before completing the full product. Ultimately, however, high costs are inevitable if we intend to enter this market seriously.

SAMPLE 2.5 continued

In your class, discuss the ethical values and policies that you think are most appropriate for doing the work planned for your course. You might begin with the values listed in Table 2.4 on page 36. What values would your class add to or omit from this list? Have some people in the course look up codes of ethics for various professional groups, such as mechanical engineers, biological researchers, and technical communicators. The easiest way to find these codes is to search the World Wide Web (see Chapter 4 for guidance). How well do these codes reflect the values listed in Table 2.4 and the values your class comes up with? On the basis of your discussion, create a code of ethics for individual and group work in your class.

EXERCISE **2.1**

Return to the audience-action profile you wrote for one of the scenarios in Exercise 1.1 on page 21. Write an extension of the profile based on the ethical, political, and cultural considerations given in Guidelines at a Glance 2.

EXERCISE **2.2**

Recommendations for Further Reading

Allen, Lori, and Dan Voss. *Ethics in Technical Communication: Shades of Gray.* New York: Wiley, 1997.

Andrews, Deborah C., ed. *International Dimensions of Technical Communication.* Arlington, VA: Society for Technical Communication, 1996.

Axtell, Roger E. *Do's and Taboos around the World.* New York: Wiley, 1993.

———. *The Do's and Taboos of Body Language around the World.* New York: Wiley, 1990.

———. *Do's and Taboos of Using English around the World.* New York: Wiley, 1995.

Brockmann, R. John, and Fern Rook, eds. *Technical Communication and Ethics.* Washington, DC: Society for Technical Communication, 1989.

Ede, Lisa, and Andrea Lunsford. *Singular Texts/Plural Authors: Perspectives on Collaborative Writing.* Carbondale, IL: Southern Illinois University Press, 1990.

Greider, William. *Who Will Tell the People?: The Betrayal of American Democracy.* New York: Simon and Schuster, 1992.

Hare, A. Paul. *Creativity in Small Groups.* Beverly Hills, CA: Sage, 1982.

Hoft, Nancy L. *International Technical Communication: How to Export Information about High Technology.* New York: Wiley, 1995.

Horton, William. "Global Graphics." In *Illustrating Computer Documentation.* New York: Wiley, 1991, 207–17.

———. "Icons for International Products." In *The Icon Book.* New York: Wiley, 1994, 241–67.

Killingsworth, M. Jimmie, and Betsy G. Jones. "Division of Labor or Integrated Teams: A Crux in the Management of Technical Communication?" *Technical Communication* 36 (1989): 210–221.

Lay, Mary M., and William M. Karis. *Collaborative Writing in Industry: Investigations in Theory and Practice.* Amityville, NY: Baywood, 1991.

Matalene, Carolyn B. *Worlds of Writing: Teaching and Learning in Discourse Communities of Work.* New York: Random House, 1989.

Singer, Peter. *Practical Ethics.* 2nd Edition. Cambridge: Cambridge University Press, 1993.

Toulmin, Stephen, Richard Rieke, and Allan Janik. *An Introduction to Reasoning.* 2nd Edition. New York: Macmillan, 1984.

Zuboff, Shoshana. *In the Age of the Smart Machine: The Future of Work and Power.* New York: Basic, 1987.

Additional Reading for Advanced Research

Blyler, Nancy Roundy, and Charlotte Thralls, eds. *Professional Communication: The Social Prespective.* Newbury Park, CA: Sage, 1993.

Dombrowski, Paul M. "The Lessons of the *Challenger* Investigations." *IEEE Transactions* 34 (Dec. 1991): 211–216.

Hall, Edward T. *Beyond Culture.* New York: Anchor, 1976.

———. *The Silent Language.* New York: Anchor, 1973.

Herndl, Carl G., Barbara A. Fennell, and Carolyn R. Miller, "Understanding Failures in Organizational Discourse: The Accident at Three Mile Island and the Shuttle *Challenger* Disaster." *Textual Dynamics of the Professions: Historical and Contemporary Studies of Writing in Professional Communities.* Ed. Charles Bazerman and James Paradis. Madison, WI: University of Wisconsin Press, 1991: 279–306.

Killingsworth, M. Jimmie, and Michael Gilbertson. *Signs, Genres, and Communities in Technical Communication.* Amityville, NY: Baywood, 1992.

Odell, Lee, and Dixie Goswami, eds. *Writing in Nonacademic Settings.* New York: Guilford, 1985.

Winsor, Dorothy A. "The Construction of Knowledge in Organizations: Asking the Right Questions about the *Challenger.*" *Journal of Business and Technical Communication* 4 (1990): 7–20.

Zappen, James P. "Rhetorical and Technical Communication: An Argument for Historical and Political Pluralism." *Iowa State Journal of Business and Technical Communication* 1 (1987): 29–44.

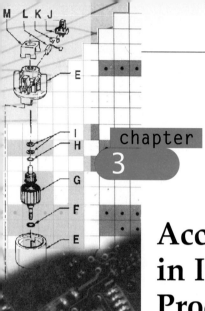

Accommodating the User in Information Products and Processes: The **CORE** Method

CHAPTER OUTLINE
▼▼▼▼▼▼▼▼▼▼▼▼▼▼▼▼▼▼▼▼▼▼▼▼▼▼▼

Anticipating User Expectations in
  Information Products: Media and Genre
Keeping the User Foremost in the Writing
  Process: The CORE Method
Writing the Management Plan

CHAPTER OBJECTIVES
▼▼▼▼▼▼▼▼▼▼▼▼▼▼▼▼▼▼▼▼▼▼▼▼▼▼▼

After you have worked through this chapter,
you should be able to do the following:

- Recognize the different expectations that
  users bring to information products in
  various media and genres
- Use the CORE method to bring trial users
  into the processes of developing
  information products in various media and
  genres
- Write a management plan that coordinates
  production efforts for an individual or
  team project

Chapters 1 and 2 offered methods of broadening your understanding of the contexts within which technical communicators work. If you are producing an information product, you should be able to imagine people receiving it in a context of use, which is embedded in a broader context of social and cultural life.

In this chapter, we ask you to return your attention to the context of production—but keep your imagination working. Now you need to envision the more immediate context for each bit of information you have to impart—the information product itself: the speech, document, or electronic file you create. As you plan the processes for developing the product, you should try to keep the user in mind, first by recognizing how users typically approach information products in different media and genres, and second by following a method that allows you to design the product for users, getting actual user feedback at every stage of production.

This chapter leads you through a discussion of typical expectations associated with the media and genres of technical communication and then recommends a method for meeting those expectations in production. We call the process the **CORE method** because it begins with having you write a short document, the *core* of your communication, a document that will differ according to the genre in which you are working. With each step of product development, you add to this core document, seeking the response of trial users at each stage, until the information product is complete. When applied effectively, the CORE method results in an information product designed for easy use, one that allows users to convert information into action efficiently.

Anticipating User Expectations in Information Products: Media and Genre

In our discussion of cultural contexts in Chapters 1 and 2, we suggested that certain cultures prefer communications to be delivered in certain ways. In fact, technical communication itself evolves only under certain cultural conditions. It typically embodies the learning styles of people in technologically developed countries, people who look favorably upon personal independence, social mobility, economic growth, and technological progress.

Anthropologist Edward T. Hall distinguishes between three general kinds of learning styles: the *formal,* the *informal,* and the *technical.* Usually all three types are present in any culture but are emphasized in varying degrees. Table 3.1 summarizes the differences in the three types of learning by which people receive, assimilate, and act on new information.

Imagine a tennis coach using all three approaches in a single lesson on how to serve. Applying the formal approach, she could say, "No, no, be sure to follow through every time." Switching to informal, she would say, "Just watch me,

TABLE 3.1

Three types of learning.

Type	Learning Style	Typical Attitudes
formal	behavior guided by precepts, warnings, and corrections ("Don't slurp your soup; it's bad manners"—i.e. "bad form.")	context-dependent, conservative, strong interest in maintaining traditions and social order
informal	behavior guided by example along with a few technical details ("Do it the way I do it; hold your spoon closer to your mouth.")	still context-dependent, but more flexible, more oriented toward individual development and social change than formal type, but not as efficient and systematic as the technical type
technical	behavior guided by specific, detailed, step-by-step instructions ("Fill the spoon with soup; bring it slowly toward your mouth; once you make contact with the lips, tip the spoon slightly and spill the soup into the mouth without sucking.")	highly mobile from context to context, oriented toward use by individuals, highly favorable to social change, efficient and systematic

Based on Edward T. Hall, *The Silent Language,* New York: Doubleday, 1973, pages 68–90.

how I follow through without looking up too early." Or applying a fully technical style, she might work through each step from how to lift the racket at a certain angle to the final point of looking up after the follow-through, perhaps even reinforcing the lesson by showing the student a computerized analysis of the serving motion.

Notice that formal and informal learning tend to be completely dependent on certain contextual conditions. The teacher must be present at the time of the lesson, and the student's understanding depends on frequent contact with the teacher. The instructions are usually delivered orally. But by the time we get to technical instruction, we begin to see that having the instructions written down— or recorded in some way—would be a great benefit to all involved. The teacher could write a book or make a video and reach more students. The students could save money by learning at their own rate and on their own schedule.

Such perceived advantages account for the growth of technical education and the growth of technical communication in a socially mobile, fast-changing technological society. This does not mean that technical learning replaces formal and informal learning entirely in a developing culture or that it is necessarily better. Even computer companies still occasionally send trainers out to teach informal lessons to big clients. Many people receive this sort of instruction with great appreciation. The same people may look with resentment upon a computer manual; they prefer a "human touch" to the seemingly impersonal, "cold" approach of the printed manual.

The point here is twofold:

1. Every shift from formal and informal to technical communication involves trade-offs. Technical communication is not necessarily more effective just because it is more "advanced."
2. Every shift in the communication medium—from speech to writing, for example—carries with it changes in the way the audience receives the communication. Writing may seem "cold" or overwhelming compared to live speech, for example, even though it can convey much more information.

Technical communication favors media that allow for greater distribution and detailed, systematic communication, such as printed writing, audio-visual recordings, and hypertextual websites. In your projects, you may have to choose between different media, or you may be assigned to adopt a specific medium of communication. Either way, you need to know the trade-offs in audience effectiveness and user expectations that confront you with each media choice.

Effects of Different Media

Technical communicators work in four general media formats: live speech, writing (usually printed writing), audio-visual communication (A-V), and computer-mediated communication (CMC). Table 3.2 summarizes the main advantages and disadvantages of each format.

In the chapters that follow in Part Two of this textbook, you will learn that in the process of developing information products in any of these media, you need to

- **capitalize** on the advantages and
- **compensate** for the disadvantages.

When you give an oral presentation, for example, you have to restrict yourself to making only a few main points, perhaps as few as three, to allow for a relatively short audience attention span. But you can use written and visual aids to increase the attention span—for example, overhead slides with lists of key points and colorful graphics that seize the attention and impress the memory of the viewer. You can even provide handouts, using written words and graphics to extend the "reach" of the oral presentation.

When you prepare written documents, you can adjust your style to make it more fluent, open, and active; this gives the reader the sense that within this "cold" text is a helpful presence interested in their successful completion of the desired actions. By using style and document design in this way, you create a working environment that simulates a training session with a live instructor.

Audio-visual presentations and computer-mediated communication (CMC) represent technologies designed to compensate for some of the weaknesses of

TABLE 3.2
Advantages and disadvantages of different communication media.

Medium	Advantages	Disadvantages
live speech	Communicators can ■ use body language, gestures, and variations in vocal delivery; ■ make adjustments according to immediate feedback; ■ respond to questions (interactive); ■ control pace; and ■ condense background and rationale.	Communicators ■ have no written record of what the audience receives; ■ must be physically present; and ■ must deal with space and time constraints. Audience has ■ no written record of what was said; ■ no way to reinforce learning without the presence of the speaker; and ■ limited attention span.
writing (or print)	Communicators can ■ provide extended rationales; ■ present extended background information; and ■ convey more information to larger, more distant audiences. Audience can ■ be selective about what, when, and where to read; ■ use at their own pace; ■ review and re-read; and ■ carry the product with them.	Communicators ■ may miss the level of the audience; ■ cannot answer questions or respond to feedback; ■ may seem distant, cold, condescending, or unresponsive; and ■ incur expenses associated with printing and distributing products. Audience may not ■ question the author; or ■ probe or expand on ideas.
audio-visual communication (A-V)	Communicators ■ have all the advantages of the written medium; and ■ can simulate the personal presence of the author. Audience ■ has all the advantages of the written medium; and ■ assimilates the communication with less effort than written material.	Communicators ■ incur greater expenses than printing; ■ require technical expertise; and ■ are limited by considerations of time and audience attention span. Audience may not be able to review the product easily.
computer-mediated communication (CMC)	Communicators can ■ enjoy all the advantages of the written medium; ■ use a variety of presentational forms (text, graphics, A-V, and interactive); and ■ receive audience feedback and responses. Audience can enjoy all the advantages of the written medium.	Communicators must have ■ access to expensive computer equipment; and ■ the power and expertise to maintain the equipment during contexts of production and use. Audience ■ must have access to expensive computer equipment; and ■ may have more difficulty "navigating" than in written text.

speech and writing. A-V makes speech recordable for easy review, and CMC gives writing an interactive potential. But the new media introduce new problems. A-V viewers tend to become passive and to lose the spirit of action essential to effective communication, and CMC users may suffer from wandering attention or may get lost in the numerous possibilities for interaction, finally deciding that navigating the possibilities is a waste of time.

Again, the key to solving these problems lies in thoughtful design. You must use what you know about the potential effects of alternate media to improve your chances of effective communication in the medium you are using at any particular moment. To make good decisions in media design, you have to be very clear about the purpose of the information product, the reason the user needs it, and what actions the user expects to accomplish from using it. The question you must ask for every product you design is one that we will repeat many times in this book:

Who is going to do what with this information?

Though the purpose and use of every document is distinct, we can discern trends of use that relate to categories of information products. These categories are called **genres.**

Try to compensate for the loss of contact you have with the users of the print documents you produce by imagining what expectations they will have and what actions they will hope to accomplish beyond reading.

Categories of User Action: Genres of Technical Communication

Genre is the French word for "type" or "kind" (cognate with the scientific word "genus" and with "generic"). When people speak of particular "genres of writing," they usually mean a type of writing recognizable to most readers, such as a novel, short story, epic poem, advertisement, or letter of introduction.

Recent rhetorical theorists have added meaning and value to the term by asking us to think of genre as a form of social action.[1] For technical communication—which we think of as information in action—this view of genre is extremely helpful. Rather than being just a collection of document features arranged in familiar ways, each genre represents a special set of expectations about how users will act upon reading the document.

The main genres of technical communication are the *report,* the *proposal,* and the *manual.*[2] Reports and proposals provide information designed to help users make decisions and solve problems. Manuals provide technical instructions to show users how to carry out actions. Table 3.3 summarizes some of the differing expectations that users bring to these documents.

TABLE 3.3

User expectations in three genres of technical communication.

Genre	Context of Production	Context of Use (Who is going to do what with this information?)
report	Expert authors, typically researchers in organizations like consulting agencies or think tanks, provide analyses of problems, activities, and options, and give recommendations.	Decision makers such as government officials or managers decide on an action plan based on the analyses and recommendations of experts.
proposal	Expert authors who have a special ability and a special understanding of how to solve a problem or fulfill a social need provide a rationale and plan of action or product design.	Reviewers who have the money or power to grant permission to proceed decide whether to fund or permit a proposed project. Often they must decide which is the best of several proposals.
manual	Expert authors impart their know-how on carrying out a technical process in step-by-step instructions.	Users (often novice users) learn how to perform actions by following technical instructions.

1. See Carolyn R. Miller, "Genre as Social Action," *Quarterly Journal of Speech* 70 (1984): 151–167.

2. See M. Jimmie Killingsworth and Michael K. Gilbertson, *Signs, Genres, and Communities in Technical Communication* (Amityville, NY: Baywood, 1992).

The genres listed in Table 3.3 are not the only ones in technical communication, but they are the fundamental ones. Two other commonly recognized genres are correspondence (letters and memos) and scientific papers. In Part Three of this book, which covers each genre in detail, we present scientific papers as a special variant of the technical report and correspondence as a general format for short communications that often do the same kinds of work done by the fundamental genres. A letter, for example, might report, propose, or instruct.

For now, you should simply recognize that different genres do different kinds of work and fulfill different user expectations in conventional ways. With some sensitivity to the kind of genre you might be working with, you can add detail and insight to your audience-action profile and management plan.

Keeping the User Foremost in the Writing Process: The CORE Method

No matter which medium or genre you employ, you will always face the problem of reconciling production and use. The product you develop must be "user-friendly," responsive to the needs and expectations of the people that will act on the information you provide. To achieve this goal, you need to follow a systematic process that maximizizes your understanding of potential users and actual users. You need to bring the context of use into the context of production as much as possible.

Toward this end, we recommend a process we call the CORE method of developing information products. In this approach, you begin by performing an audience-action analysis—taking account of users from the very start—and then write short core documents. These documents give the core of the information product, the main argument the user needs to understand a problem and decide on a solution, or the central tasks the user needs to complete. You can present the core document to trial users to get feedback on what you need as you proceed. Since it is short, it does not make great demands on the trial user, and it is easier for you to revise. Imagine waiting until the product is nearly complete and finding out from users that it is not what they expected! Will you have to start over or will you try to pawn the thing off as best you can? Unfortunately many technical communicators face this dilemma, the result of bad planning and lack of foresight.

The CORE method involves the following steps:

1. Conduct an audience-action (or rhetorical) analysis. In a written profile, analyze the communication situation, and define the capabilities of the document producers and the needs of the document users. Stipulate the kind of document to be produced and the kind of information needed.
2. After initial research, write a core document of no more than two pages. Share it with trial readers to get a sense of the kind of information your

audience will need. The more readers, the better. The closer the readers are to your actual audience, the better. Continue the research process.

3. Develop graphics to support the main points in your core document. Try to cover as much of your information as you can in graphical form. Adapt some of these displays for an oral presentation.
4. Present an oral report, or briefing, to the trial or actual audience.
5. Develop the full written document, using your core document and the graphics you designed for an audience of users.

The CORE method has three notable advantages:

1. It is a realistic approach. A common practice in industry and government is for managers or other decision-makers to ask for a short paper—sometimes called a position paper, a concept paper, or a white paper—before requesting a full document. This practice allows the audience to direct the author toward the most crucial dimensions of the issue under investigation. It also gives the audience a chance to cut the project off before either the writer or the reader wastes valuable time (or money) on a full report.
2. It is a practical approach. Whether used in the workplace or in class, the core document requires a clear understanding of purpose, audience, and context right from the start. It lets you focus on these problems without having to expend too much energy in writing. By engaging the audience early, you can avoid costly wrong turns and false starts. Moreover, trial readers (including teachers and peer reviewers) are usually more willing to help if they don't have to read too much.
3. It is a collaborative approach. It facilitates interactions among multiple authors and editors throughout the writing process.

Figure 3.1 gives an overview of the writing process as conceived in the CORE method.

The flexibility of the CORE method—its adaptability for various writing tasks—makes it a good management tool. But a few points will not vary:

- *The need to enlist trial readers throughout the process.* Trial readers may be coauthors, editors, peer reviewers, sample users, or your instructor.
- *The need to employ a variety of communication media.* Technical communication is increasingly oriented toward a mix of visual, oral, written, and electronic discourse. You will need to develop skills in all of these areas.
- *The need to review and revise frequently.* The exercises and writing assignments in this textbook represent an attempt to get you accustomed to the standard rhythm of writing, reviewing, and revising in the production process.

The shaded box in Figure 3.1 indicates the part of the process introduced already in Chapters 1 and 2: audience-action analysis and profile writing. The chapters in Part Two help you to develop skills keyed to other phases in the process. Chapter 4 covers research; Chapter 5, writing core documents for re-

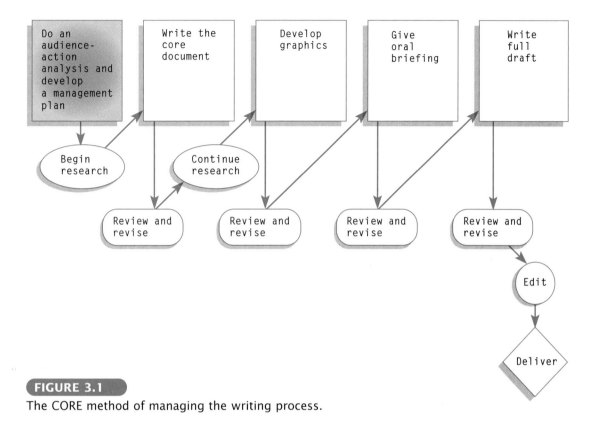

The CORE method of managing the writing process.

ports, scientific papers, proposals, and instructional manuals; Chapters 6 and 7, graphics and design; Chapter 8, oral presentation; Chapter 9, writing style; Chapter 10, editing. Then Part Three addresses specifics for each genre.

Before you go further in the process of developing information products, however, we recommend one final step in the prewriting phase: the management plan.

Writing the Management Plan

Writing the management plan completes your preparations for launching into the CORE method, or any other process to develop information products. Addressed to the supervisor of your project (your manager or teacher), the plan is particularly important if you are working in a team. All members of the team should consider it to be a binding agreement.

To present your management plan, prepare a short document set up in memorandum format with a date and "To," "From," and "Subject" lines at the

GUIDELINES AT A GLANCE 3
Writing a Management Plan

- Use memorandum format (see Chapter 11 if you need further guidance).
- Analyze the tasks required by the project (for example: researching, drafting, illustrating, editing, document testing, printing, and copying).
- Discuss which team members are best qualified to perform the various tasks and decide on a basic approach for the team.
- Choose a method for integrating each team member's contributions (division of labor, integrated team method, or a combination).
- Set a production schedule, assigning tasks to each team member.
- Use the project's audience-action analysis and draft profile (see Chapters 1 and 2) as the foundation for your management plan, adding a specific schedule and assignment of tasks. A clear image of what will happen in the project should arise in the mind of the management plan's reader.
- Submit the management plan to your supervisor (or instructor).

top (see Chapter 11 for a discussion of memo writing). Your memo should present a schedule of production, and in the case of a team project, the roles of each team member and a method for integrating contributions from different members. As background for the schedule of production, you should begin with an audience-action profile like the ones you developed in Chapters 1 and 2.

Guidelines at a Glance 3 summarizes the key steps in producing the management plan. Sample 3.1 shows a management plan for the insurance software case from Chapters 1 and 2. Sample 3.2 shows a management plan written by a student for a technical manual project (the completed project is discussed in Chapter 15).

EXERCISE 3.1

In class or in small groups, evaluate each of the two management plans (Samples 3.1 and 3.2). How could they be improved? How would a management plan for your own project be different or similar?

EXERCISE 3.2

Return to the audience-action profile you wrote for one of the scenarios in Exercise 1.1 on page 21 and the extension you wrote in Exercise 2.2 on page 49. Use these drafts as the foundation to write a management plan for the project.

Memorandum

January 10, 1999

To: William White, Director, Technical Writing Division

From: Jim Smith, Team Leader, ClientBase Documentation Project

Subject: Management Plan for Version 2.0 Documentation

The Technical Writing Division has the assignment of producing the user manuals and online help for version 2.0 of ClientBase. Used exclusively by insurance offices, this software allows agents to collect, arrange, and access client information and file claims for their clients. Agents and clerical workers with no computer experience beyond word processing and often little more than a high school education maintain client files and enter claim information via computers and modems. They are networked with their home offices, which receive the claims, process them, assign adjusters or make payments as required. So the documentation must be accurate, easy, and fast to use and should help the users begin their own work quickly.

To best meet the needs of the user, we will follow a team approach. The team will include two writers (Smith and Vallejo), an interface designer (Jones), an illustrator (Bernard), and an editor (Whister). The documentation team will use the engineering specifications for the new version (2.0), copies of the old version (1.2) and its documentation, and surveys returned by users of version 1.2. The team has also researched the documentation practices of our two leading competitors. We will create a plan and a draft copy of the new manuals and prototypes of the new screens and online help. Then we will test the prototypes with trial users and revise the draft as needed.

The problem we face is to make the computer screens and the support documents as simple to use as possible while allowing the users to take full advantage of all the software options. The interface of version 1.2 walks the user through each step of each function ("create client file," "file claim," etc.). The user selects functions by pull-down menus, each of which includes a help session. When the user activates the selected function, the program asks the user to enter each item of information in a pop-up box. The surveys returned by users of version 1.2 found the program so easy to get started on that they never used the step-by-step tutorial provided in the manual for beginning users. They said they only opened the reference section when they had problems, and they complained that the manual didn't provide enough help with how to handle problems. So we have decided on a three-part approach to revision:

- build as many instructions as we can into the interface,
- omit the tutorial in the manual,
- and include more information and instructions on troubleshooting.

The team that created version 1.2 of ClientBase has been reassembled for the development of version 2.0. In that project, the engineering representative in charge of interface design and programming (Jones) was at first hesitant to make changes suggested by the technical writers (Smith and Vallejo) in the user interface of the program itself. For efficiency's sake, Jones wanted to keep programming changes to a minimum and consign rhetorical improvements to the print document and online help sessions. But user confusion in preliminary testing convinced Jones of the need to keep

continued

SAMPLE 3.1   Management plan for a software documentation project

the interface consistent in approach with the documentation. We are glad to have Jones back on the team so that we do not have to fight this battle again.

The team is fully integrated and ready to work together. The only possible problem is Vallejo's commitment to continue intensive work on the QuestTime project during the beginning of the ClientBase revision. She has agreed to devote Tuesdays and Wednesdays to the ClientBase project except during July when all her time must be given to QuestTime. We will request additional staff during that month to compensate for Vallejo's absence. Smith will assume training responsibilities during that time.

In the new version, we will try to solve some ethical problems we encountered in version 1. We were convinced by our marketing representative (a visiting member of the team) to "go easy" on the number of warnings and cautions we included in the instructions. The argument was that too many "negative messages" would hurt sales. But our surveys revealed a high level of user frustration over not being warned of the sensitivity of the program during certain data entry routines. Unwary users who had lost data felt betrayed by the documentation. Our new approach, oriented toward minimal documentation with maximum troubleshooting instructions, should solve the problem for users. But we fear a hard sell will be needed for marketing. So we are requesting that a marketing representative be added to our team early on to gain an insider's view of our approach and to discover a marketing strategy that makes the most of the approach.

To ensure effective communication among British users, we will take the same approach we took in version 1.2, creating a "B" version for use in Great Britain. After developing the American version, the team will create version 2.0B by screening the core information to be sure no Americanisms exist in the instructions and by changing date, time, and currency formats in both the interface and the documentation.

We are also aware that, if version 2.0 succeeds in our American and British markets, we may explore German markets as early as version 2.1. Initial cultural analysis suggests that mere translation may not be sufficient to satisfy German users. They tend to prefer generous documentation with great technical detail (see Hoft, *International Technical Communication,* New York: Wiley, 1995, pages 336–37). Very likely, our minimalist approach will not work for this market. We may be able to write an expanded and updated version of the documentation for version 1.2 and then have it translated into German for testing. We recommend that this effort begin as soon as possible after testing the American and British versions. To avoid high costs, we can find German users to test a prototype segment of the German version before completing the full product. Ultimately, however, high costs are inevitable if we intend to enter this market seriously.

We now have our production schedule ready for your approval:

Feb. 1   Smith and Vallejo submit a revision plan to the full team. Jones and Bernard begin work on the interface with Smith as advisor.

Mar. 15   Jones and Bernard present the revised interface for team testing. Vallejo and Whister serve as test subjects. Smith observes.

Apr. 1   Work on the draft interface is completed. Smith, Vallejo, and Bernard begin work on manual set, with Jones as advisor. Bring the marketing team member on board.

May 1   First draft of the full documentation is ready for team testing. Whister and the marketing rep serve as test subjects. Jones observes.

SAMPLE 3.1 continued

June 1   Alpha version of the interface and manuals, each keyed to the other, is
            reading for testing with trial users. Smith, Vallejo, and Jones carry out
            testing.

July 1   The Beta version is ready for user testing. The substitute for Vallejo comes
            on board.

Aug. 1   Usability testing ends, final drafts begin. Vallejo rejoins the team. Whister
            takes charge of revision process.

Sept. 1   Version 2.0 documentation is delivered to production. Work begins on
            version 2.0B. Smith and Vallejo represent the team on work with the
            international office.

As soon as we get your approval on the schedule, we can proceed with the first
draft. Thank you for your continuing support of our work.

SAMPLE 3.1 continued

MEMORANDUM

September 25, 1997

To: Professor M. Jimmie Killingsworth

From: Norm Woody, Heather Parsons, Eve Rickenbacker

Subject: Management Plan for Windows® Quick Start Project

    As we decided in our meeting with you this week, we will
be creating a simple users guide for Microsoft Windows®. Such
a manual is needed in the university's microcomputer center,
and we have been commissioned to write and test the manual
by the Director of Computing Services, with whom we spoke
last week.
    The users of the manual will be students, mostly first-
time users of Windows® who don't have time to read the full
documentation and need a quick way to get started. Our
research shows that students learn best when they can work
on their own projects as soon as possible, so we don't waste
any time in getting them typing in their own papers and
incorporating their own graphics. The user assistants at the
Micro Center also tell us that students don't have much of

SAMPLE 3.2   Management plan for a student project   *continued*

an "attention span" when it comes to reading documentation, so we have to keep this short. If we keep it brief, we can keep costs low, too. We will use "screen dumps" as our main graphics, which are inexpensive as well as useful for guiding students from one step to the next.

The manual will be totally "localized." The language and motivational material (even a joke we are planning to use) are designed only for the primary context of use, our university. We hope to motivate students by making this "their own" manual. We will do our best to keep them out of trouble with good warnings and brief troubleshooting guides. We will address the problem areas we discover by user-testing the core document (a task analysis) and two subsequent drafts.

Since Norm knows the most about the software, he will direct the project and be the lead writer. Eve and Heather will focus on editing, testing, and production.

Below is our plan of work and our schedule for completing the project:

October 1:   Norm delivers the first draft of the task analysis to Eve for editing.

October 5:   First user test (using the edited draft of the task analysis). Heather, who knows the least about the software, will be the test subject. Norm will serve as her guide, answering questions and getting her out of any trouble that develops. Eve will observe the transaction, making notes on a separate copy of the task analysis to use in the next revision. We will also videotape the session.

October 5–10:   Norm and Eve will redraft the task analysis based on the results of the user test. Heather will edit their work. The whole team will work on the first expansion of the draft, adding graphics and warnings.

October 10:   Second user test: Using the draft of the task analysis with rudimentary graphics and warnings, two novice users will be selected from the class. Again, Norm will serve as user guide, with Eve and Heather observing and making notes. Once again, we will videotape the session.

October 10–15: Using the results of the second user test, Eve and Heather will produce the next draft, the first version of a fully completed manual.

SAMPLE 3.2 continued

```
October 15:    Training session in class, led by the full
               team. Norm will operate the overhead
               projection of the computer screen. Eve and
               Heather will provide an overview of the
               operating procedures.
October 15-20: Using feedback from the training session,
               the full team will work on a revision of
               the manual.
October 23:    The completed manual will be submitted for
               your evaluation and comments.

     We understand that, after you return the manual, we will
probably need to make another revision.
     We also understand the importance of sticking to our
commitments. Our initials on this memo indicate our
willingness to work as a team. We understand that, if any
of us falters in our responsibilities, that person may be
removed from the team.
     We look forward to your comments and suggestions. Thank
you for your attention to our project.
```

SAMPLE 3.2 continued

Recommendations for Further Reading

Hall, Edward T. *The Silent Language.* New York: Anchor, 1973.

Palmer, Richard Phillips, and Harvey Varnett. *How to Manage Information: A Systems Approach.* Phoenix, AZ: Oryx, 1990.

Schriver, Karen. *Dynamics in Document Design: Creating Text for Readers.* New York: Wiley, 1997.

Additional Reading for Advanced Research

Killingsworth, M. Jimmie, and Michael Gilbertson. *Signs, Genres, and Communities in Technical Communication.* Amityville, NY: Baywood, 1992.

Miller, Carolyn R. "Genre as Social Action." *Quarterly Journal of Speech* 70 (1984): 151–167.

Processes of Technical Communication

▼▼

Treating knowledge as design treats it as active, to be used, rather than passive, to be stored.

—*D. N. Perkins,* Knowledge as Design

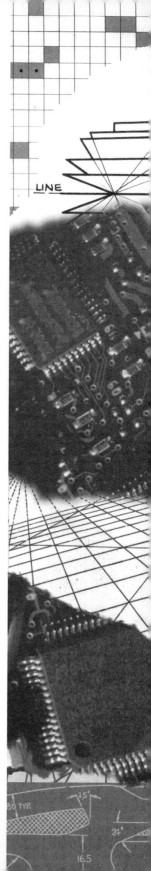

▼▼▼

Readers deserve documents that meet their needs.

—*Karen A. Schriver,* Dynamics in Document Design

▼▼▼

Dazzled by flashy technology, habituated to rock-music videos, and pressured for ever-greater productivity, users of technical products just don't read documents the way we writers wish they would. Documents that are merely polite or friendly don't do the job for audiences who are bored, busy, and sometimes hostile. These audiences must be *seduced* into reading.

—*William Horton,* Secrets of User-Seductive Documents

Research: Discovering and Developing Information

CHAPTER OUTLINE
▼▼▼▼▼▼▼▼▼▼▼▼▼▼▼▼▼▼▼▼▼▼▼▼▼▼

Researching Your Audience
Researching Your Topic
Creating a Usable Base of Information

CHAPTER OBJECTIVES
▼▼▼▼▼▼▼▼▼▼▼▼▼▼▼▼▼▼▼▼▼▼▼▼▼▼▼

After you have worked through this chapter, you should be able to do the following:

- Research and analyze the needs of prospective users—your audience
- Locate the information you need to produce technical documents by reviewing printed documents and electronic resources
- Develop additional information resources through interviews, archival research, correspondence, and surveys
- Recognize occasions suitable for original empirical research, and apply quantitative and qualitative techniques where training permits
- Create a usable base of information, smoothly integrating the process of inquiry and the process of writing

T he CORE method introduced in Chapter 3 begins with the need to gather reliable data. As Figure 4.1 suggests, you must have something to say before you can begin to write—and that means researching your topic. In addition, for a document to "grow" into a usable product, it must be nurtured and tested with substantive research at several stages of the process.

This chapter covers traditional methods of library research associated with standard textbooks and courses in English. It also examines methods associated more directly with technical communication—the use of interviews, surveys, systematic observations, and experiments, for example. Finally, it explains how to turn your raw research into a usable form for developing an information product.

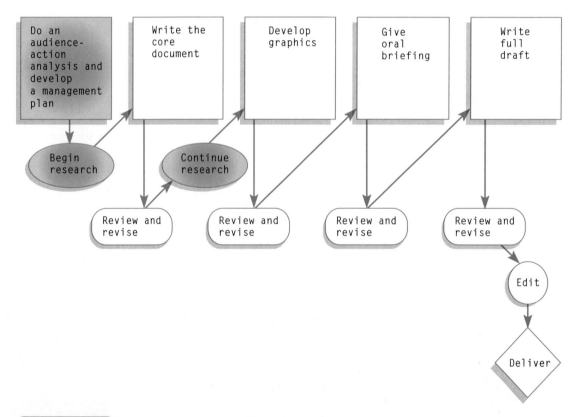

FIGURE 4.1

Data-gathering activities in the CORE method. The method begins with researching audiences and topics.

Researching Your Audience

During the research process you must focus on both your own needs (getting enough information to meet your assignment) and those of your intended audience from the very beginning, recording and developing the information that will make the reader productive. Only in this way can you make effective decisions about what information will be most useful when it is time to write. The audience awareness that you need for research requires an extension of the audience-action analysis we discussed in Chapter 1.

As Figure 4.2 suggests, you can turn the labels of the contextual model we developed in Chapter 1 into research questions about the context of use:

- Who are the users? Exactly who is the audience?
- What information is already available to these users? What are their databases?
- What kinds of actions will the users take based on the information you provide?
- What kind of document will best serve the users' needs? Do they need a report? A proposal? A manual? Should it be brief and general or long and detailed?

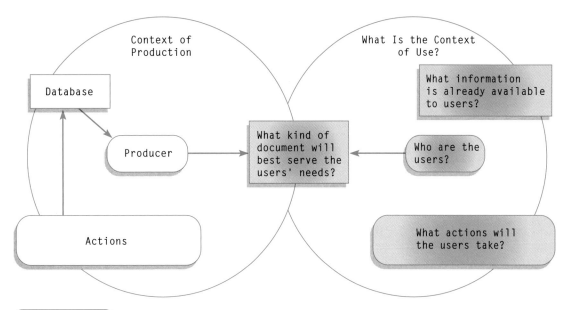

FIGURE 4.2

A model for defining the context of use. Before you can effectively gather data for a document, you need to answer questions about your audience and its concerns.

Two comprehensive approaches to audience analysis can help you answer these questions:

- *The market research approach* is a realistic method that involves locating and testing an actual group of users.
- *The rhetorical approach* stresses creativity and imagination, urging that you do more than *find* an audience—you must *create* your audience.

Chances are, you will need to draw heavily on both approaches, allowing the strengths of one to compensate for the other's weaknesses.

The Market Research Approach to Audience Analysis

Essentially, when using the market research approach, you should follow these steps:

1. *Decide on a target audience.* It must be narrow enough that you can identify real people in real situations who can serve as representatives of your market. In narrowing your target, of course, you will leave out other possible users. Maybe you can afford to miss them entirely, or maybe they will come along despite your efforts. Authors of manuals for personal computers, for example, often target business users because experience shows that, though many different kinds of people will *buy the computer,* business users are most likely to *read the manual.* Most hackers, college students, and game players don't bother with the tutorial anyhow, so they won't feel slighted if all the examples in the book refer to business applications (such as writing memos or keeping office accounts).
2. *Select representatives of the sample market.* In business these people may be reliable customers or customers of your competition—whom you can reach by purchased mailing lists, mailed questionnaires, or (with more chance of irritation) telephone surveys. Sometimes, the selection process is highly systematic, following established rules for sample selection; other times, the process is rushed and random, with the researchers hastily scrambling for feedback. In academic technical writing, authors often ask colleagues or students—representatives of expert and novice audiences in their field—to read and comment on early drafts of their work. The members of your own technical communication class may represent your target audience if you are writing to a student readership or to a well-educated mass audience from your region of the country.
3. *Gather information about the representatives of the target audience.* This information may be filed internally within your organization. Computer companies, for example, place response cards in their manuals, asking customers to inform them about problems they encounter or about useful features of the product. The responses are analyzed and classified in some kind of docu-

ment. Other kinds of surveys may be carried out by telephone or electronic bulletin boards, over user hotlines, or in person.

4. *Observe representatives of the target audience in action.* You can gather qualitative information about your audience, using ethnographic, case study, or clinical techniques. One such technique is the "talk aloud protocol." Market researchers ask the test subject to use the product while making a steady stream of comments about what's going on. The researchers record the activity and the comments, then analyze them for as many subjects as possible, sometimes supplementing the action session with interviews and follow-up questionnaires. Technical writers often develop *usability tests* for their draft manuals, observing readers as they attempt to follow a manual's instructions for using the product.[1]

In sum, the aim of all market research is to define the context of use so that the writer can integrate it into the context of production.

The Rhetorical Approach to Audience Analysis

The rhetorical approach relies on the author's ability to predict (or imagine) audience reactions on the basis of the following information:

- universal human characteristics (such as attention span or need for feedback)
- limits imposed by role or background (such as educational level or performance needs)
- expectations arising from previous reading or experience (such as habits and conventions for making sentences, formatting reports, or addressing superiors in a company hierarchy)

Authors cannot always gather these data through research, so they must apply analysis, critical insight, and principled design. In Chapters 5–8 we will explore the process of rhetorical analysis in more detail. For now, remember that as you gather data and make notes or comments about your audience, you should recognize that your audience's needs will be affected by their education, culture, and other "human factors."

Collaborating with Other Researchers

Case 4.1 shows the importance of collaboration—working with others not only to write, but also to gather information. Research and development projects rarely involve a single researcher. Usually, there are multiple contributors, so everything in this chapter about individual research habits must be considered

---

1. See Roger Grice and Lenore Ridgway, "Information Product Testing: An Integral Part of Information Development," *Perspectives on Software Documentation: Inquiries and Innovations,* ed. T. T. Barker (Amityville, NY: Baywood, 1991), 209–228.

in this light. The first audience for your rough information will be your closest colleagues, your coauthors, editors, or managers.

case 4.1   A SIMPLE USABILITY TEST

A group of students in one of our classes wrote a short manual for the graphics software in our computer production lab. One of the four students, the least experienced in using the software, was asked not to participate in writing the first draft. When the draft was completed, she became the first test subject. She sat down at the computer and worked through the tutorial of the draft manual. One team member sat beside her to answer questions while another made notes about the questions, the answers, and the steady stream of talk—"Let's see, it says here to place the pointer on the box at the left of the screen. Do I do that with the mouse?" Another student videotaped the transactions.

The four students studied the recordings and the notes and made their revisions with problems and solutions in mind. Then they tested their next draft on two more subjects: an experienced user and a novice.

They discovered that the expert user had no real need for the tutorial. He could handle all tasks simply by referring to the menus and help sessions provided in the user interface of the computer. But novice users needed to walk the familiar ground of an instructional manual before they were comfortable working only with the information available on the screen. For them, the manual was a bridge to total reliance on the computer.

On the basis of these results, the authors decided to ignore expert users and to pitch their manual toward first-time users. They revised with this in mind and announced their decision in the manual's preface.

Researching Your Topic

Once you have conducted your audience-action analysis by defining the context of use, you can begin gathering data on your topic, which will become the content of your document. Thorough researchers use many resources—public and university libraries, archives, company files, research centers, corporate offices, think tanks, and laboratories. None of these, by itself, is likely to provide all the information you will need, so you must know how to find your way around a variety of information-rich places.

You can access material in most locations through a fairly standard set of information resources that fall into two large categories:

1. Published resources, or public information, available through
   - print media (published literature, catalogs, indexes, bibliographies, etc.) and
   - electronic media (databases, the Internet, etc.)

2. Unpublished resources, or original research, available through
   - oral and written records (interviews, correspondence, archival and file re-search) and
   - empirical studies (surveys, field observations, experiments)

In the research stage of any project, you build your own database by mining these information resources. From this base of information, you can launch a variety of projects. You can interpret old information for new purposes and audiences, or you can do original research to fill the information gaps you discover in the published literature.

Most public information resources—whether print or electronic—are organized for *tiered access.* This is the organizing principle used in newspaper publishing. Consider, for instance, Sample 4.1, a rather typical news story.

The article is organized into three distinct structural units—headline, first paragraph, and remainder of article. This structure allows for selective reading.

At the most general level, all readers browse the headlines, at least on the front page. If the headline grabs you, you will probably move to the next level, the first paragraph, which summarizes the main points of the story. If you are still interested at that point, you can find further details in the remainder of the story. Figure 4.3 shows this tiering scheme.

---

3 Oil Spills Unleash Millions of Gallons

Newport, R.I. — Major oil spills threatened Rhode Island, Delaware and Texas Saturday after three tanker accidents in just over 132 hours unleashed an estimated 2.55 million gallons of oil.

The extraordinary sequence of events began Friday afternoon in Narragansett Bay, not far from here, when the Greek-registered tanker *World Prodigy* ran aground on Brenton Reef. The reef tore holes in the vessel and released up to 1.5 million gallons of light heating oil into Narragansett Bay, the Coast Guard said.

About 12 hours later, up to 800,000 gallons of heavy crude oil oozed into the Delaware River after the Uruguayan-registered *Presidente Rivera* ran aground near the town of Claymont, upriver from Wilmington.

Sandwiched between those huge spills was the estimated loss of 250,000 gallons of heavy crude oil into the Houston Ship Channel near La Porte, Texas. That spill occurred late Friday when a tug-driven barge was punctured and lost its load after colliding with a Panamanian-registered tanker.

Unlike the situation in Valdez, Alaska, where 11 million gallons of crude oil were spilled in the grounding of the *Exxon Valdez* on March 24, government and environmental workers moved quickly in an effort to combat all three spills.

The White House, which had been criticized for its slow response to the Alaska disaster, dispatched William K. Reilly, administrator of the Environmental Protection Agency, and Manuel Lujan Jr.,

(continued on page 11)

SAMPLE 4.1   News story with tiered access to information
Source: Bob Drogin and William J. Eaton, *Los Angeles Times,* printed in *The Commercial-Appeal,* (Memphis), June 25, 1989, page 1.

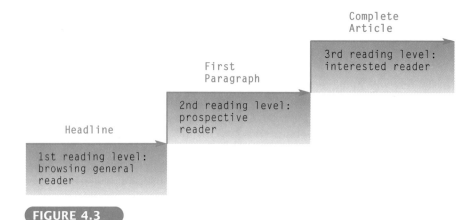

FIGURE 4.3

Tiered access in a news story. This organizational scheme accommodates three levels of readership, ranging from least interested to most interested.

Access to the information becomes more challenging at each tier. The greatest number of readers will read the large print, boldface headline—which is both easy to find and easy to read. A smaller number will read the first paragraph, which is easy to find but, because of the small print, requires greater effort to read. The smallest number of readers will take the trouble to read further and turn to page 11 to get all the details of the story.

Researching Print Resources: The Literature

Even highly technical information is organized for tiered access. If you flip through an academic journal such as *The Physical Review* or *Ecological Monographs,* you will find that, like newspaper headlines, the article *titles* are written to give a strong sense of the topic of each piece. That's one reason why technical articles often seem to have jaw-breaking titles such as "Fish Predation in Size-Structured Populations of Treefrog Tadpoles" or "Effects of Food Supplementation, Song Playback, and Temperature on Vocal Territorial Behavior of Carolina Wrens."

These titles may not be particularly elegant, but they do tell readers glancing through a bibliography or table of contents exactly what the article covers. The authors maximize the number of potential *keywords,* recognizable terms that will help fellow researchers gain access to information needed for further research. If you're working in the field of population ecology, for example, words such as "predation," "size-structured," "population," "food supplementation," and "territorial behavior" will cue your further reading. If you saw the titles above and your field was either amphibian research ("treefrog tadpoles")

ABSTRACT—We found both the song rate and the rate of song-type change of male Carolina Wrens (*Thryothorus ludovicianus*) in winter were positively correlated with ambient temperature. When the effect was controlled statistically, food supplementation significantly increased both the song rate and the rate of song-type change. Song playback did not significantly increase either song rate or rate of song-type change, however. Because foraging and singing are mutually exclusive behaviors in Carolina Wrens, the increase in vocal territorial behavior associated with warmer temperatures and food supplementation may reflect a decrease in the time required for foraging. The rate of vocal territorial behavior in winter may be more dependent on the amount of food available to wrens than on the presence of intruders.

SAMPLE 4.2   Abstract for "Effects of Food Supplementation, Song Playback, and Temperature on Vocal Territorial Behavior of Carolina Wrens"
Source: Joy G. Strain and Ronald L. Mumme, *The Auk* 105 [1988], 11.

or ornithology ("Carolina Wrens"), you would probably move on to the next reading level.

Here you encounter the *abstract*—a short, usually paragraph-length summary of the article. Sample 4.2 is the abstract for "Effects of Food Supplementation, Song Playback, and Temperature on Vocal Territorial Behavior of Carolina Wrens."

If you are still interested after reading the abstract, you can move to the third tier and take on the whole article. Figure 4.4 gives a schematic view of the specialized journal's use of tiered access.

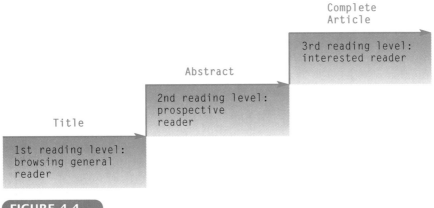

FIGURE 4.4

Tiered access in a specialized academic journal.

Of course, even with this standardized organization to serve you, leafing through publications is hardly the best method for finding your way into the specialized literature on your topic. Fortunately, *bibliographies and indexes*—both printed and online—also offer tiered access. First, you can use keywords in your subject area—such as "population size," "predation," or "territorial behavior"—to locate article titles from journals listed in an index. Some indexes provide a second tier in the form of abstracts. At the very least, a bibliography or index will lead you to the right place in the right journal. Then, after you read the abstract, you can study the whole article if it's on target.

case 4.2 FINDING LITERATURE ON AN ASSIGNED TOPIC

Imagine that you work as an engineer at an electrical cooperative in a rural county. The chairman of the board, your boss, saw a public television show about global warming and wants you to write a short information sheet about how ordinary consumers can adjust their energy-consumption patterns to help keep the greenhouse effect in check.

Because your specialty is solar energy, you don't have an expert's understanding of research on global climate change. But in such a small office, you have to do the job of a generalist. As an experienced researcher, you know how to find out what you need to know. Here's what you do:

1. At the local university's library, look through the subject index of the card catalog (or online guide to the collection) to access textbooks or general interest books that you can use to get a handle on your topic. The bibliographies and footnotes in these books will lead you to others. If you have trouble, consult your most dependable information resource—the reference librarian.

    You find a textbook and two general interest books on your topic: Penelope ReVelle and Charles ReVelle, *The Environment: Issues and Choices for Society* (3rd ed., Boston: Jones, 1988); Stephen Schneider, *Global Warming: Are We Entering the Greenhouse Century?* (San Francisco: Sierra Club, 1989); Jonathan Weiner, *The Next One Hundred Years: Shaping the Fate of Our Living Earth* (New York: Bantam, 1990).

2. Next, you search the *Reader's Guide to Periodical Literature* and the library's periodical list to locate articles in popular, widely circulated magazines.

    You find three useful articles in popular magazines: Barry Commoner, "A Reporter at Large: The Environment" (*New Yorker*, June 15, 1987: 46–71); David Brand, "Is the Earth Warming Up?" (*Time*, July 4, 1988: 18); Michael D. Lemonick, "Feeling the Heat" (*Time*, Jan. 2, 1989: 36–41).

3. Then you check the *Monthly Catalog of U.S. Government Publications* for any recent information distributed by the federal government. You consult the li-

brary's holdings of government documents, which are usually contained in a special room or section. You will definitely have to consult the reference librarian in this pursuit, for the range of government publications deposited in different libraries and the method of cataloging them vary widely from place to place. If your library does not have a document you need, you can request it directly from the agency that produced it. Usually, the agencies respond quickly to requests.

The librarian helps you find a promising government publication: *Policy Options for Stabilizing Global Climate,* Office of Policy, Planning, and Evaluation, Environmental Protection Agency, 1989.

4. Next, to get a sense of expert opinion on the topic without taking on the specialized literature, you look into the science news and essay sections of more specialized journals written for educated but nonexpert audiences. If the *Reader's Guide to Periodical Literature* doesn't serve you well enough, you check the more specialized indexes such as the *General Science Index* and the *Applied Science and Technology Index.*

You find a few articles from the general interest sections of science journals: V. Ramanathan, "The Greenhouse Theory of Climate Change: A Test by an Inadvertent Global Experiment" (*Science* 240 [1988]: 293–299); Stephen Schneider, "The Greenhouse Effect: Science and Policy" (*Science* 243 [1989]: 771–781); Eliot Marshall, "EPA's Plan for Cooling the Global Greenhouse" (*Science* 243 [1989]: 1544–1545).

5. Since you have some advanced training, you can go a step further and explore the specialized technical literature.[2] To locate articles on your topic, you look into specialized indexes such as *Biological Abstracts, Chemical Abstracts, Bibliography and Index of Geology,* and the *Science Citation Index.*

This search yields one specialized article that you can understand in a general way: Hansen, James, et al., "Global Climate Changes as Forecast by Goddard Institute for Space Studies Three-Dimensional Model" (*Journal of Geophysical Research* 93 [1988]: 9341–9364).

Printed information resources usually present neatly tiered information. Of all the sources mentioned, general books and articles in popular magazines are the easiest to find and to read. They are widely circulated and easily accessed at most public libraries. They are also written in a style that is suitable for a general audience, delivering technical terminology and specialized concepts only in small doses. Beyond this point of access, we find increasing specialization of information and decreasing accessibility and readability (Figure 4.5).

---

2. Most undergraduate students in technical communication courses will find it tough going beyond this point.

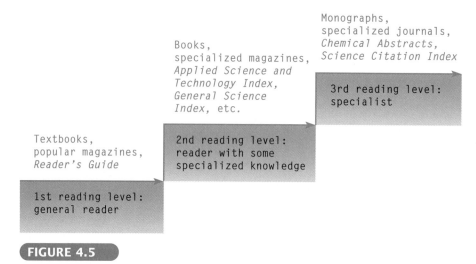

Books,
specialized magazines,
*Applied Science and
Technology Index,
General Science
Index,* etc.

Monographs,
specialized journals,
*Chemical Abstracts,
Science Citation Index*

3rd reading level:
specialist

Textbooks,
popular magazines,
Reader's Guide

2nd reading level:
reader with some
specialized knowledge

1st reading level:
general reader

FIGURE 4.5

Tiered access for print resources. As specialization of information increases, accessibility and readability decrease.

Researching Electronic Databases

Though electronic databases are growing more powerful and more available for use by students and the general public, they are still a relatively new information technology. They are changing so rapidly that any general statement we make about them here may not be true by the time you read this book.

Not having benefited from the centuries of trial and error that have shaped the print information system, electronic databases are neither as well organized nor as reliable. But they are the research tools of the future and even the present, so you will not be able to avoid them. Chances are, your own library has already replaced its card catalog with a local electronic database—an online catalog.

As you approach these tools, remember, first of all, that electronic databases are not as carefully tiered as printed resources. One of the great advantages of computer storage is that huge amounts of information can be contained in a small space. You've no doubt heard that a single microchip can store libraries full of information. But this advantage leads to a compensating disadvantage: Because the capacity of electronic databases is so large, the temptation is to put *everything* into them. Consequently, users may be swamped with lists of titles on any given subject and forced to perform their own tiering to get at relevant information.

The best way to do your own tiering is to study the art of effective descriptors. A *descriptor* is a keyword or group of related words that are used to initiate your search of the database. To discover the right descriptors, you need to define your topic in terms that are widely used in indexing and organizing information in your field of study.

In our library, for example, when we wanted to read about electronic databases, we came up empty-handed in our search of the online catalog by entering "electronic information bases" as the descriptor. We had to seek help from a librarian, who told us to try the phrase "online searching," which brought up about a hundred titles on the screen.

You can also refine your choice of keywords by referring to standard reference books that give lists of subject headings that indexers and researchers frequently use. One of the best and most widely available is *Subject Headings Used in the Dictionary Catalogs of the Library of Congress,* known by the short title of *Library of Congress Subject Headings.* Though this book was designed for use in researching printed resources, it works pretty well for choosing descriptors, too, since most electronic databases are designed by librarians to index printed resources.

To supplement this tried and true source, you will find an increasing number of descriptor guides to specialized fields of research. Most disciplines have begun to develop a *thesaurus of descriptors.* Ask your reference librarian to help you find the best one for your field of work.

Here is another hint on building a good combination of descriptors for online searching: Find a few articles on your topic and notice typical combinations of words, especially words that are used in titles and major section headings. One good reason for starting early to read and take notes on the sources you discover is that this initial research will help you become familiar with key terminology in your topic area. This will help you extend your research beyond the superficial level.

Finally, we recommend that you resort to specialized online searching—the searches that you usually have to pay for—only after you have exhausted more general online searches and print resources. In undergraduate work, you will probably need to search only the database maintained by your own university's library.

However, if you are working toward publication or doing graduate research, your literature review will need to be exhaustive. You should read everything on your topic published in the last ten years or so and much of the older work as well, so you will need to work through the best specialized databases in your field.

Many of the old printed indexes, such as *Chemical Abstracts,* are now available online. In some libraries these offer considerable advantages for users who are interested in conducting extensive literature searches. You may, for example, not only view the title and abstract of articles in your field but even order a printout of the article itself.

Conducting Internet Research

To conduct research on the Internet, you will need:

- A computer with a telephone or modem link to the Internet or to a host computer designated as an Internet server,
- Software programs that allow you to perform specified operations (such as browsing, searching, or sending and receiving electronic mail), and
- A basic working knowledge of how the Internet is structured and the mechanics of moving through it.

Hardware and Site Identification   To access the Internet, your computer is probably linked to a server (a computer that stores files) through a telephone line. While large main frame computers were the original servers, powerful home computers can now act as servers. Large servers are generally housed in educational, government, or commercial organizations, although other groups have formed local area network servers (LANs). Each server is identified by an address, usually a series of abbreviations that reflect the name and type of organization. For instance, <oeri.gov> is a government server operated by the Office of Educational Research and Improvement (oeri); <tenet.edu> is a server operated by the state of Texas for educational purposes. In this instance, tenet stands for Texas Education NETwork. Other common types of servers include noncommercial organizations (.org), commercial entities (.com), networks (.net), and military institutions (.mil).

A specific person or site is identified by prefacing the server name and type with the person's identification name or assigned code and separating the two with the symbol "@". Together, the user name and server name and type make up the user's Internet address. For example, <jack&jill@aol.com> tells us that a user named jack&jill can receive information on the commercial server operated by America Online; <jofutbal@tamu.edu> tells us that a user named jofutbal can receive information on the educational server operated by Texas A&M University. To contact either of these users, you must have appropriate software programs that allow you to perform different tasks (e.g., send and receive e-mail, join discussion groups, or participate in listservs). With the Internet addresses of specific people or sites, you can contact them for information. (See Chapter 11 for a further discussion of e-mail.) However, much more information is available on the Internet.

Beginning in the 1960s as a government research tool, the Internet has expanded to encompass vast quantities of information in both text and graphic form. In addition to communication functions (see Chapter 11), the Internet allows researchers to browse and search through documents posted by individuals and organizations. The documents may be available as text only or they may consist of a combination of text, full-color graphics, and even video and/or audio functions.

With Internet access, you can search the World Wide Web (also referred to as the web), an enormous collection of millions of posted pages of information accessible through software programs called browsers, search engines, or transfer protocols. Web sites have addresses called URLs (Universal Resource Locators) that include the mode of access (usually "http://," which stands for HyperText Transport Protocol), followed by the type of file accessible (usually either "www" for World Wide Web or "gopher"), followed by the server name and type of organization operating the server. For example, <http://www.sedl.org> means there is an organization named sedl that posts information on the web. By typing this URL into a browser (see below), you will reach the organization's web page.

Since files may be nested (placed within other files), addresses may need to be expanded to find the exact location. For example, a technical writing class syllabus can be found at <http://www-english.tamu.edu/pers/fac/Killingsworth/engl.210.html>. The URL tells us that a file in html format (HyperText Markup Language) entitled engl.210 can be found nested in the Palmer file, within the faculty file, within the personal file, that can be accessed through the World Wide Web by way of the English department server that is operated by Texas A&M University, an educational institution.

Of course, just as soon as you become familiar with recognizing this form of address, a new type will probably emerge. We can only hope it will be simpler.

Software Programs   Three general types of software programs allow you to perform specific tasks using the Internet:

- *Browsers* allow you to identify and view posted information.
- *Search engines* allow you to look for specific key words either in the documents or their titles.
- *Transfer programs* allow you to transfer information from a host computer to your home, classroom, or office computer.

Browsers   Currently, browsers are the preferred programs for navigating through the Web. They may be text-based or graphics-based, although to make full use of the graphics potential, a high-speed computer and modem (your Internet connection) are needed. However, don't be discouraged if you are using a slower computer and modem. The text you access is generally the same; the fancy bells and whistles will be missing, which may not be a drawback if you are conducting academic research.

Browsers are basically software programs that display information and contain buttons for performing a range of functions, such as moving to other Internet locations or linking with search engines, graphics, or other software programs. Some well-known browsers include Microsoft Internet Explorer, Mosaic, and Netscape.

Search Engines   There are numerous programs available that allow you to search electronic sources for information. Just as you can tell a word processing program to locate each incidence of a specified word in a document, you can direct different search engines either to look for general information using broad subject areas or to find very specific information using key words and sophisticated boolean cues. Boolean logic reduces all values to TRUE or FALSE, allowing for simple (digital) choices in a series of steps.

Search engines differ in their approaches to 1) conducting the search itself, 2) limiting or controlling the number of sites that are searched, 3) indexing sites, 4) ranking sites, and 5) allowing complex boolean searches. For this reason, it is a good idea to try several different engines before abandoning a search.

In conducting searches, engines may proceed linearly using categories, following one link to the next, or in spider web fashion, branching out in multiple directions simultaneously. The latter type accesses more sites within a specified time frame, but if you are searching for something very specific rather than trying to find *all* information on a topic, this may not be desirable. Samples 4.3 and 4.4 display web pages for two popular search engines. Sample 4.3 offers search options under multiple categories, while Sample 4.4 allows only a key word search.

Most engines limit the pool of sites they sort through. Some require registration of proposed web pages and other Internet documents before including them in the pool to be searched. Others may access documents not available to the general public. For example, a search engine operating at an academic institution may accesses sites within the institution that have not been posted on the Internet.

Not all search engines index sites, but those that do generally develop categories of sites. This means that instead of searching through the entire Internet for information related to a specific scientific discovery, by clicking on the "science" category, the search engine searches only those sites identified by the developers of the software program to be science-related. Indexing can speed things up considerably since the search engine may sort through only a few thousand sites rather than millions. The drawback, however, is that things may not always be categorized appropriately or as you would expect. For instance, a particular article you want on string theory may have inadvertently been categorized as "arts and crafts" rather than as "science."

Those search engines that rank results of searches do so using different formulae that may reflect the frequency with which a key word appears in a title or in the text of a document, or how often it appears near the beginning of the document. Some search engines limit access to only the highest ranked results

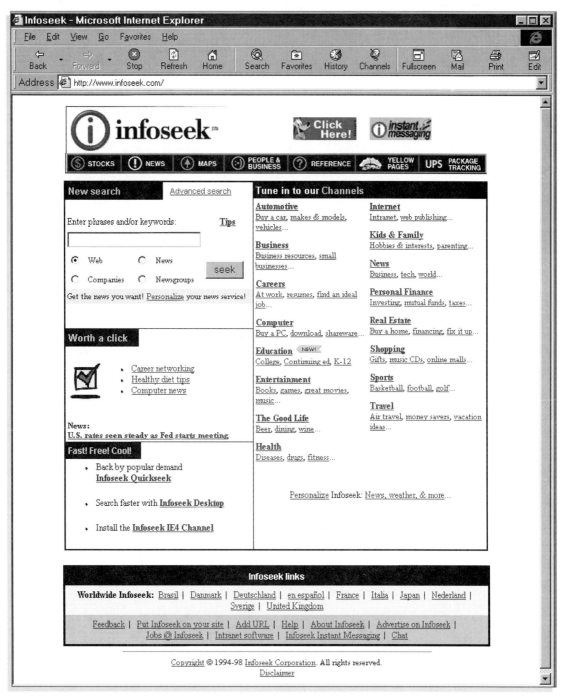

SAMPLE 4.3 Popular search engine with multiple category and key word search options.

Source: http://www.infoseek.com/

SAMPLE 4.4   Popular search engine with key word search option.
Reproduced with the permission of Digital Equipment Corporation. AltaVista, the AltaVista logo, and the Digital logo are trademarks of Digital Equipment Corporation.
Source: http://www.altavista.digital.com

(e.g., the top 50 or 100 selections), while others list literally thousands of sites, conveniently prioritized by your computer.[3]

A rudimentary knowledge of using booleans will also help you in conducting electronic searches. For instance, if you type "bucky balls" as key words, some search engines will find every site containing the word "bucky" (probably not very many) and every site containing the word "balls," while

3. See Eric Branscomb's *Casting Your Net: A Student's Guide to Research on the Internet* (Boston: Allyn and Bacon, 1998) for a chart comparing strengths of different widely used search engines.

GUIDELINES AT A GLANCE 4
Tips for using search engines

- Be careful to not to pick a topic that is too broad or a key word with multiple meanings.
- Try different key words.
- Different combinations of key words can get different results.
- Know the syntax of the search engine you use (e.g., proper use of booleans).
- To get more in-depth information, learn how to use the advanced search mechanisms on the different search engines.
- The same search path done on different computers may get different results, depending on the defaults set on the hardware and software.
- Search engines from different universities may obtain different results.

other engines will recognize this as a single term and find only sites that list the two words together. Obviously, the latter is preferable, yet the first search engine might have achieved the same results had you simply inserted the word "and" between the two words or used quotation marks or italics, according to the guidelines offered by the search engine. These guidelines generally appear on the search engine's home page in the form of either a pull-down menu or a button. Some search engines allow very complex advanced searches and provide short tutorials to help you master the appropriate boolean and style requirements.

Guidelines at a Glance 4 lists some additional tips our students offer for using search engines.

Transfer Programs   File transfer protocols (ftp) are software programs that transfer files from one computer to another. Most of the current search engines now have this ability so that you all you need to do is click on some highlighted text to download a file.

The process of moving information between computers has become easier over the years, but you still may have problems. You may lose formatting or find yourself unable to open documents you have transferred (downloaded) to your computer. For example, if you type a document using a specific software program, there is no guarantee that the person trying to read your document has the same program. As new alliances are made between major hardware and software manufacturers, software programs are being developed to translate from one format to another. However, if you don't have the appropriate translation programs, you still may not be able to read a downloaded document.

Be cautious and selective when downloading files. Some may carry computer viruses; others may take literally hours to download; still others (especially graphics) may take up more memory than your computer can handle.

There are ways to move text between computers without uploading or downloading them. For instance, rather than attaching a file to an electronic mail message, you can save a document as a text only file and embed it within the text of the e-mail message. That way, if recipients can read the message, they can also read the attachment. If the document is short, you may also be able to simply highlight and copy it to a word processing program on your desktop. Or, you can highlight it, select "new mail message" from the menu bar and send the information to yourself via e-mail. A third option is to select "quote document" from the menu bar.

You can also copy graphics from the web by holding the mouse button down (the right one, if there are two) while pointing to the desired graphic. A pull-down menu will give you several options, including "copy image." You can then paste the image into a regular graphics program. In fact, most word processing programs now have built-in graphics capabilities that allow you to paste graphics directly into your text document.

Even though it may be tempting to cut and paste information from the web into your own document, be careful not to plagiarize; give credit for quotations, paraphrases, ideas, and graphics.

Credibility and Reliability of Electronic Sources   Most books and periodicals undergo a strenuous process of editorial and/or peer review before publication and distribution, a process that sometimes results in delays of a year or more before the information is available. The Internet offers the advantage of speed, making information available literally minutes after web pages are posted. But there are trade-offs.

The Internet is a largely unregulated living testimony to humankind's freedom of speech. Anyone with electronic access can post web pages. The challenge for both the technical writer and researcher, then, is to be certain that posted information is accurate and reliable. How can we do this?

First, acknowledge that, unlike print publications, most information on the Internet has not been critically reviewed for content or scrutinized for grammatical errors. Be skeptical. Assume the posted information is bogus and search for clues to establish its credibility and reliability. Guidelines at a Glance 5 offer some questions you might ask yourself when reviewing electronic documents.

There is much reputable information available electronically. For instance, many professional electronic journals now available on-line have been subjected to a streamlined review process similar to that of print publications. But you must check diligently the accuracy and reliability of your sources. Whenever possible, apply a basic tenet of qualitative analysis when conducting research on the Internet: *triangulate data*—that is, confirm information you find

GUIDELINES AT A GLANCE 5
Questions for reviewing electronic resources

- Is an author listed? If so, are credentials provided? Documents that give no author names may be less reliable or credible than those that do.

- Who allowed the posting? That is, who runs the server that posts the information? Is it a reputable government, military, or educational organization that might enforce strict guidelines for posting? Or is it a commercial provider that may be trying to sell something? Just because a site is sponsored by a commercial server, however, does not mean it is not worth accessing. For instance, in pursuing the search for bucky balls, we accessed this commercial web page: <http://www.htp.com/ ARTLEX/Hobbies/Science/Index.htm>, which included a link to a bucky ball home page sponsored by the Department of Physics and Astronomy of SUNY at Stony Brook, <http://buckminster.physics.sunysb.edu/>.

- If the server is reputable, does it restrict, screen, or endorse postings? An institution may endorse some web pages, but allow students or employees to post unscreened personal pages. (For example, a large research institution was recently embarrassed when it learned that web surfers were accessing its home page from the home page of a woman who had posted nude photos of herself at various institutional landmarks.)

- Does the information appear accurate? Even if the server and author are reputable, some files on the Internet are not locked and can be altered by users. Be wary of false information.

- Does the information appear to be biased? Is a specific product or cause being promoted? If so, will this bias hinder your research?

- Is there a date indicating when the information was posted or updated? Much information on the Internet is already outdated. A current date may be an indication that the site is reviewed or updated regularly. However, a disadvantage to using Internet resources is that web sites can be withdrawn from public access without notice, with the result that another researcher (or teacher) is not able to confirm your source.

from at least three unrelated sources, some of which are *not* electronic (e.g., books, articles, interviews, etc.).

Citing Electronic Sources   Since electronic postings are not subject to the same restrictions as print materials, finding enough information to cite your source may be a challenge. Nevertheless, copyright laws also apply to material posted on the web. Thus, whether you are acknowledging the source of text or graphics, it is important that you provide enough information for someone who wants to track your research and locate the same sites you found. At a minimum, this would include the complete web address and the date on which you accessed it. You should try to include the same information you would if the document were available in print form, but this is not always possible.

GUIDELINES AT A GLANCE 6
Citing electronic sources

- Check recent editions of your preferred style manual for specific directions on how to cite electronic sources.
- If no author or sponsoring agency is listed on the web page itself, check the web address for the name of the server posting the information.
- If the source is not on the page you are viewing, backtrack through the web addresses to find the source.
- For important or frequently used web addresses, use the menu bar to make bookmarks.
- Copy the URLs from the web pages and paste the information into an ongoing reference list (usually a word processing file).
- When you print important information from the web, be sure to record the URL.

As electronic research becomes more widespread and acceptable, authors of the major style manuals have added guidelines for citing electronic sources. The University of Vermont has posted abbreviated guidelines for students using either APA (American Psychological Association) or MLA (Modern Language Association) styles at <http://www.uvm.edu/~xli/reference/apa.html> and <http://www.uvm.edu/~xli/reference/mla.html>.

Guidelines at a Glance 6 offers additional guidance on citing electronic sources.

Developing Unpublished Information Resources

So far, we have covered the kinds of public information resources that are generally available to any student researcher. You may be able to complete most of the assignments in your technical communication class by accessing only the printed resources that are indexed in standard bibliographies and in the most accessible electronic databases. The reports that you write, though, may be little more than reviews of the relevant literature.

Most professional researchers would be only beginning their work when they had exhausted these resources. In the information economy, the research with the highest value is *original research.* Original researchers open the door to previously inaccessible information, or they create new knowledge by revelations or insights that enhance the reader's understanding of previously known facts and phenomena.

In this section we take a brief look at several means of developing original research:

- researching files and archives
- interviewing

- using correspondence and surveys
- doing field studies and clinical observations
- performing experiments

You may already be doing lab research, field work, and archival research. You may also have access to files and to experts in your school or in the company where you work. Or you may have expertise in techniques that are poorly documented in the public literature. By making the most of these opportunities as a researcher, by bringing new information to the public, you will take a big step toward professionalizing your research and your writing.

Researching Files and Archives: Access and Ethics Technical documents tend to be "active" only as long as they are immediately useful. But where do old reports, memos, letters, and other outdated documents go? If they survive housecleaning, they usually find their way into the archives, which all companies and governments keep. Though this "old" information may hold limited interest for technical researchers and readers, it should not be overlooked. Research in files and archives is especially useful in the preparation of documents that call for evaluation and review. When the past performance of an office or a project is an issue—in annual reports, for example, or during a reorganization, or in proposals for future funding—an organization's internal files are an important resource.

Documents that involve project planning (such as proposals) frequently refer to old projects that failed because of clearly identifiable reasons, such as lack of sufficient funding, inadequate personnel, equipment, or facilities. Access to files and archives can give you a view of history that published research has overlooked or ignored, actions and events that have been recorded but have not been open to public view.

The issue of access involves a number of *physical problems,* such as how to find the right files and how to find the energy to plow through them. It also involves *legal and ethical questions.* Legal access can be a problem, since a company's files are its own property. Even if you are an employee of the company, you may need permission to use old files for any report that will be circulated among nonemployees. Even for internal communications, opening access to certain files may violate policies of privacy or privileges of rank.

Of course, if you are not an employee of an organization, you do not have physical access to its files unless the company grants it to you. Then you are obligated not only to get permission to use the information, but also to reveal in some detail how you intend to use your findings in any work you plan to publish.

The same is true for personal files. If a professor or a fellow student shares notes and unpublished manuscripts with you and you want to use that information in something you are writing, you must first get permission and then acknowledge the source of the information, giving credit where credit is due.

Corporate files and archives may contain key information for
projects that preceded yours and yield important clues about ways
to succeed where others have failed.

The products of other people's research are construed as their *intellectual property* by the courts. Even if this weren't the case, good research ethics would require you to seek permission for use, reveal your purposes in advance, and
acknowledge the favor in print.

Interviewing   Journalists learned long ago how to capture history in the making. They interview experts on important subjects and agents of important actions, creating a public record of what their informants say about the past, the present, and the future.

Professional technical writers, in similar fashion, consider the expertise of their colleagues and the know-how of major players in their organizations as part of the company database. They call on members of the technical staff to explain difficult concepts and new procedures; they ask technicians to demonstrate the use of new products; they go to management and marketing experts for hints about future plans and development trends. Their notes from these conversations become the subject matter of countless user manuals, sales brochures, and annual reports.

If, as a student researcher, you know experts on the subject you are researching, you may consider including an interview as one of your information resources.

Telephone interviewing can be an efficient way to query experts on a technical topic. Good preparation is essential to compensate for the normal give-and-take of face-to-face interaction.

Here are some guidelines for conducting interviews:

1. In approaching potential sources, *be courteous and even deferential.* Such behavior is certainly good manners, and it may help to loosen the tongue of your informant.

2. From the start, be sure to make clear how you intend to use the information. *Get permission up front* and *acknowledge the helpfulness of your source later* when you write your document.

3. *Be prepared when you enter the interview.* Never go in cold; never "wing it." You should be as thoroughly informed as possible when you talk to others about your topic. Don't waste their time by asking them to explain something that you could have learned from a quick reading of a widely accessible published source, such as a textbook or an article in *Time* magazine.

4. *Let the informant talk, but gently keep turning the conversation back to your topic.* From journalists we can derive a simple method of interviewing:[4] Always keep a mental checklist of the questions you used initially to analyze your research topic: *Who? What? When? Where? How? Why? With what results?* By keeping these questions in mind—or by actually keying pages of a stenographer's pad to each question—you can ensure coverage of the topic. Also, have ready any questions about special problems you encountered when reading company reports or published literature. With luck, your source can help you solve such problems.

5. *Take good notes, but realize that you will be able to get down only the main points, along with a few quotes and key phrases.* I always try to get to a computer immediately after an interview to record a longer version of the notes I take in the interview. This allows me to add my own reflections while my informant's words are still fresh in my mind.

6. Many people tape record their interviews, but this makes some informants nervous and overly cautious. It also sets you up for a laborious task later when you have to transcribe your tapes. Despite these shortcomings, tapes allow greater accuracy. If you do several interviews in quick succession, you will surely have to depend on your tape recorder. *Before recording an interview, ask for permission to do so, and prepare the machine and the tapes so that you can record discreetly during the interview. Then transcribe the tapes carefully after the interview.* You can treat the transcription, the written version of the interview, as a regular source in your bibliography.

Another thing we can learn from journalists—especially *bad* journalists—is that interviewing has shortcomings. When you depend on information provided in an interview, you place yourself at the mercy of your informant. Can

---

4. A more elaborate model for interviewing is given an exhaustive treatment in Earl McDowell's book *Interviewing Practices for Technical Writers* (Amityville, NY: Baywood, 1991).

you be sure that this unpublished information is reliable? After all, it has not been subjected to editorial and critical review in the public research forum. And what about conflicting information from two sources?

In a technical setting, the information gained through interviews usually directs you back to published sources, which can help you resolve conflicts and establish the authority of your sources. Does the information provided by an informant seem to follow a trend in the public literature? Has your expert established his or her own authority in approved channels of publication?

We all have a natural tendency to favor a source who has taken the time to help us. But when it is time to write, regardless of how much you appreciate and acknowledge your sources, you must use your own knowledge and judgment to *maintain your critical perspective.* As a researcher, you are ultimately responsible for what you write. You have to mediate among your sources, present your own version of the truth, and be prepared to defend it.

Using Correspondence and Surveys

Using Correspondence and Surveys Social scientists often try to corroborate information provided by informants by increasing the number of people responding to the research questions. They do this by developing written resources, usually in the form of surveys and questionnaires, to supplement print resources and conversations with knowledgeable people.

As a student researcher, you should consider extending your research along these lines. You can simply write to knowledgeable people, inviting them to respond to a general question. Or you can do a survey, either by telephone, in person (at a mall or in the university center), or by mail, based on a questionnaire asking sample populations about your topic. If you decide to conduct a survey by mail, be prepared to spend some money (for stationery, postage, and return postage) and to take some time—at least a month—to gather such information.

Be aware of hidden difficulties in designing and interpreting questionnaires. Inexperienced surveyors tend to ask questions that are ambiguous, wordy, or too complex; seek information that the respondents cannot deliver; and seem suspicious to the respondents, or hint at purposes beyond the one stated by the researcher.[5]

A bad survey may also produce data that are difficult to interpret. Most social scientists prefer answers that can be counted and analyzed quantitatively by established statistical methods. Such methods provide good rules for how to make generalizations from answers given by samples of a general population. You should avoid questions like "Do you enjoy reading novels?" It would be better to ask, "How many novels do you read each month?" or "On a scale from 1 to 5 (1 = not at all, 5 = very much), how would you rate your enjoyment of reading novels?"

---

5. Janice Lauer and William Asher, *Composition Research: Empirical Designs* (New York: Oxford University Press, 1988), 66.

Methods of interpretation are fuzzier for open-ended questions, and respondents are less likely to answer them. Open-ended questions require the respondent to reflect on the question and write more than a numerical or yes–no response. But they also allow researchers to delve into more complex issues and accommodate individuality among respondents.

Survey methodology has developed into a rather esoteric field.[6] Experienced researchers rely on an elaborate system of rules and policies to conduct and interpret surveys with good statistical procedures. It is well beyond the scope of this text to deal with these extensive methods. If you are not trained in statistical methodology, you should collaborate with someone who is before you try a survey. Otherwise, you could waste a lot of time and money doing an amateurish survey that yields worthless results. But if you are trained in statistics, psychology, sociology, education, or the other social sciences, you have probably learned at least the rudiments of survey methodology. If so, you can probably manage a survey that would add a dimension of original research to your work in technical communication.

For example, you can use a survey to update, verify, or disprove information that is available in the public literature on your topic. This approach is especially useful for determining whether a generalization holds up for a local population. Say, for example, you read an article that says that college students tend to prefer detective fiction to popular romances, but you notice that the sample population was drawn mainly from East Coast and West Coast colleges. You might test the generality of the claim by surveying students in Midwestern and Southern schools. Or perhaps you are writing a proposal for solving a problem on your campus. You could survey the people you think will benefit from your plan and include the data to support your arguments.

Samples 4.5, 4.6, and 4.7 illustrate how one technical communication student carried out a survey of experts in his field. Notice how the researcher thoroughly explains in the letter both the overall project and the questionnaire, keeping the explanations brief and placing action items in a clear position of emphasis. The first sentence of the fourth paragraph, for example, says precisely what the researcher wants most urgently from the subject: "Please complete Part One." He courteously tiers the assignment so that the busy expert can skip the less important parts, but he also provides an opportunity to supply additional information if the time and inclination are there. He not only gives an optional Part Two and a reply card but also mixes yes–no and short answer questions (easy and fast to answer) with open-ended questions that would require a more considered response. Finally, he follows good research ethics, assuring the subject that "anonymity will be honored."

6. A good general introduction is given by D. R. Berdie and J. F. Anderson, *Questionnaires: Designs and Use* (Metuchen, NJ: Scarecrow Press, 1974). The American Statistical Association's *Survey Manual* is also useful. For using survey methodology in technical communication, see Paul V. Anderson, "Survey Methodology," *Writing in Nonacademic Settings*, Lee Odell and Dixie Goswami, eds. (New York: Guilford, 1986), 453–498.

PO Box 155
English Department
Memphis State University
Memphis, TN 38111
20 August 1991

M. Jimmie Killingsworth
Director, Writing Program
Department of English
Texas A&M University
College Station, TX 77843-4224

Dear M. Jimmie Killingsworth:

I am writing to ask your help in supplying information for
my master's thesis research project. As a technical and
professional writing major at Memphis State University, I am
researching the interaction between authors and editors in
manuscript preparation for publication. To learn more about
the general behavior of editors I am surveying the writers
who contributed to six technical and business communication
journals during 1990.

I hope that you can take a few minutes to complete the
enclosed questionnaire in regard to your article "A Grammar
of Person for Technical Writing" published in The Technical
Writing Teacher in 1990 and mail it to me in the enclosed,
stamped envelope.

The questionnaire is in two parts. Part One is labeled
"Quick Answer" and can be completed easily in about two to
three minutes. Part Two is labeled "Short Answer" and asks
you to provide some specific examples from your experience
with your editor. Part Two can be completed in less than
fifteen minutes, depending on whether you answer from memory
or look up your manuscript to jog your memory.

Please complete Part One. If you are willing to take a few
minutes to provide more details, complete Part Two also--
your anecdotes will be helpful and appreciated. Please be
assured that no answers that might reveal your identity
will be included in my report.

Finally, I am also interested in interviewing some authors
about their interaction with their editors. If you are
willing for me to phone you with some follow-up questions,
please provide your name and phone number on the enclosed,

SAMPLE 4.5 Student survey of experts—cover letter   *continued* stamped card. Again, be assured that your anonymity will be honored.

I would appreciate your returning this survey by 14 September so I can begin evaluating survey results soon; however, if you take a bit longer, I will still be glad to get your response.

Thank you very much for your time and as much information as you are able to give.

Sincerely,

Noel Clements

Enclosures

SAMPLE 4.5 continued

WRITER/EDITOR INTERACTION QUESTIONNAIRE

Part One: Quick Answer

SUBMISSION/ACCEPTANCE INFORMATION

1. Was your article solicited by an editor or guest editor?   Y   N
2. Had you previously submitted this article to another journal?   Y   N
   If yes, how many others?
3. Before you submitted the article to the journal in which
   it was published, had you published in this journal before?   Y   N
4. Approximately how long after you submitted the article did you receive either an acceptance or suggestions for revision?
5. Was the acceptance conditional on major revisions?   Y   N
6. How many versions did you submit before the editor was satisfied?
7. How many reviewers read your article?
8. How many reviewers were in favor of acceptance?

SAMPLE 4.6   Student survey of experts—questionnaire

REVISION INFORMATION

1. Did you receive feedback from a colleague before this submission?   Y   N

2. Did the editor make changes in your article without consulting you?   Y   N

3. Beyond copyediting changes, were any substantive changes made that you did not make?   Y   N

4. What major revisions did the journal editor ask you to make? Check all that apply.

   ___ organization   ___ cut length

   ___ title change   ___ cut specific material

   ___ documentation   ___ add specific material

   ___ word choice   ___ other (please specify)

   Other: _____

5. Was your journal editor specific enough about the revisions you were required to make so that you knew what to do?   Y   N

6. Were all requests for revisions reasonable?   Y   N

7. Did you protest or argue any suggested changes?   Y   N
   If so, were you successful in persuading the editor to accept your point of view?   Y   N

8. Were there some revision suggestions that you felt were unreasonable or unhelpful, but not worth arguing about?   Y   N

9. Now that you look back, how satisfied are you with the editing of your article?
   Highly Satisfied   Satisfied   Unsatisfied   Highly Unsatisfied

Part Two: Short Answer

1. Please explain any unreasonable requests for revision: _____

   _____

   _____

2. If you protested against any suggested changes, what was your argument? _____

   _____

   _____

   _____

3. Please describe the journal editor's reaction to your point of view. _____

   _____

   _____

   _____

4. Please give an example or two of revision suggestions that you felt were unreasonable or unhelpful but not worth arguing about. _____

   _____

   _____

5. Please tell us anything particularly interesting or important about your experience with the journal editor. _____

   _____

   _____

   _____

SAMPLE 4.6 continued

INTERVIEW CARD

I would like to learn more about the interaction between writers and editors in technical and business communication journals by hearing from you personally. If you are willing to answer follow-up questions in a brief phone interview, please complete this preaddressed, stamped card and send it today.

Name _____

Phone _____   _____

Best time(s)/day(s) to call _____

_____

SAMPLE 4.7   Student survey of experts—return card

Conducting Field Studies and Making Clinical Observations   Another research method that has gained wide acceptance in the social sciences is known as "qualitative" or "naturalistic" research. With roots in medical and natural history research, this kind of study involves systematic observation of phenomena in natural settings. Health researchers study patients in hospitals and at home; ethnographers study indigenous cultures in remote settings; sociologists study inner-city youths in ghettos; educational researchers observe children at work in schools; and so on. On the basis of the notes they take on their observations, they write case studies that students and practitioners in these fields depend on to supplement their own practical experience.

Qualitative researchers try to minimize their influence on events and to serve mainly as recorders and interpreters of their observations. Since the presence of researchers always has some influence on what happens, they try to account for, or "correct for," this influence. They interview people they observe or people who have observed the same phenomena they are studying. They compare and contrast what various informants say, and they compare their findings to reports in the literature. They may even supplement their observational data with quantitative findings drawn from surveys and demographic studies.

As a student researcher, you will have opportunities to carry out potentially useful, and even publishable, observations. You will have access to sites that are ripe for such studies—the workplace, the lab, and the clinic, not to mention the dorm, the locker room, and the cafeteria. But take warning: You cannot substitute casual observations for systematic research. Furthermore, legal and ethical restrictions are even stronger in case studies involving human subjects than in matters related to intellectual property. Qualitative research deals with people's lives, not just their "data."

Though most basic textbooks in anthropology, psychology, and sociology cover the methodological and ethical issues involved in naturalistic research, you should seek a mentor—a trained professional researcher—before you undertake such research. Your mentor can help you develop a rigorous system for recording and interpreting your results so that your efforts will not be wasted. Your mentor can also advise you as to how to take every measure to avoid offending your informants.

Here are a few hints for conducting an effective study:

1. *Select a site where you can blend into the regular context of work,* as an employee, a fellow worker, a manager, or a student intern, for example. Avoid being a researcher with a notebook peering into other people's business.

2. *Take sparse but careful notes while you are in the field, supplementing them quickly and fully when you return to your office, study, or workstation.*

3. *Verify your observations with multiple sources.* Interview many people involved in your project. Do multiple interviews with informants who are particularly important and willing to talk, but be careful not to bias your data by being overly sympathetic to the most accessible informants. Supplement your observational data and interviews with a thorough review of the literature that treats similar situations, as well as with full historical or archival research, including work in files relating to human subjects. Whenever possible, also collect quantitative data from surveys and other empirical research.

4. *Make photographs, videotapes, or audiotapes only with permission.* Make clear, from the start, the purposes of your research and the possible outlets for publication.

5. *Try to interpret your observations within the limits of the situation you are observing, usually without resorting to cause-and-effect reasoning.*

This last hint suggests many of the difficulties involved in this kind of research. If you depend totally on what you see, not on what you suppose or intuit, you generally have to stick to the questions of *what, who, where, when* and *how.* Avoid speculations about *why* unless you report them as data drawn from interviews or surveys. For example, you can say that *you saw* a man scream at his computer, and you can say that *he said* that he screamed at it because he was having a bad day. You can also report that the man's file reveals a history of calm behavior at work. But you cannot say *why* he *really* screamed at it. Nor can you say that the computer itself, despite others' testimony about its notorious difficulty, is the sole cause of this interruption in the normally serene atmosphere of the office. After all, the man may have recently been under more stress than you know about, and the others in the office may merely be sympathizing with him.

So much for the rigors of observational analysis. The discussion of note-taking and coding later in this chapter returns to qualitative research methodology to discuss the very useful systems of information analysis that have evolved in ethnography and the case study method.[7]

Performing Experiments   If field observations are "naturalistic" in their effort to describe things precisely as they are found, the methodology of experimental research prefers to introduce variables into controlled situations that are modeled on natural conditions. Experimental scientists don't just observe things as they find them; they "twist the lion's tail," as Francis Bacon once said. Instead of waiting for something to happen, they create the happening and then extrapolate from their findings to generalizations about natural occurrences.

This kind of reasoning is governed by rules and standard procedures developed over centuries of scientific practice. The elaborate methods of scientific research cannot be covered here. But you may spend much of your undergraduate career refining your understanding of the scientific method as you do experiments for chemistry, biology, psychology, or physics courses or test engineering designs in labs. What you learn in technical communication classes will help you to present the experimental findings you develop in your other courses of study.

Be aware, however, of the differences between the kind of experimental results you have on hand and the results that count as original research. In searching the literature, you will notice gaps in the knowledge base of your field. These gaps entice professional researchers to take on new projects in original research. By contrast, many of the experimental projects you undertake in your classes may not be original in this sense. They are more likely designed to teach you fine points of methodology or to demonstrate the means by which old discoveries were made. In professional research, the experimental design, as well as the results of the experiment, is usually original.

Complex questions about the originality, objectivity, factuality, and validity of scientific research are well beyond the scope of this text. We can only em-

---

7. If you are interested in pursuing this fascinating field of research on your own, you can find good introductions in R. K. Yin, *Case Study Research* (London: Sage, 1984); M. H. Agar, *Speaking of Ethnography* (London: Sage, 1985); and, for technical communicators, Stephen Doheny-Farina and Lee Odell, "Ethnographic Research on Writing: Assumptions and Methodology," *Writing in Nonacademic Settings,* L. Odell and D. Goswami, eds. (New York: Guilford, 1986). The spirit as well as the letter of this approach are best communicated in Clifford Geertz's now classic book *The Interpretation of Culture* (New York: Basic Books, 1973), especially Chapter 1, "Thick Description: Toward an Interpretive Theory of Culture."

phasize that the results of experiments done with scientific rigor carry perhaps more weight than any other form of evidence in our culture. As a researcher, use this kind of information if you have access to it. For now, we must leave it at that, with the promise to return to questions about scientific writing in Chapter 13.[8]

Creating a Usable Base of Information

Once you have assembled a critical mass of data, you should move quickly beyond the idea that research is a simple matter of *collecting* information. It also involves mental processing, the *development* of information.

Early in your research, you should start processing information and shaping it into a usable form. The best way to do this is to get the information into writing. The very act of writing is a concrete way of thinking through topics and facts and getting them straight in your head. With help from a photocopy machine and a highlighting pen, you have to take fewer notes than students of yesteryear did, but don't give up note-taking entirely. Jot down important ideas and talk them over with others. Attempt some trial drafts of sentences and paragraphs for the document you are working on.

The rest of this chapter presents a few techniques you can use to get your research into a usable written form.

Start with a Working Outline

A working outline presents a simple way of organizing your notes as you take them. Some authors prefer to take notes in a fairly random way, not bothering to organize them in a notebook, card file, or computer file keyed to subtopics or other categories. This approach is fine if you have plenty of time later to digest and process your notes. Anticipating time crunches, though, you will be better off to begin with an analysis of what you already know about your topic, looking for patterns and issues.

You can do this by creating a code of keywords. Use these to make tabs for your notebook or card file or headings for each page or card; or write the keywords in your notebook margins as you make notes, taking time later to cut and paste. The keywords can help you in structuring your research now—in searching electronic databases, for example—and in organizing your writing later.

---

8. Even Chapter 13 will offer only a cursory glance at important topics in the history and philosophy of science. Students who are serious about scientific research should read Ian Hacking's book *Representing and Intervening: Introductory Topics in the Philosophy of Natural Science* (Cambridge, England: Cambridge University Press, 1983).

You can use topical keywords to code ideas, themes, or groups of facts. The project on global warming in Case 4.2, for example, could be developed around this short working outline:

1. Scientific theory and controversy
2. History of global warming
3. Global warming now
4. CFCs, rainforests, etc.
5. Energy use, especially consumption of fossil fuels

You can add new topics, dividing and rewriting the old ones, as the project progresses. This method approximates the kind of content analysis used by ethnographers.[9] They lay out an initial set of themes, issues, or hypotheses—interpretive categories corresponding to explanations or patterns of behavior and key events. Then they expand, review, and revise their topics as they go. When the time comes to write the research report, the analysis is complete; they are ready to get their thoughts on paper.

Alternatively, you can base your working outline on topic analysis, using the key questions of *what, who, when, where, why, how,* and *with what results.* These could generate a somewhat different set of categories:

1. What is global warming, the greenhouse effect, ozone depletion, and so forth?
2. Who is responsible for the theory, and who is responsible for the effect itself?
3. When and where can we find evidence of the greenhouse effect?
4. Why should we worry?
5. How can we counteract the effects of global warming?
6. What will be the results if we take these steps? What if we don't?

The drawback of using a working outline is that you can limit your understanding of the topic by neglecting a key issue that you didn't notice the first time around. Just remember to be open to new topics, or even to a total revision of your outline, as you do your research.

You can also record your data as scientists and field workers do, in a lab or field notebook. In these notebooks, researchers create their own blank tables and fill them in as they go. As the data are produced, a neat place in the notebook awaits them. The numbers immediately fall into an interpretive category (dictated by the experimental design) and have a special relation to other data.

Keep a List of Sources and a Full Running Bibliography

You must carefully keep track of where you get your information. Devote a section of your notebook to a running list of sources, including books, articles, in-

---

9. The method is often called "iterative content analysis." See David M. Fetterman, *Ethnography Step by Step* (Newbury Park, CA: Sage, 1989), Chapter 5.

terviews, and observations. Record enough information about each source to enable another researcher to retrace your steps.

For interviews, observations, and experiments, keep all the pertinent information about people, places, times, and circumstances. Create a full heading in your notes, something like the following:

- Interview with John R. Smith, Vice-President of Finance, Atlas Corporation, Greenville, Mississippi, 9/9/91, Atlas Corporate Offices, by previous appointment.
- Personal Communication with Dr. Cynthia Jones, Senior Research Scientist, MaBell Labs, Poughkeepsie, New York, at Conference on Electronic Communications, New York, New York, 10/14/90, working lunch.
- Field observations, Neutron Test Site, Tonapah, Nevada, with Bill Jacobs, Toni Richardson, Hal Ortega, 6/9/89, 10:00 am. Sunny sky, no clouds. Dry, no rain this month.

For books, articles, and other published documents, write down the complete publication data in a running bibliography. In fact, we recommend that, from the very start, you write out the full citation in the documentary style you will use in your finished product, as in Sample 4.8.[10]

Electrify Your Notes

If you keep your notes on a computer, you can easily alphabetize your running bibliography. When it is time to produce the paper, you can simply copy the full

---

Anderson, W. T. (1990, July/August). Green politics now come in four distinct shades. The Utne Reader, 52–53.

Brown, L. R., Durning, A. B., Flavin, C., French, H. F., Jacobson, J., Lowe, M. D., Postel, S., Renner, M., Starke, L., & Young, J. E. (1990). State of the world 1990: A Worldwatch Institute report on progress toward a sustainable society. New York: W. W. Norton.

Commoner, B. (1971). The closing circle: Nature, man, and technology. New York: Knopf.

Comp, T. A. (Ed.). (1989). Blueprint for the environment: A plan for federal action. Salt Lake City, UT: Howe Brothers.

SAMPLE 4.8   Running bibliography in APA style

10. For information on documentation styles, see Chapter 10 on editing.

list and edit out all the titles you did not refer to. Your list of references—a chore for most writers—can be quickly finished.

This is just one of the advantages of keeping notes on a computer. It allows you to cut and paste using interview transcripts, author quotations, your own observations and comments, and previous documents you've created. If you have a laptop computer, your writing may never touch paper until the last draft of the finished document. If your information is written or typed on note cards or paper, highlighted on photocopies, or recorded on audiotapes or videotapes, you will have to add an extra step (or steps) to the writing process: transcription of notes and/or rewriting of the notes in the first draft. Even so, it's worth the effort. Electronically stored notes are ready to be processed into the document.

In addition, you can make numerical data easier to handle by employing spreadsheet programs and other database software. Since such programs can receive and incorporate new data at any time, why not enter them frequently as the project progresses? By doing so, you can see patterns as they develop and get a head start on interpreting your results.

Make Different Kinds of Notes—Direct Quotations, Summaries, Paraphrases, Comments, and Interpretations

As time and the situation allow, vary the type of notes you take from the literature you read, the people you talk with, and the events you observe. The more kinds of material you have when you start to write, the more likely you will be to find the information that suits the task. How you take notes should depend on

- the nature of the talk, writing, or action you are dealing with and
- the use to which you will put the information.

Direct quotations are best for passages of reading or conversation that you may want to reproduce later as a direct quotation in your document or for text you want to question and review carefully later, in relation to your other notes. For example, you may want to quote passages that you find difficult to understand on first reading. Then you can go back and puzzle over the passage when you have more time. The note you take may look something like the card in Sample 4.9.

You may also want to use a direct quotation for passages that display striking language that you may want to borrow later. See Sample 4.10 for an example. Also, use direct quotations for passages that may be controversial, as in Sample 4.11.

When you take the note for a direct quotation, check what you wrote carefully, word by word. In interviews, ask the informant for permission to quote directly. Read back sentences that you plan to use verbatim or that you don't

Source: Schneider, Stephen H. (1989). _Global Warming: Are We Entering the Greenhouse Century?_ San Francisco: Sierra Club Books.

pp. 84–86:   "Despite the pitfalls and the fact that records in various countries contain individual fluctuations from year to year, both the GISS and CRU reconstructions, even after recent corrections for urban heating, agree on the two salient features: an approximately 0.5°C (1°F) warming trend over the past century and several warm years bunched in the 1980s (1988, 1987, and 1981 being the warmest years in the record)."

SAMPLE 4.9   Research note card with a direct quotation that may be difficult to understand on first reading

understand. Allow the informant the opportunity to correct or modify the sentence.

Partial quotations are also useful if you want to reproduce striking or controversial language. You could, for example, work parts of the full quotations above into your own sentences, as in Sample 4.12.

Source: Schneider, Stephen H. (1989). _Global Warming: Are We Entering the Greenhouse Century?_ San Francisco: Sierra Club Books.

p. 200:   "Nowadays, everybody is doing something about the weather, but nobody is talking about it."

p. 206:   "The public's perceptions, of course, are primarily shaped by the next category, the media, which thrive on the four Ds: drama, disaster, debate, and dichotomy."

SAMPLE 4.10   Research note card with direct quotations of striking language

Source: Schneider, Stephen H. (1989). _Global Warming: Are We Entering the_
          Greenhouse Century? San Francisco: Sierra Club Books.

pp. 235: "The greenhouse effect and ozone depletion teach the same lesson:
          we cannot continue to use the atmosphere as a sewer without expecting
          substantial and potentially irreversible global environmental disruption."

SAMPLE 4.11   Research note card with a direct quotation of a possibly
controversial statement

Summaries or paraphrases are preferable to direct quotations in most technical writing. When you summarize in documents, you can work information
from other sources smoothly into your own paragraphs, maintaining a stylistic
consistency and a readable fluency. When you summarize during note-taking,
you claim the information as your own by putting it in your own words. You
can also take only the information you need.

Source: Schneider, Stephen H. (1989). _Global Warming: Are We Entering the_
          Greenhouse Century? San Francisco: Sierra Club Books.

Schneider complains about the media's addiction to "the four Ds: drama, disaster,
debate, and dichotomy" (p. 206).

He says we should stop using the atmosphere as a "sewer" (p. 235).

SAMPLE 4.12   Research note card with a partial quotation

From the long and difficult example in Sample 4.9, for instance, you might extract only the extent of the warming and the number of hot years, as in Sample 4.13.

The downside of summarizing and paraphrasing is that you might misinterpret your source's meaning or lose nuances of meaning. Some of this is inevitable, but most can be avoided. Never let your guard down as a researcher. Be on the lookout for the traps of misunderstanding and misreading.

Remember also that you will be guilty of *plagiarism*—intellectual theft—if you do not use citations to credit the source of your information, even when you are summarizing and paraphrasing instead of quoting directly. You would obviously be plagiarizing if you borrowed a sentence from the example in Sample 4.10—*The public's perceptions, of course, are primarily shaped by the next category, the media, which thrive on the four Ds: drama, disaster, debate, and dichotomy*—without using quotation marks and without citing your source. But it would also be wrong to borrow a piece of the sentence—for example, the phrase *"the four Ds: drama, disaster, debate, and dichotomy."* Even to criticize the media for its constant emphasis on these four areas could be construed as plagiarism, since the author of this passage was the first to propose the fourfold critique (though hardly the first to complain about each item individually). Always cite an author's ideas or interpretations as well as any information that is not available in a general reference book, such as a dictionary or encyclopedia. Such common information is frequently reproduced without citation unless it involves illustrations, which are always strictly copyrighted.

If you have any doubt about whether to cite something, go ahead and give credit. It is better to be safe than sorry. In your notes, always keep a careful

Source: Schneider, Stephen H. (1989). Global Warming: Are We Entering the Greenhouse Century? San Francisco: Sierra Club Books.

pp. 84–86:   Most scientists agree that the earth has warmed by a half degree Celsius or a full degree Fahrenheit over the last one hundred years and that three of those years—1981, 1987, and 1988—are the hottest on record.

SAMPLE 4.13   Research note card with a paraphrase

record of your information sources even if you don't need to cite them in the final document. By doing so, you can always retrace your steps and check your information or decide later what to cite and what to consider part of the general body of knowledge.

Comments usually involve evaluations of your source made in the heat of reading. For example, on the note in Sample 4.11, we might write "controversial" beside Stephen Schneider's contention that the same lesson is to be learned from the greenhouse effect and ozone depletion. Or we might create a cross reference by writing, "See the different viewpoint in Ramanathan." Comments also can be notes directing your further research, such as "look into this later," or "reevaluate earlier position," or "compare to data from Tom's lab."

Interpretation is the process of showing significance, of saying what is important and what is not. In technical communication, interpretations usually involve the tough little question "So what?"[11]

If we read, for example, that the cost of housing will go up 20 percent in the next two years, we may ask, "So what does this mean for our plans? Should we stay in our rental house, waiting for the cloud of recession to lift, or should we try to buy a new home now while the prices are still relatively low?"

Likewise, a mechanic reading a repair manual may ask "so what?" when reading "Part C is designed to aid the action of Part D." Does this mean that the mechanic should have Part C examined if Part D fails instead of just replacing the failed part? Or is the statement merely an explanation of operation theory, signifying no real need for action? If it is the latter, perhaps it is not even needed in a manual devoted to servicing the product.

It may be dangerous to dismiss and discard any piece of information early in your interpretive process, but in general, your final notes should include only significant information that points directly to action outcomes. If you can connect your information to your reader's "so whats," then you have effectively interpreted your data. Now you can help your reader create meaning in the context of use. Meaningful information in technical writing usually does one or more of the following:

- shows the reader how to *act*
  Example: If you want to do your part to slow global warming, ride a bicycle or carpool to work.
- helps the user to make a *decision*
  Example: On the basis of the information provided by Professor Schneider, our committee urges the immediate development of alternative energy techniques that reduce emissions of carbon into the atmosphere.
- reveals the likely *outcome* of an action
  Example: If we don't develop the alternative sources soon, our coastlines may be threatened as seriously as our city air already is.

11. We are indebted to our friend and former colleague, Professor Donald Cunningham, for impressing us with the importance of this question. See his useful book, *Creating Technical Manuals* (New York: McGraw-Hill, 1985).

If you interpret your information as you record it, you have started to make your way up the final peak of information development. You are processing the old and creating new information. You have begun to produce knowledge that will make a difference in the lives of those who use it.

Part 1: Case study. Select one of the following situations and list the steps you will take to analyze the audience. For instance, you may begin with a market research analysis, asking questions such as "Who is the target audience?" Determine who the audience is as best you can with the information provided. Finally, explain the major differences between the context of production and the context of use and how you plan to resolve them.

EXERCISE **4.1**

1. Your company has decided to develop and market a complex but useful irrigation system for single-lot, landscaped yards and lawns. This computerized unit will not only enable users to program watering times for various locations in their yards but also allow them to electronically adjust the amount of water they use. After the installation the entire system may be controlled from indoors, and the actual watering apparatus will be concealed in various parts of the landscape. The projected retail cost of the unit will be $1800 plus an installation fee of $500.
2. Your company has developed a multimedia package to take advantage of the growing trend among affluent homeowners to purchase and use a wide variety of electronic components.[12] The package includes equipment to translate normal electronic signals into radio frequencies that are then translated into digital data that is read by a computer. With this package, the homeowner can program and control electronic components singly or collectively. Each component (e.g., televisions, video camcorders, compact disc players, turntables, cassette tape players, receivers, speaker systems, answering machines, computers, printers, fax machines, radios, ovens, yard sprinkler systems, heating and cooling systems, or home security systems) can be linked to a computer software program included in the package, which allows the homeowner to program components to go on or off at specified times. As the company's technical writer, you are assigned the task of developing a document to convince the leaders of a very strong, culturally diverse neighborhood association that your company's request to move its plant to their predominately poor urban area should be approved.

Part 2: Evaluation. Each of the following is a group project, requiring four or five students in each group. Group members should work according to the guidelines for collaboration established in Part One.

1. Find a manual for a computer program (such as WordPerfect® or Excel®) and perform a user test on a specific section or function (such as page numbering) using the talk-aloud method described earlier in this chapter. Write a short report on the usability of the manual, stressing the interaction between the context of production and the context of use that you see between the

---

12. The idea for this exercise came from Adele Duran, one of our technical writing students.

manual and the sample user. What are the differences in the databases between the two, for example, and how are these resolved?

2. Each group should write instructions for how to use one of the electronic retrieval systems in the college library (such as First Search or the ERIC database). This short manual should direct its users to find a few very specific pieces of information in each system. When completed, the manuals will be traded with the other groups in the class, which will perform user tests at the library and evaluate the instructions in terms of market research and rhetorical analysis.

EXERCISE 4.2

Conducting an Internet search.   This exercise requires computer Internet access.

Use a search engine to find out about employment opportunities in your selected field of study. You can do this by either entering key works that describe the job, task, or profession, or by entering the name of a specific company. If you choose the key word option, keep modifying your search terms until the number of "hits" is reasonable. Try to find specific information about the job (e.g., education, skills, experience, salary range).

(*Tip:* If you are dissatisfied with the operation or results of the search engine you accessed, try a different one. Some offer additional key words or search categories to assist in employment searches. While some may take longer than others, the results may be worth the extra wait time.)

EXERCISE 4.3

Part 1. Compiling a bibliography.   In the early stages of research, it's helpful to compile a bibliography, particularly an annotated bibliography, which lists the works you have found and includes a brief summary of each. Working with your group, compile a two- to three-page list of sources relevant to one of the following topics. Try to get sources from more than one tier of readership, and be sure to use a standard citation style (see Sample 4.8).

- municipal regulations concerning bicycling on city streets
- alternatives to the nine-month school calendar
- food irradiation
- financial aspects of community recycling programs
- local gun ownership laws
- recent developments in automotive emissions control
- "video dial tone" technology
- nutritional information labeling accuracy

Part 2. Management plan for original research.   Consider how you might conduct original research to expand your information base for the topic you chose in Part 1. For example, to learn more about bicycling regulations, you might survey local residents about their satisfaction with road access and safety for cyclists, conduct a field study by bicycling to class for a week and recording your experiences, or interview local police officers on accident rates involving bicycles.

Once you have decided how you would proceed, compose a *management plan,* in which you outline the types of research and the breakdown of responsibilities among your group members. See Chapter 3 for a management plan format.

Treat the following passages as though they were sources for a report. For each, provide a summary of the passage and a paraphrase of its main point(s), comment briefly on its importance or value to a study of the topic, and give your interpretation of the passage by telling what makes the passage significant.

EXERCISE
4.4

1. The age of the Renaissance was an age of contrasts and contradictions. It was an age of powerful personalities, cruel military men, clever and ruthless statesmen, but also of exquisite artists, gentle poets, and dedicated scholars. There were men of enormous wealth, but multitudes who suffered abject poverty. It was a time when nights were consumed in debauchery, but also devoted to vigils and prayer. It was a time for display and pomp, but a time also for preachers of penitence, humility, and withdrawal to a solitary life. It was a day for progress coupled with retrogression. It boasted of the dignity of man but bewailed his misery. It could be humanistic and yet act totally inhumane. It coupled a pronounced interest in man with a weariness with life and a longing for a celestial home. The Renaissance was colored in many hues, changing sometimes subtly, sometimes sharply, often swiftly. Contrasts and contradictions are basic to human nature and present in nearly every man; what brings them so clearly into focus in the Renaissance is the fact that history was in rapid movement and Italy was in a state of accelerated transition. (17)

   Spitz, Lewis W. *The Renaissance and Reformation Movements.* Vol. 1. St. Louis: Concordia Publishing House, 1971.

2. At some of the nation's most prestigious schools, the SAT scores of the student body are as follows: at Amherst, 66 percent of the students score above 600 on the verbal and 83 percent above the 600 on the mathematics; at Bryn Mawr, 70 percent above 600 on the verbal and 70 percent over 600 on the mathematics; at Haverford, 67 and 86 percent; at MIT, 72 and 97 percent. The median student SAT scores for the verbal and mathematics portions are 600 at Brown, Columbia, Cornell, Dartmouth, Duke, Georgetown, Harvard, Oberlin, Princeton, Williams, Yale, and other colleges ranked as most competitive, such as Franklin and Marshall, Lafayette, Brandeis, and Lehigh, range in the high 500s and low 600s. (452)

   William, Walter E. "Campus Racism." *Cultural Tapestry: Readings for a Pluralistic Society,* ed. Faun Bernbach Evans, Barbara Gleason, and Mark Wiley. New York: HarperCollins, 1992: 450–455.

3. The next target for productivity enhancement is the systems analyst, data-base specialist, expert systems specialist, et al. Unfortunately, the goal of improving productivity in systems work has been a two-edged sword. On the one hand we are asked to be more *productive*. On the other hand, we are frequently criticized for not producing *quality* systems. Historically, we have viewed productivity and quality as equal and opposite forces. In other words, efforts to improve quality (e.g., *Structured Analysis*) have resulted in reduced productivity (and vice versa). . . . Indeed, data processing and systems development managers find themselves walking a tightrope, trying to find the optimum balance between productivity and quality. But it doesn't have to work that way. We can eat our cake and have it too—enter Computer Aided Systems Engineering (CASE). Improvements in productivity and quality are the prime goals of CASE. (3)

Whitten, Jeffrey L. and Lonnie D. Bentley. *Using Excelerator For Systems Analysis and Design.* Homewood, IL, and Boston: Irwin, 1987.

Recommendations for Further Reading

Berdie, Douglas R., John F. Anderson, and Marsha A. Niebuhr. *Questionnaires: Designs and Use.* 2nd edition. Metuchen, NJ: Scarecrow Press, 1986.

Branscomb, H. Eric. *Casting Your Net: A Student's Guide to Research on the Internet.* Boston: Allyn and Bacon, 1998.

Clark, Carol Lea. *Working the Web: A Student's Guide.* Orlando, FL: Harcourt, 1997.

Fetterman, David M. *Ethnography Step by Step.* Newbury Park, CA: Sage, 1989.

McDowell, Earl E. *Interviewing Practices for Technical Writers.* Amityville, NY: Baywood, 1991.

Meadow, Charles T., and Pauline Cochrane. *Basics of Online Searching.* New York: Wiley-Interscience, 1981.

Odell, Lee, and Dixie Goswami, eds. *Writing in Nonacademic Settings.* New York: Guilford, 1986.

Palmer, Richard Phillips, and Harvey Varnett. *How to Manage Information: A Systems Approach.* Phoenix, AZ: Oryx, 1990.

Perkins, D. N. *Knowledge as Design.* Hillsdale, NJ: Lawrence Erlbaum, 1986.

Redish, Janice C., and David A. Schell. "Writing and Testing Instructions for Usability." *Technical Writing: Theory and Practice,* ed. B. Fearing and K. Sparrow. New York: Modern Language Association, 1989: 63–71.

Additional Reading for Advanced Research

Allen, Jo. "Breaking with Tradition: New Directions in Audience Analysis." *Technical Writing: Theory and Practice,* ed. B. Fearing and K. Sparrow. New York: Modern Language Association, 1989: 53–62.

Grice, Roger A. "Document Development in Industry." *Technical Writing: Theory and Practice,* ed. B. Fearing and K. Sparrow. New York: Modern Language Association, 1989: 27–32.

Grice, Roger A., and Lenore Ridgway. "Information Product Testing: An Integral Part of Information Development." *Perspectives on Software Documentation: Inquiries and Innovations,* ed. T. T. Barker. Amityville, NY: Baywood, 1991: 209–228.

Keene, M., and M. Barnes-Ostrander. "Audience Analysis and Adaptation." *Research in Technical Communication: A Bibliographic Sourcebook,* ed. M. Moran and D. Journet. Westport, CT: Greenwood, 1985: 163–191.

Selzer, Jack. "Composing Processes for Technical Discourse." *Technical Writing: Theory and Practice,* ed. B. Fearing and K. Sparrow. New York: Modern Language Association, 1989: 43–50.

Writing Core Documents

CHAPTER OUTLINE
▼▼▼▼▼▼▼▼▼▼▼▼▼▼▼▼▼▼▼▼▼▼▼▼▼▼▼

Choosing a Primary Genre: Report, Proposal, or Manual

Designing the Core Document

Writing the Core Document

CHAPTER OBJECTIVES
▼▼▼▼▼▼▼▼▼▼▼▼▼▼▼▼▼▼▼▼▼▼▼▼▼▼▼

After you have worked through this chapter, you should be able to do the following:

- Differentiate between characteristics of the three primary genres in technical communication
- Recognize the appropriate genre for your information product
- Design and develop a core document in any of the genres

Once you have completed your audience-action analysis and profile and conducted preliminary research, you are ready to develop the core document. In the CORE method described in this textbook (see Figure 5.1), the core document is written early in the process and is both preceded and followed by research and revision. This chapter will guide you through the processes of identifying the appropriate genre for your core document (modifying your projected audience or purpose as needed) and designing a position paper, concept paper, or task analysis.

Each of the three primary genres of technical communication has its own kind of core document. Thus, the first step in writing the document is to decide which genre will best communicate your information.

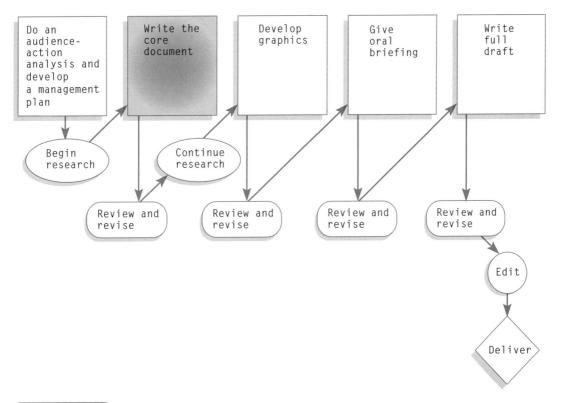

FIGURE 5.1

The place of the core document in the CORE method.

Choosing a Primary Genre:
Report, Proposal, or Manual

Though there are many variations in format, the key genres are the report, the proposal, and the manual. All of these genres deal with problems and actions, but at different stages. Reports usually review and evaluate past actions, then recommend general courses of action for the future. Proposals consider a present problem and offer specific solutions for the future. Manuals help users solve specific problems in the present.

If we give each genre a voice,

- The report says, "I did this."
- The proposal says, "I will do this."
- The manual says, "Do this."

Notice that the report "voice" uses the past tense and the first person. The typical author of a report is an expert using experience (usually research or special knowledge) to advise an audience of decision makers. Of course, supervisors and observers also write reports; in these instances a better summation would be "He or she (or they) did this." Also, in some instances the action is more important than the performers (as in scientific papers), so an accurate summation would be "This was done."

In all of these instances, the report deals with actions that have already occurred. But a report usually presents past actions with an eye to shaping the future. This commitment to the future is clearest in recommendation reports, which often borrow elements from manuals and proposals. The "voice" of a recommendation report would say, "I (an expert observer of past events and present conditions) think you (my client) should do this." An environmental engineer, for example, might recommend that the city of San Diego adopt a waste disposal plan that worked well for New York City in the past.

Notice that the proposal sentence is in the future tense and the subject of the sentence, the author, is clearly involved in the planned action. The identity of the author may change from *I* to *we* in corporate or collaborative proposals. In this instance, the reader is an observer or judge—probably a funding agent—or someone who controls the outcome of the project. But in this genre the proposers are the actors. They must prove that they are qualified to carry out the plan they are proposing—a set of actions that they must clearly specify.

Finally, notice that the manual sentence is in the present tense and imperative mood. The author's aim is to control action that takes place at the time of reading or just after reading. The active agent this time is the reader, not the author. The sentence says, in effect, "I (the author) am showing *you* (the reader) how to *do this*." By addressing the reader directly, the manual builds confidence, encourages and sustains interest, and guides the way through difficult procedures.

Table 5.1 compares the three genres of technical communication. The column for technical reports has been subdivided to reflect differences between technical reports on options or recommendations (Chapter 12) and scientific or experimental reports (Chapter 13). Notice that the genres differ in purpose, in-

TABLE 5.1

A comparison of characteristics of the four major genres of technical writing.

| | Technical Reports | | | |
	Reports on Options or Projects	Scientific or Experimental Reports	Technical Proposals	Technical Manuals
Purpose	To provide *information* directing a specific *audience* toward a decision about future *action* within a specific *context of use*	To advance a research agenda by contributing new information to a specialized field of scientific study	To convince an audience of evaluators that the author(s) can solve a problem by taking a specified future action; to demonstrate that something *is* broken and that *you* can fix it	To provide a *guide* that leads a *user* (usually a novice) through a technical *procedure*
Desired Audience Action	To make a decision or perform an action recommended in the report	To engage in research and discourse on a specific topic	Approval of funding or other support for carrying out the action plan	Carrying out the detailed action plan
Style	Active voice, knowledgeable, informative, to-the-point	Passive voice commonly used, with emphasis on the action; impersonal; reproducible	Narrative, with emphasis on future action; positive, hopeful tone that inspires confidence	Accessible prose, active voice, simple imperative, easy-to-use format, authoritative, helpful tone, efficient, use of graphics to display procedures
Tense	Past ("I did this")		Future ("I will do this")	Present ("Do this")
Core Document	Position paper		Pre-proposal or concept paper	Task analysis
Structure	Introduction and Background, Discussion, Conclusion	IMRaD: Introduction, Methods, Research, and Discussion	Detailed *plan of action* keyed to *a specific set of objectives* based on *projected outcomes*	Tutorial or reference

tended audience action, style, and tense. (Differences in document design will be addressed in Chapter 7.) Each also requires a different type of core document that reflects basic structural differences between the genres.

You can now return to your audience-action profile and determine the appropriate genre for your information product, based on its purpose and intended audience action. For example, if you want your audience to write letters to their state legislators protesting genetic discrimination by health insurance companies, a technical report with recommendations would be appropriate. If, however, you want your audience to provide computer equipment and space for you to pursue research on a new software program, you would write a proposal. A technical manual would be a more appropriate genre for guiding your audience in constructing and maintaining an outdoor ornamental pond.

Once you have determined the appropriate genre, you can use Table 5.1 to develop a core document that effectively incorporates the indicated tense and style preferences. Core documents are short, concise narratives (1–2 pages) that address all the major points you intend to present in your final product—at least those that you have uncovered at this stage of your research.

Designing the Core Document

In business and professional settings, core documents give managers brief overviews of projects so they can decide whether to commit the time and resources needed to complete the projects. In fact, it is not uncommon for a project to be scrapped after a core document review. Thus it is crucial that the important points you hope to make in the report, proposal, or manual are not only included in the core document, but are also presented in a form and style that will convince your readers that they *do* want to see more. To accomplish this, you need to show your readers that:

- You have done your research (you know what experts in the field have to say on the topic)
- You know what you are talking about (you have developed some expertise in the area)
- It is worth your readers' time and effort to read the full document

In some ways, your core document is similar to a preview for a movie. It allows you to showcase the best and most important ideas you intend to present. If your core document uses excessive jargon that is unfamiliar to your readers or comes across as dictatorial or dry (styles not particularly preferred in most technical communication genres—see Chapter 9), your project could very well be dead in the water. On the other hand, if your audience is a group of enthusiastic racing bicyclists and your narrative uses no racing or bicycling terminology, you might insult your readers' intelligence or give the impression that you don't really know much about this specialized topic, with the possible result that your readers would have absolutely no interest in finishing the core document, much less in reading a full written report, proposal, or manual.

Matching Purpose, Audience, and Intended Action with Genre and Style

Technical communication products are more effective if they do not try to do too much. The role of journalism is to inform the public, but technical communication (which may also involve journalism) is more focused. Not only do you want your audience to know about a specific issue, but you also want them to take very specific actions after reading your document.

For instance, a magazine article informing the public about the slaughter of starving buffalo by Montana state officials in Yellowstone National Park might use disturbing photographs and make passionate emotional appeals to elicit a public outcry. But the journalist has no real interest in whether the audience takes any action as a result of reading the article. The journalist's goal has been accomplished once the reader has read the article. On the other hand, a technical report on the same subject might involve exactly the same types of research and might even use some of the same journalistic techniques to garner public support. It will differ from a journalistic piece, however, in that:

- it will target a specific audience rather than the general public, and
- it will be written for the purpose of getting that audience to make a decision or follow a specific action that you recommend.

The two key factors, then, are audience and purpose. These factors can be manipulated so that you can select the appropriate genre and find the right style to reach your intended audience effectively. Let's first consider audience.

Revisiting Audience

When developing your audience-action analysis, you identified a target audience for your information product. You may even have conducted market research to determine the needs and peculiarities of your particular audience. However, at this point, you may need to re-examine and narrow your audience, focusing on a smaller group than you originally intended. Ask yourself: Who needs to read this information product so that the actions or decisions I recommend will be carried out? A rhetorical analysis of the communication situation may help you eliminate some groups not immediately excluded.

For example, let's reconsider a scenario from Chapter 1:

> Your company has decided to develop and market a complex but useful irrigation system for single-lot, landscaped yards and lawns. This computerized unit will not only enable users to program watering times for various locations in their yards but also allow them to adjust electronically the amount of

water they use. After the installation, the entire system can be controlled from indoors, and the actual watering apparatus will be concealed in various parts of the landscape. The projected retail cost of the unit will be $1,800 plus an installation fee of $500.

While it may appear, at first glance, that the projected audience for this scenario is anyone interested in buying an automatic sprinkler system, the following shows how one student was able to narrow the projected audience considerably by analyzing the scenario rhetorically.

Rhetorical Analysis of Projected Audience
by Mark Hosack

The product being sold is a complex computerized irrigation system for a single-lot yard. This eliminates those who live in apartments, town houses, or all other members of the human population who do not live in the stereotypical urban or suburban family home.

This system is also labeled "complex" and "computerized." This nuance leads me to feel the product is targeted toward those who put time and effort into the landscape of their yards, not those who simply mow their lawns. The terms *complex* and *computerized* also imply an audience that thinks of themselves as "complex" and capable of using a computerized product. This same audience, if they were to see an irrigation system labeled as *simple* and *manual*, might believe themselves to be "above" the product. The labels *complex* and *computerized* are designed to attract an audience that believes they are "sophisticated" and beyond the simple layperson. However, the term *computerized*, in this day and age, also implies a machine that works for you, a nonthinking product—push the button and it works. This may be more attractive to a "lazier" audience or one that is too busy to maintain their lawns with a simple system.

The apparatus is concealed in the landscape. Thus the audience must be concerned about the landscape's appearance. The irrigation system appeals for aesthetic reasons.

The last detail mentioned is the price. Since I have no prior experience pricing irrigation systems, I do not know if the $1,800 cost and the $500 installation fee are high, but from the description, I will assume that they are. Thus, the target audience must also be either middle to upper class, or possess a great love for their lawn and believe more money creates more beauty. To summarize, the target audience consists of those people who

- own a single-lot yard,
- exhibit a sense of aesthetics,
- live in relative wealth or possess great appreciation for landscaping, and
- believe themselves to be capable of operating a "complex," computerized machine.

As you begin to develop the core document, you may need to modify your target audience so that the genre and style you select will most effectively reach the readers and persuade them to do what you want them to do. It will be much easier now for Mark to design an information product suitable for this more specific audience.

Revisiting Purpose and Intended Audience Action   You may also need to revise your intended purpose, which should involve audience action. You have probably written academic reports throughout most of your school career—reports written for your teacher to prove that you could conduct research, write, and follow a specified format. Thus, when researching topics for your technical communication products, it is easy to fall back into old habits and lose sight of your purpose.

Academic writing allows you to prove or demonstrate knowledge to your readers. The information you present is of paramount importance, while other issues (such as reader enjoyment or increase in reader knowledge) are less important. However, *technical* writing focuses on what the audience is supposed to *do* with the information you present. The information itself is flexible, in that you may need different kinds of information to influence different audiences. As you conduct your research and begin drafting your core document, try to keep your intended audience action foremost in your thoughts.

Technical communication products are *not* exhaustive, extensive documents that cover every aspect of a topic. Rather, they strive to present *only* the most important information to users—the information that will help them make the decision or perform the action you suggest. Your task, then, is to sift through and condense the information so that users can quickly grasp all the main points and issues (without losing sight of the complexity of the topic) and find out exactly what you are recommending or instructing them to do.

Writing the Core Document

The core document represents your first attempt to move from structured information—the working outline and notes developed in your research—toward an argument that contains actual positions and recommendations. Some writers find it useful to construct an outline at this point, but an outline is a structure with only an implied argument. The core document should make the argument explicit.

If you do use an outline, remember that it is for *your* use. This outline may be the same one you use for your complete information product. However, it is more likely that it will change somewhat based on new or pertinent information you uncover as you continue research, or on feedback you receive from your instructor or peers after completing your core document and oral presentation.

Remember that your core document is *not* the final product but only a summary of the main points to inform your manager, employer, or funding agency

GUIDELINES AT A GLANCE 7

Writing a position paper for a technical report

First paragraph (Introduction and Background):
- State your purpose.
- State your position or recommendation on the issue.
- Provide brief background information, explaining key terms and orienting the audience to the project or problem.

Second paragraph (Discussion):
- Bring the audience up to date by briefly describing the current status of the project.

Third paragraph (Conclusion):
- Explain your position briefly, giving special attention to the benefits of the decision you recommend.

of the nature of the project you are working on. As such, it should consist of only 1–2 pages.

Drafting a Position Paper

Technical reports can be categorized into several types: reports on options, reports on recommendations, and reports on original scientific or experimental research. The last type uses a unique IMRaD structure discussed in Chapter 13 and is thus listed as a separate category in Table 5.1. However, the first two types essentially follow the same generic structure, with only slight differences (discussed in Chapter 12). As indicated in Table 5.1, the structure of a technical report includes three sections:

   I. Introduction and Background Information
  II. Discussion
 III. Conclusion

Guidelines at a Glance 7 shows how to develop a *position paper*—the core document for a report.

Sample 5.1 is a position paper for a student project—a technical report recommending a specific approach for recovering oil from old wells.

Remember from Figure 5.1 that the core document is also subject to the ongoing process of review and revision. As such, the quality of your draft position paper can be significantly improved if you submit it to your instructor or peers and incorporate the feedback into the final position paper.

Enhanced Oil Recovery (EOR) Techniques

by Lee Thomas

The purpose of this report is to provide information about possible options for an enhanced oil recovery program to a managerial team that is in the process of initiating a program to increase the production in the company's oil field. Several options are possible, but the most effective course of action is the combination of a steam flood and a well stimulation program.

Enhanced oil recovery techniques are employed when an oil field reaches a point where the rate of production significantly slows or decreases. EOR techniques can boost the production rate and increase the amount of recoverable oil. Secondary recovery techniques facilitate the flow of oil to the well bore using several basic concepts: restoring the pressure lost in a field due to production, pushing the oil through the formation using a variety of fluids, decreasing the forces between the porous formation and the oil, and repairing damage done to the formation due to oil production. A wide range of techniques is available, from the use of microbes to flooding methods. Each is suitable for different oil types and formation characteristics.

The field under consideration has reached maturity and production has begun to decline. Secondary recovery techniques should be employed to increase the life of the field and the amount of oil that can be produced. The combination of a steam flood and well stimulation program will increase the rate of production and extend the life of the field. Steam generated using a cogeneration power plant, supplied by produced water, can serve a dual purpose. The steam will be injected into the formation and the electricity generated by the power plant can be sold to a utility company. The steam flood portion of the operation will generate money and offset the cost of the well stimulation program. An outside contractor will conduct the well stimulation program. This approach will limit liability considerations and reduce the problem of a large start-up expense.

The well stimulation program should help remove problems with the older well caused by blockages in the formation surrounding the well bore. The recommended enhanced recovery program should result in increased production with a limited amount of liability and a minimum cost.

SAMPLE 5.1   Student position paper for a technical report

Drafting the Concept Paper or Pre-Proposal

In writing proposals, not only must you demonstrate that you have done your research, know your topic, and have something you want your reader to do, but you must also convince the readers that the problem is a serious one with practical solutions and that you are the person who can best solve the problem, with their support. The structure for proposals (from Table 5.1) thus calls for a detailed plan of action keyed to a specific set of objectives based on projected outcomes. The core document for this genre is a *concept paper*, also referred to as a *pre-proposal*.

The writing process for research proposals, as for all proposals, begins with the discovery of a problem or need. Sometimes the funding agency identifies a problem, then seeks a qualified applicant to handle the solution; sometimes an applicant determines the problem and seeks funding from an agency.

In the former situation, the agency generates a *solicited proposal*. It has the money but lacks the expertise or facilities to solve the problem, so it issues a *request for proposal*, or *RFP*, to find applicants who believe they can solve it. Then the agency, assisted by a board of experts, evaluates the proposals and awards a project contract to the proposal writer who presents the most persuasive plan. Some agencies also issue *requests for bids*, which provide specifications for solving the problem and simply request a price tag.

Though some funding agencies are corporations and businesses, most typically they are government bureaus (the armed services, the Forest Service, the Department of Agriculture, etc.). Through RFPs published in outlets such as *Commerce Business Daily* and *The Federal Register*, such groups seek temporary partnerships with other organizations or individuals who have access to special information or techniques.

A solicited proposal is a responsive document. It says, "I (or we) can provide the help you need in solving the problem." The proposal writer promises to deliver the best product or service available. Such proposals usually involve objectives connected with technological research and development, but they may be written to win everything from defense contracts for building missiles to education contracts for workshops on children's literature.

If the proposal writer is the one who recognizes the problem and will ultimately solve it, then he or she submits an *unsolicited proposal* to a funding agency that maintains a budget for innovative and creative work defined within fairly general guidelines—usually presented as "mission statements." Groups of this sort include government organizations such as the National Science Foundation, the National Institutes of Health, the National Endowment for the Humanities, and the Fund for the Improvement of Post-Secondary Education as well as private foundations such as the Rockefeller Foundation, the Guggenheim Foundation, and the Smithsonian Institution. These groups offer grants to researchers, mainly from universities, whose studies can contribute to a global program of systematic research.

Unsolicited proposals are research initiatives. They claim, in effect, that with the help of the funding agency, you can deliver results that will make a difference in your field of study. Ultimately, your work should be publishable. When your book or article appears in print, it will acknowledge funding from the agency, so the agency's mission will be fulfilled in a visible way. The agency can then show its supervising institution (for example, Congress) a list of important publications that report work that it funded. In this way, agencies and authors support and perpetuate each other's work. Their collaboration is another form of action partnership.

The difference between solicited and unsolicited proposals relates to who initiates the work. This difference can have a profound effect on the research. For solicited proposals, the funding agency sets the goals; for unsolicited proposals, the proposal writer establishes the need for research and carries it out.

However, the distinctions between solicited and unsolicited proposals are not always so clear-cut. Many RFPs let proposal writers decide what they want to do, and many agencies that accept unsolicited proposals regularly issue *calls for proposals* on specific topics.

There is a range in the amount of responsibility and freedom that a proposal writer has in writing both kinds of proposals. Figure 5.2 shows a continuum on which one extreme is represented by requests for bids, for which proposers have no responsibility for problem analysis. At the other extreme, represented by the free grant, or award (such as the Nobel Prize), the researcher has full responsibility for defining and implementing the project. Between these two extremes lies the *proposal zone,* wherein the funding agency and proposal writer bear varying degrees of responsibility for problem analysis.

The structure for this genre is similar to that for technical reports in that both include introduction and background sections. However, because the

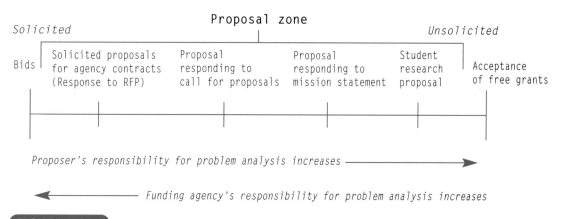

FIGURE 5.2

The proposal zone.

GUIDELINES AT A GLANCE 8

Writing a concept paper or pre-proposal for an informal technical proposal

---

Paragraph 1 (Introduction):
- Briefly state the problem that the proposal addresses.
- Briefly explain the purpose of the proposal in terms of the intended audience's interests.
- Briefly describe how you plan to solve the problem.

Paragraph 2 (Background):
- Analyze the problem by describing the old way of doing things.
- Explain key terms and orient the audience to the problem.

Paragraph 3 (Plans and Requirements):
- Describe how your project will solve the problem.
- Describe what funds, equipment, supplies, or services you will need to solve the problem.

Paragraph 4 (Benefits):
- State the benefits of the project for all parties involved.

---

proposal genre goes beyond recommending a decision or action by requesting support for a project that you identify, describe, and purport to solve, you must explain not only how you intend to solve the problem, but also what you will need from your readers to do so and how they will benefit from supporting your effort. The structure looks like this:

I. Introduction
   A. Brief statement of the problem
   B. Brief statement of the solution
   C. Brief statement of the project needs
II. Background on the problem
III. Project plans and requirements
IV. Benefits of the project

You can use this structure effectively for informal proposals directed to managers, employers, or panels of judges. Guidelines at a Glance 8 lists steps in writing a concept paper or pre-proposal.

More formal proposals, such as research and contract proposals, may require a more detailed core document that includes any or all of the following sections:

- Literature review
- Problem analysis
- Rationale for the project
- Objectives
- Project management and personnel

- Timeline
- Products and their dissemination
- Budget
- Cost–benefit analysis
- Evaluation of effectiveness
- Guarantees or assurances that you can do what you say you can

Instead of guessing which sections, if any, the funding agency may require, we recommend that you call the agency and request a copy of the RFP or any guidelines the funding agency can provide.[1]

Whether you are developing a formal or informal proposal, the core document will need to provide all of the information a funding agency needs to know before supporting your project. Once it has been drafted, get feedback from your instructor or peers so that you can revise it. Sample 5.2 is an example of a completed concept paper.

Ozonation as a Primary Water Treatment Process:
A Concept Paper

Current techniques used by the city of Bryan in the treatment of drinking water involve the use of very large amounts of extremely hazardous and environmentally unsafe chemicals. To meet new federal regulations and to create better environmental standards for the future, the city's water treatment plant needs to implement an optional method of treating the community's water supply. This proposal introduces an improved process called ozonation.

The current system relies too heavily on chlorine. Chlorine is known to be an effective antibacterial chemical and has enjoyed wide use. But recent studies question the heavy use of chlorine. They show it to be more hazardous to human health than previously believed, as well as environmentally unsafe and increasingly expensive.

By contrast, ozone used as a source of water treatment is a safe and effective means of complying with regulations and ensuring safe water resources for the future. This proposal asks for funding to generate a technical and cost proposal to be submitted to the Environmental Protection Agency as a first step toward implementing an ozone-based treatment system for the city of Bryan.

Because the plan is incremental, going one step at a time, it avoids the high costs usually associated with re-engineering water treatment facilities. But it moves the city in the right direction, toward safer and more future-oriented water engineering.

SAMPLE 5.2   Concept paper for an engineering proposal (see the full proposal in Chapter 14, Sample 14.7)

---

1. For more information on developing formal research and contract proposals, we recommend Herman Holtz and Terry Schmidt, *The Winning Proposal—How to Write It* (New York: McGraw-Hill, 1981).

Drafting a Task Analysis for a Technical Manual

The procedure for designing technical manuals is very different from that used in planning technical reports and proposals. Rather than trying to persuade readers to act, you will guide them through a series of actions they will perform. Thus you can abandon the background section, because there is no need to convince the reader that there is a problem. If not already convinced that a problem exists, he or she would not be using your manual.

Just as there are several types of technical reports, manuals can also be categorized into two types: tutorial or reference. Tutorial manuals are designed to be used once by an individual, who, after following the procedure, does not need to refer to it again. Of course the manual can then be passed along to other individuals for subsequent use, but its purpose will have been accomplished after a single use. An example would be a manual on installing a fireplace screen. Once installed, there is no need to take the screen out and repeat the procedure. A reference manual, on the other hand, is designed as a desktop companion to be used again and again by the same individual or group of individuals whenever problems develop. An example would be a troubleshooting manual for a computer's hard drive. You may never need it, but should you have difficulties with your machine, you could use the manual to identify a specific problem and try to fix it yourself.

There are slight differences in the overall structure of a tutorial and reference manual, but both follow this basic format (remember, the emphasis is not on the information, but on the actions readers are to perform):

I. Introduction
    A. Purpose and Audience
    B. Objectives
    C. Overview of Manual Contents
II. Identification of first task or problem
       Step-by-step solution to first task or problem
III. Identification of second task or problem, etc.

Tutorial manuals should focus on solving a single problem while reference manuals may discuss multiple problems that must first be identified and explained. Thus, under section II, your tutorial core document will simply state the problem and divide it into a series of tasks, each of which will then be subdivided into steps your reader can follow. However, reference manuals must go into more detail about each problem, perhaps describing multiple symptoms to help the user determine exactly what needs to be fixed. Once the problem is identified, the procedure is the same, with each task subdivided into steps. Chapter 15 provides a schematic for helping you break tasks or problems into concrete, feasible steps. Once this has been done, you will need to draft a com-

GUIDELINES AT A GLANCE 9

Writing a task analysis for a technical manual

Paragraph 1 or Heading (Introduction):

- Give the title of the manual.
- Briefly describe the intended audience.
- Briefly state the purpose of the manual.
- If you are writing a tutorial manual, provide a brief overview of the procedure your reader will follow.
- If you are writing a reference manual, provide a brief overview of the manual, describing its organization.

Paragraphs 2, 3, etc. (Task Analysis):

- If you are writing a tutorial manual, simply transfer the lists of tasks and steps from your outline to your memorandum or letter. (Be certain that each step is an imperative sentence.)
- If you are writing a reference manual, explain the problem and how to identify it as the problem; then transfer from your outline the lists of tasks and steps for the problem you have identified, guiding your reader through the tasks and steps in solving it; repeat as necessary, until all problems have been addressed.

plete imperative sentence for each step, telling your reader what to do. Try to limit the number of steps to seven, breaking down tasks into sub-tasks if necessary. The sentences about the first task or problem will make up section II. For reference manuals, sentences about the second task or problem will make up section III, and so forth.

Guidelines at a Glance 9 lists the steps in developing a *task analysis.*

Sample 5.3 is a student task analysis for a tutorial manual.

EXERCISE 5.1

For this exercise, form a technical writing team with three or four classmates and discuss how you would approach the following problem.

After reading Lee's position paper on EOR techniques (Sample 5.1), your manager or employer assigns your technical writing team the task of modifying it so that it falls into a different genre. Instead of a recommendation report, she asks you to develop a concept paper to solicit funding for the project. What changes will your team need to make to the purpose, audience, and intended audience action? What other changes will be required for the team to complete the assignment?

Title: Basic CPR

Audience: Anyone who needs to learn cardiopulmonary resuscitation (CPR)

Purpose: To give the user a general familiarity with the processes of CPR, as a foundation for a "hands-on" course

Task Analysis:
Preliminaries
1. Check for unresponsiveness.
2. Shout for help.
3. Position the victim.
4. Remember your ABCs: Airway, Breathing, and Circulation.

Airway: Open the victim's airway by using the head-tilt/chin-lift method.
1. Tilt the head.
2. Lift the chin.

Breathing: Check for breathing using the look/listen/feel method.
1. Look at the chest to see if it is moving.
2. Listen for breathing noises.
3. Feel for air from the victim's mouth.

If the victim is not breathing, perform the mouth-to-mouth technique.
1. Pinch the victim's nose with the hand holding the forehead.
2. Sealing your lips tightly around the victim's mouth, give four full breaths.
3. Watch for the chest to rise with each breath.
4. Pause long enough between breaths to completely refill your lungs.
5. After four full breaths, take the victim's pulse.
6. If the victim has a pulse but is not breathing, give a full breath every five seconds.
7. If the victim has no pulse, begin compressions to restore circulation.

Circulation: Compress the chest to get the heart pumping before continuing mouth-to-mouth.
1. Kneel beside the victim.
2. Locate the correct hand position.
3. With both hands, apply pressure to the victim's sternum and then release.
4. After 15 compressions, give two breaths.
5. Check again for pulse and breath.
6. Maintain 15 compressions per two breaths until breathing and pulse are restored or until help arrives.

SAMPLE 5.3   Task analysis for a CPR manual (tutorial)

EXERCISE
5.2

Identify the major funding agencies or organizations that support research projects in your field. Contact two or three of them and request that they send you sample RFPs or funding guidelines. When you have received them, determine what additional information they require beyond that which you are instructed to address in Guidelines at a Glance 8.

For this exercise, form a technical writing team with three or four classmates.

Think about how three or four favorite businesses in your community operate. If any perform less efficiently than you would like, identify those areas that could be improved. Then do a task analysis for each problem to determine whether a technical manual could be used to solve the problem. Consider presenting your task analysis to the manager of the business and using this as a project for developing a full manual (see Chapter 14).

EXERCISE
5.3

Recommendations for Further Reading

Schriver, Karen A. *Dynamics in Document Design.* New York: Wiley, 1997.

Additional Reading for Advanced Research

Killingsworth, M. Jimmie, and Michael Gilbertson. *Signs, Genres, and Communities in Technical Communication.* Amityville, NY: Baywood, 1992.
Miller, Carolyn R. "Genre as Social Action." *Quarterly Journal of Speech* 70 (1984): 151–167.

chapter

6

Developing
Purposeful Graphics

CHAPTER OUTLINE
▼▼▼▼▼▼▼▼▼▼▼▼▼▼▼▼▼▼▼▼▼▼▼▼▼

Planning Graphics for Specific Functions
Producing Report-Quality Graphics

CHAPTER OBJECTIVES
▼▼▼▼▼▼▼▼▼▼▼▼▼▼▼▼▼▼▼▼▼▼▼▼▼▼▼▼

After you have worked through this chapter, you should be able to do the following:

- Use graphics as a primary means of developing technical information
- Plan and design graphics that enhance the usability and rhetorical effectiveness of technical documents
- Recognize and implement various methods for producing report-quality graphics

The graphics-oriented (GO) method of document development was conceived by a group of technical writers who had to produce multivolume proposals with tight deadlines and many contributing authors.[1] Their method challenges the old notion that graphics are "visual aids" that merely decorate or slightly enhance a document that is primarily textual. Instead, the GO method treats graphics as a primary communication tool. This outlook is fast becoming the norm in technical communication.

Realizing the preference for visual representation among both technical professionals and ordinary users in the video age, the GO method urges authors to present the most important information in visual form. Ideally, the graphics should be almost independent of the written text. The advantages in production—in which visual displays work better than words as the basis for discussions among coauthors—are matched by the advantages in use for readers who face the daily task of evaluating hundreds of proposals. Reviewers of submissions for federal grants may spend as little as twelve minutes on each proposal they read, and good graphics allow them to grasp the "big picture" at a glance.[2]

As Figure 6.1 shows, the GO method's philosophy is integral to the CORE method for producing technical documents. As soon as you establish your main points in a short core document you should immediately begin producing graphics to summarize and display the content you've discovered in your research. This practice will help you to further develop and refine the information. It will also start you on your way to producing the kind of visually rich documents that busy information users prefer.

If you are not used to thinking visually during the writing process, you may have to train yourself to adopt this new orientation. But you do not have to become a great artist. For the technical writer, understanding the functions and rhetorical qualities of graphics is more important than being able to create fancy drawings. As a technical professional, you will probably be able to hire a graphic artist when you need one; as a student, you can create report-quality graphics using readily available technology, such as the photocopy machine, the 35mm camera, and the graphics packages that are available for personal computers. The purpose of this chapter, then, is to help you *conceive* good graphic plans and to implement those plans, at least in a rudimentary form.

Planning Graphics for Specific Functions

In technical communication, graphics combine words, diagrams, pictures, and other design elements to allow the reader to take in information at a glance. In

1. R. Green, "The Graphic-Oriented (GO) Proposal Primer," *Proceedings of the International Technical Communication Conference* (1985) VC30.

2. See Robert Lefferts, *Elements of Graphics: How to Prepare Charts and Graphs for Effective Reports* (New York: Harper & Row, 1981) 5. See also Rudolf Arnheim, *Visual Thinking* (Berkeley: University of California Press, 1969) 249.

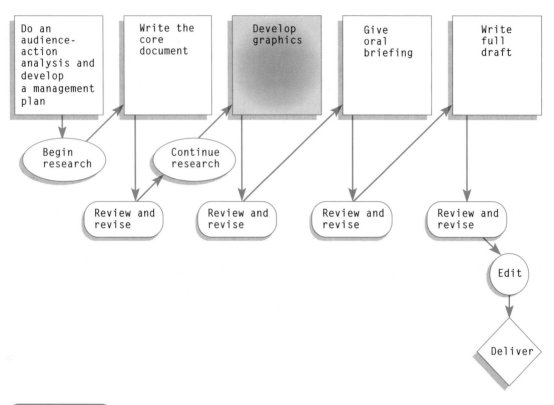

FIGURE 6.1

The place of graphical development in the CORE method for producing technical documents.

oral reports, these graphics (including outlines and lists of words, phrases, or short sentences) are the only text the audience will read; the rest of the information is delivered orally by the presenter. In written reports, graphics support and complement the written text in special ways.

To see how a graphical display can do things that writing can't (and vice versa), compare the paragraph in Sample 6.1 to the table in Sample 6.2. How well does the table convey the information in the italicized portions of the paragraph as compared to the nonitalicized portions?

Obviously, the table cannot make the kinds of comparisons and correlations that are stated in the italicized phrases of the paragraph. And the data in the table are awkwardly represented in the paragraph. So the table cannot completely replace the paragraph, nor can the paragraph effectively replace the table; verbal and visual elements are not interchangeable but *complementary.*

Bill weighed 200 pounds in January, 180 in February, and 170 in March. *Like Bill, Joe and Harry lost 20 pounds in the second month and 10 in the third month of the program.* Joe weighed in at 190 in January, dropped to 170 in February, and finished at 160. *In similar fashion,* Harry fell from 180 to 160 to 150.

SAMPLE 6.1   Paragraph that needs a table

Source: M. Jimmie Killingsworth and Michael K. Gilbertson, *Signs, Genres, and Communities in Technical Communication* (Amityville, NY: Baywood, 1992) 45.

Different categories of graphics complement the text in different ways—that is, they serve specific, conventional functions within the context of technical communication. Before you can develop effective graphics, you must ask yourself a series of analytical questions:

- What are the most important bits of information I have to convey to my audience?
- What is the rhetorical purpose of that information? In other words, what function does each bit of information serve in light of the overall purpose of my information product?
- What kind of graphic would best suit this function? In other words, which kind will enable my audience to understand key information, remember it, and process it as part of my overall message?

The following sections will help you answer this last question and provide examples that you can use to begin your planning.

	Jan	Feb	March
Bill	200	180	170
Joe	190	170	160
Harry	180	160	150

SAMPLE 6.2   Table showing weight record (by pounds) in a three-month program

Source: M. Jimmie Killingsworth and Michael K. Gilbertson, *Signs, Genres, and Communities in Technical Communication* (Amityville, NY: Baywood, 1992) 46.

Use Graphics to Preview and Review Key Points

A graphical display helps an audience sort information and perceive the structure of the information.[3] In Sample 6.2, for instance, the table does not make *statements* about the weight watchers' losses, but it does enable the reader to see those changes and relationships very quickly.

In an oral report, you can use outlines and lists in a similar way—as forecasting devices to show your audience the points you plan to cover in your report and to review points you have already made. By previewing and reviewing the information in this way, you help viewers to form a receptive mental structure for your report—an empty array to be filled up as they listen. Ideally, this will enable them to perceive your report as structured information rather than as random thoughts and impressions.

Outlines such as the one in Sample 6.3 are particularly effective for previewing and reviewing whole reports, whereas *lists* are more useful for shorter sections, as shown in Sample 6.4.

In both outlines and lists, *keep the number of words to a minimum*. Also, *lists should be grammatically parallel*. That is, every element should have the same grammatical structure—nouns with nouns (Sample 6.4), phrases with phrases (Sample 6.3), sentences with sentences (Sample 6.5).

Interventions in Early Childhood

<u>Purpose</u>: To provide information on reducing the risk of violent behavior among adolescents.

<u>Position</u>: Intervention in adolescents is too late to prevent violent behavior; intervention must begin in childhood.

1. Factors Leading to Violent Behavior
   1.1   Family and peer relationships
   1.2   Living environment
   1.3   Media violence
2. Methods of Early Childhood Intervention
   2.1   Care giving and discipline skills
   2.2   Anger management
   2.3   Communication
   2.4   Conflict resolution and problem-solving skills
3. The Need for Further Research and Evaluation

SAMPLE 6.3   Outline as a previewing device for an entire oral report

3. Carolyn Rude, "Format in Instruction Manuals: Applications of Existing Research," *Journal of Business and Technical Communication* 2.1 (January 1988): 63–77.

```
Critical Issues for Further Research
  • cultural norms
  • socioeconomic status
  • media violence
  • children's resiliency
  • social policy
```

SAMPLE 6.4   List as a previewing device for one section of an oral report

Finally, use *bullets (•)* or other such symbols for lists in which the order of the items is determined by importance or simply by the sequence of the report (as in Sample 6.4). Use *numbers* in chronological lists or lists in which the sequence of items represents a special order of actions (Sample 6.5).

Flowcharts are also good devices for preview and review in both oral and written reports, especially in those that deal with processes or sequences of events. Lists work fine for showing a linear sequence of events, as in Sample 6.5. Flowcharts work better for showing interrelated (multilinear) events that do not always "flow" in a straight path but instead branch, overlap, or are recursive.

Like tables, flowcharts can greatly simplify expressions that would be tortuous in prose. Compare, for instance, the paragraph in Sample 6.6 to the flowchart in Sample 6.7, shown on page 142.

It is best to keep flowcharts simple and readable, especially in oral reports. If you want to pack in a lot of complex information, use more than one flowchart to illustrate it.

```
Problem-Solving Method
  1. Define the problem.
  2. Analyze the problem.
  3. Consider options for solving the problem.
  4. Select an option.
  5. Implement the selection.
  6. Develop criteria for evaluating the process and results.
```

SAMPLE 6.5   Numbered list of chronological events

Once the project begins, the product team creates version 1 of the software, while the research team gathers the initial data. When they have completed these tasks, each team inputs their results, and the testing team evaluates their work. If the testing team finds the input of either team inadequate, they can return the material to either the product team or the research team. The team that gets the information returned then must revise and submit to another evaluation by the testing team. Once the testing team validates the input, it authorizes initial release of the product.

SAMPLE 6.6   Paragraph (in need of a flowchart) describing interrelated multilinear events

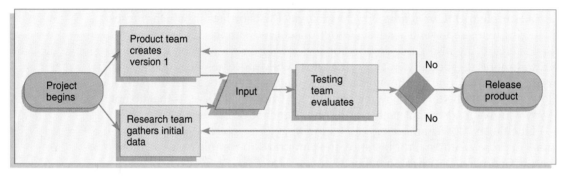

SAMPLE 6.7   Simple flowchart showing interrelated, multilinear events

Use Graphics to Orient the Audience in Space and Time

Maps, flowcharts, and other kinds of charts and diagrams can be more effective than verbal texts in showing readers *where* and *when* actions take place.

Maps are cognitively useful in both oral and written reports; they allow the reader to locate the project geographically. They can point up advantages such as proximity to resources (airports, cheap power, etc.) or show how benefits will be spread over a large geographical area (Sample 6.8). Or they can display research results, such as distribution and frequency of a disease (Sample 6.9, page 144).

Flowcharts are as effective in orienting the audience as they are in previewing information. With a simple color highlight, the flowchart in Sample 6.7 can be used to orient the reader to the specific part of the process under discussion

SAMPLE 6.8   Map showing geographical distribution of regional housing offices

at a particular point in the report. Sample 6.10 on page 144 highlights the evaluation task of the production process depicted in the original graphic.

Gantt charts orient readers by showing the relation of activities to each other in time. Like flowcharts, they indicate when certain events will occur in relation to other events. As Sample 6.11 on page 145 suggests, they are especially useful for illustrating projects in which the timelines of different activities overlap.

Diagrams and line drawings can visually orient the audience to a place or part of a mechanism under discussion. Technical instructions frequently use "maps" of equipment and processes to allow the user to see how parts relate to the whole. Sample 6.12 on page 145, for example, uses *callouts* (words connected to the diagram by lines) to name the parts of a machine.

Sample 6.13 on page 146 uses a modified organizational chart to relate tasks and steps hierarchically to an overall set of functions in a word-processing program.

Diagrams can also be extremely useful in making abstract relationships seem more concrete. A simple picture can be a powerful way of showing the relationship of parts of a theory to the whole conception. Sample 6.14, on page 147, is a computer-generated visual representation of an abstract mathematical

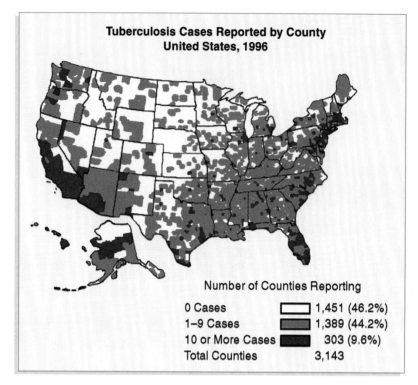

SAMPLE 6.9 Map showing frequency of cases of tuberculosis in the United States in 1996, reported by county

Source: National Center for HIV, STD, and TB Prevention, Center for Disease Control, Washington, DC [Online]. Tuberculosis Cases Reported by County, United States, 1996. Updated October 3, 1997. Available http://www.cdc.gov/nchstp/tb/surv/surv96jpg/surv3.jpg. Accessed April 12, 1998.

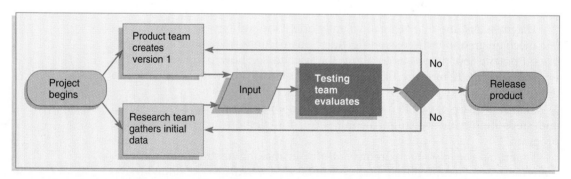

SAMPLE 6.10 Flowchart emphasizing part of process under discussion (darkened, with white type)

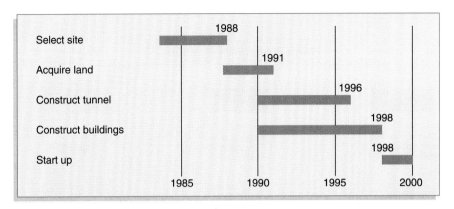

SAMPLE 6.11   Gantt chart

problem. As such, it does not represent a concrete event. However, Sample 6.15, on page 148, is a computer-generated model of what could happen if a specific large object of known dimensions fell from space into the ocean.

Sample 6.16 on page 148 is a different kind of orientational graphic. Instead of depicting a set of theoretical constructs, it presents a simplified version of visible reality. This one shows the proper outcome of a procedure for setting up a

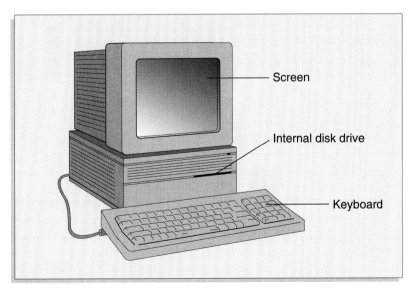

SAMPLE 6.12   Orientational graphic with callouts showing the relation of parts to the whole

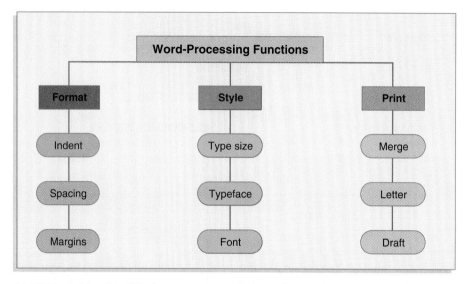

SAMPLE 6.13 Modified organizational chart showing how tasks and steps fit into an overall functional scheme

Source: M. Jimmie Killingsworth and Michael K. Gilbertson, *Signs, Genres, and Communities in Technical Communication* (Amityville, NY: Baywood, 1992) 57.

computer. Graphics such as this can be very important in providing *feedback* to let the reader know whether the task has been correctly completed.[4] They essentially say, "When you're finished, you should see something like this."

Sample 6.17 on page 149 is another kind of outcome-oriented, "realistic" graphic commonly used in computer manuals. It shows a computer screen just as users should see it if they have correctly followed instructions up to a specified point.

Photographs can also be useful as realistic orientational graphics. But the great strength of photographs—their ability to realistically display scenes and objects in great detail—can also be a weakness. To be effective as orientational graphics, photographs must be carefully composed to distinguish the subject under discussion from the surrounding details. Good captions and labels can help. On the whole, however, there's no substitute for good composition.

The photo in sample 6.18 on page 150, for example, would not be very effective in delivering the local Humane Society's recommendation to license and tag all outdoor pets—Sparky's tags are all but lost in the clutter of background

4. For an excellent treatment of feedback as a principle of design, see Donald A. Norman, *The Psychology of Everyday Things* (New York: Basic Books, 1988) 27–29. (After the first printing, the book was retitled *The Design of Everyday Things*.) See also Norman's new book: *Turn Signals Are the Facial Expressions of Automobiles* (Reading, MA: Addison-Wesley, 1992).

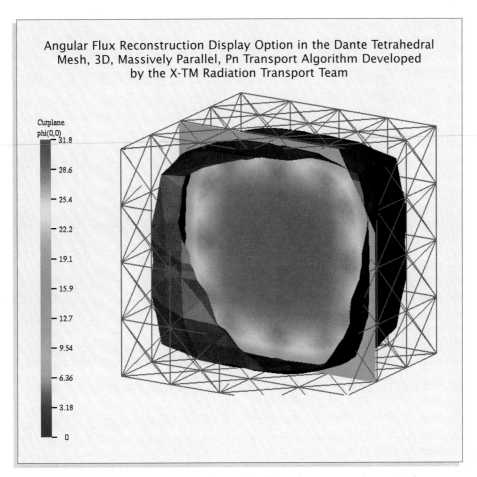

Angular Flux Reconstruction Display Option in the Dante Tetrahedral Mesh, 3D, Massively Parallel, Pn Transport Algorithm Developed by the X-TM Radiation Transport Team

SAMPLE 6.14   Three-dimensional model of an abstract mathematical problem

Source: X-TM Radiation Transport Team, Applied Theoretical and Computational Physics Division, Los Alamos National Laboratory, Los Alamos, NM, http://www-xdiv.lanl.gov/XTM/radtran/graphics_lib/afp/afp.html. Accessed April 12, 1998.

details. In sample 6.19 on page 151, the same photo has been manipulated to highlight the intended message by screening out the background noise.

Use Graphics to Indicate Functions and Direct Action

Functional graphics direct the user toward action. They not only locate objects in space (orient the reader) but also indicate what to do with the objects through

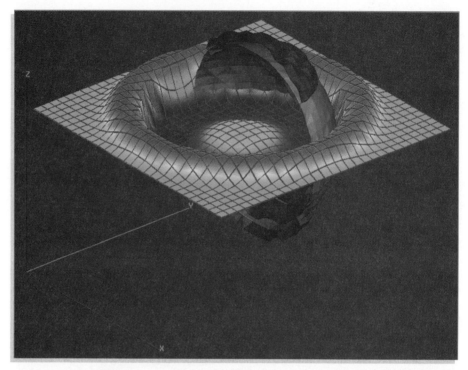

SAMPLE 6.15   Hi-density graphic depicting results of a hypothetical space power reactor water reentry accident

Source: X-TM Radiation Transport Team, Applied Theoretical and Computational Physics Division, Los Alamos National Laboratory, Los Alamos, NM [Online]. Space Power Reactor Splash. Available http://www-xdiv.lanl.gov/XTM/radtran/graphics_lib/space_power/space_power.html. Accessed April 12, 1998.

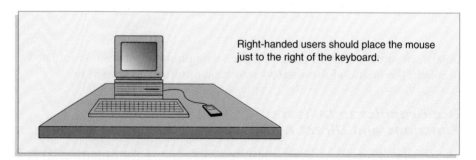

SAMPLE 6.16   Orientational graphic with caption showing outcome

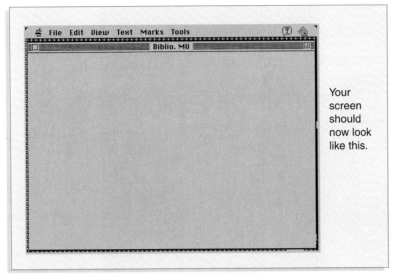

Your screen should now look like this.

SAMPLE 6.17   Graphic with caption showing outcome of completed instructions

the use of arrows, human figures in motion, or other conventional signs of activity. Most often, functional graphics are realistic drawings or photographs. The line drawings in Sample 6.20 on page 152, for example, show an adult the proper hand placement and action when expelling an object lodged in a child's windpipe.

The goal of such pictures is to all but eliminate the need for reading, allowing the user to move quickly through the procedure—a definite plus in contexts such as medical emergencies where speed is crucial.

Orientational and functional graphics like these may seem simple, but in documents designed to accompany user actions, the need to orient the reader correctly and to illustrate functions accurately is crucial—perhaps even a matter of life and death.

Use High-Density Graphics to Summarize Complex Actions, Show Relationships, and Interpret Findings

High-density graphics—complex tables, line graphs, and other data displays—condense and classify huge portions of experience in the space of a single diagram. Such graphics are especially useful in project reports and proposals. They may even be required. Some funding agencies, for example, ask proposal

SAMPLE 6.18   Photograph with too much background detail

authors to make a table that summarizes their project in a single page of text. Sample 6.21 on page 153, from a project in science education, is an example of such a table. Sample 6.22 on page 154 gives an overview of the same project, this time in the form of a diagram. As a complement to the table, the diagram gives a better sense of the relationships between the various components.

As with photographs, the summarizing power of high-density graphics is both a strength and a weakness. These complex displays save time and space, but they may also saturate the audience with too much information distributed

SAMPLE 6.19   Effective orientational photograph that highlights the subject

over too small an area of text. Saturation is especially likely in oral reports, so use high-density graphics sparingly and always with patient, clear explanations. In written reports for specialized audiences, such graphics work better, but you'll still need to use your written text to indicate the most important data points and the most significant relationships among them. In other words, you must take care to interpret complex graphics.

Sample 6.23 on page 155, a very complex table from a scientific article, offers a good example of how researchers use high-density graphics to summarize

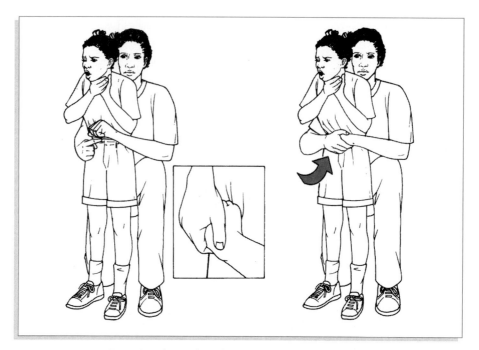

SAMPLE 6.20   Functional graphic depicting proper placement and action of hands to prevent a child from choking
Source: Courtesy of the American Red Cross. All rights reserved.

and classify experimental findings. With this mound of data placed before the audience, however, the pressure is on the written text (or the speaker in an oral presentation) to interpret the findings.

A thorough interpretation would proceed on three progressive levels:[5]

1. *The elementary level:* Every single entry (or *element*) in the table has a meaning that you could address in the discussion. You could point out, for example, that 20 tadpoles in sibship 4 survived among the large fish. It is probably not necessary to interpret every entry; rather, you should highlight a few representative ones.
2. *The intermediate level:* At this level, you guide the reader in an exploration of relationships among the individual elements. You can move vertically

5. The three levels were first analyzed by Jacques Bertin, *Graphics and Graphic Information Processing,* trans. W. Berg and P. Scott (New York: Walter de Gruyter, 1981) 12–13; see also M. Jimmie Killingsworth and Michael K. Gilbertson, "How Can Text and Graphics Be Effectively Integrated?" *Solving Problems in Technical Writing,* ed. L. Beene and P. White (New York: Oxford University Press, 1988) 135–137.

Science Futures Alliance Project

Needs and Objectives	Activities	Evaluation	Dissemination
Need: To enhance communication and cooperation among science teachers, researchers in science and education, and leaders in business and industry **Objective:** To organize and operate a community-based Science Education Alliance Network	1. Assemble an advisory board of community representatives who are interested in science education. 2. Develop a Science Teacher Corps and present in-service training sessions for cooperating school systems, involving recognized lead teachers from local schools as well as faculty researchers and teachers. 3. Organize and implement a Research Partnership Program for science teachers and student researchers. 4. Provide mini-grants to assist science teachers who are interested in participating in research and professional activities. 5. Develop an equipment grant/loan program for science teachers in community schools. 6. Develop and provide access to a database of science information (text and website) for all educational levels. 7. Produce a regional newsletter to provide a forum for project announcements, descriptions of activities, and the presentation of opinions and essays on scientific topics to stimulate ideas, participation in science, and communication among interested parties.	An ethnographic study will analyze and evaluate the effectiveness with which information is exchanged and cooperation is enhanced in the project's effort to overcome barriers in its promotion of scientific education, research, and communication. Documentation techniques will include audiotaping and videotaping of project activities, interviews with participants, and surveys taken through the Science Futures Alliance newsletter.	Public meetings, on-site in-service training sessions, newsletter, reports at professional meetings (such as NSTA)
Need: To increase the number of science majors and the number and quality of science teachers to ensure the continuation of the scientific enterprise **Objective:** To organize and implement a Science Teacher Recruitment Program	1. Organize an interuniversity/college planning committee to encourage and assist students in selecting science teaching careers. 2. Establish a Cadet Science Teacher Program that would train high school students in their respective schools and provide an opportunity for them to participate in pre-teacher preparation activities.	Ethnographic study	Brochures, newsletter, reports in teaching journals and at professional meetings
Need: To understand the efficacy of networking procedures in promoting scientific activity and in recruiting science teachers **Objective:** To produce an ethnographic study of a pilot networking project: the Science Futures Alliance Network	1. Document project, using audiotaping and videotaping, interviews with participants, and surveys attached to quarterly newsletters. 2. Analyze the data, taking into consideration the institutional, social, cultural, and personal goals that attach to each communicative interchange among the various participants. 3. Evaluate the effectiveness of the activities according to how well the objectives have been accomplished. 4. On the basis of the evaluation, make recommendations and present a model for science educators who are interested in developing similar networks.	Peer review for publication in established journals and/or scholarly presses	Scholarly articles and/or books

SAMPLE 6.21 Table summarizing a proposed project in science education

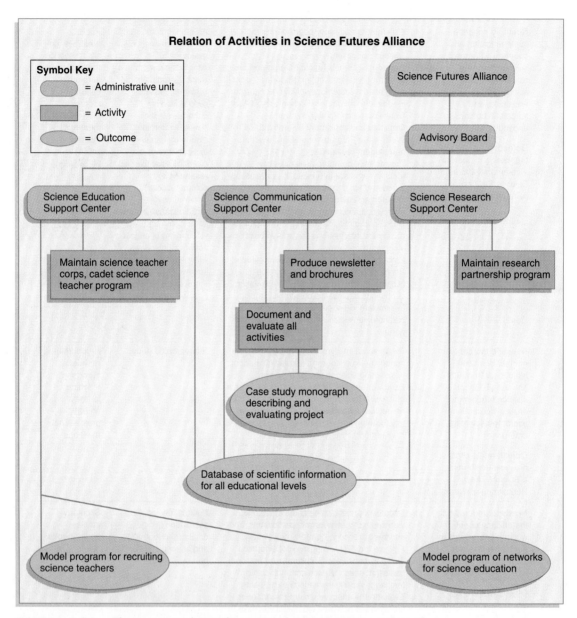

SAMPLE 6.22 Chart summarizing the same project in science education

(within a single column) and show, for example, that sibship 4 had the greatest number of survivors in the presence of the large fish. Or you can move horizontally and show that the survival rate of the large tadpoles in sibship 4 was the same regardless of the size of the predator fish.

Tadpoles			Fish			
			Large (9.8 ±0.3g)		Small (2.2 ±0.1g)	
			Number Surviving	Proportion Surviving	Number Surviving	Proportion Surviving
Small	Sibship	Mass				
	1	28	0	0.00	0	0.00
	2	23	0	0.00	3	0.15
	3	28	0	0.00	0	0.00
	4	30	0	0.00	0	0.00
	5	26	0	0.00	0	0.00
		X = 27 ± 1mg	0	0.00	0.6 ± 0.6	0.03 ± 0.03
Medium	Sibship	Mass				
	1	115	4	0.20	16	0.80
	2	114	0	0.00	6	0.30
	3	112	4	0.20	10	0.50
	4	124	12	0.60	12	0.60
	5	112	2	0.10	10	0.50
		X = 115 ± 1mg	4.4 ± 2.0	0.22 ± 0.10	10.8 ± 1.6	0.54± 0.08
Large	Sibship	Mass				
	1	351	14	0.70	20	1.00
	2	220	12	0.60	16	0.80
	3	313	13	0.65	20	1.00
	4	314	20	1.00	20	1.00
	5	351	16	0.00	19	0.95
		X = 310 ± 24mg	15.0 ± 1.4	0.75 ± 0.07	19.0 ± 0.8	0.95 ± 0.04

SAMPLE 6.23   Complex table showing tadpole survival rate relative to size and predator size
Source: Raymond Semlitsch and J. W. Gibbons, "Fish Predation in Size-Structured Populations of Treefrog Tadpoles," *Oecologia* 75 [1988]: 323. Copyright © Springer-Verlag.

3. *The macro level:* At this level, you draw conclusions. The table as a whole shows clearly that the size of the predator fish in relation to the size of the tadpole prey is a significant factor in the survival of the tadpoles. Evolutionary theory would suggest that natural selection favors large tadpoles.

Another kind of high-density graphic, the double line graph in Sample 6.24, on page 156, offers similar opportunities for interpretation. You can discuss a single data point at the elementary level (showing, for example, that admissions reached their peak in 1990, when 3000 students entered the university). At the intermediate level, you can demonstrate that admissions increased dramati-

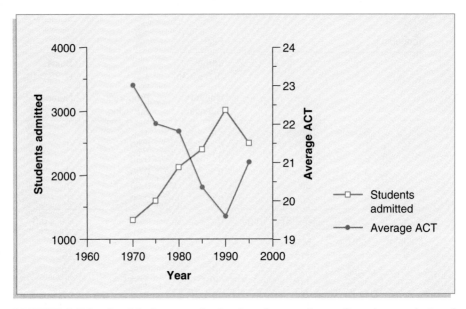

SAMPLE 6.24 Double line graph plotting the numbers of students admitted to a state university against average ACT scores (fictional data for illustration only)

cally between 1970 and 1990. Finally, at the macro level, you can argue that the number of students admitted and the average ACT score of the incoming class are inversely proportional. Any information that is developed graphically along two axes (an "X axis" and a "Y axis") would lend itself to these three levels of interpretation.

Use Low-Density Graphics to Display and Emphasize Key Points

Graphics that show simpler relationships and strong contrasts are often more useful in emphasizing very important points. They are likely to be among your most important graphical tools in all oral reports and in written reports directed at nonexpert audiences.

The simple table in Sample 6.25, for example, summarizes the outcomes of a research project in which students at various levels were exposed to a special program in health education. The researchers used different methodologies (experimental and naturalistic) for two different populations. The table compares, in a simple format, the overall attitudinal changes in the two groups. The check marks indicate significant degrees of change. The senior high populations in both the experimental group and the naturalistic group showed significant

	Experimental		Naturalistic	
	Jr. high	Sr. high	Jr. high	Sr. high
Knowledge of health	✓	✓	✓	✓
Attitude changes		✓	✓	✓
Drug use		✓		✓
Cigarette smoking		✓		✓
Alcohol consumption				✓

SAMPLE 6.25   Simple table summarizing the results from a research project in education

change on the topic of drug use, for example, whereas the junior high groups showed only insignificant attitudinal change on this topic.

Low-density bar charts and column charts are also excellent devices for emphasizing strong contrasts. Sample 6.26 on page 158 shows the difference in energy use per capita among developed and developing nations. Although the main point of this graphic is to illustrate the difference in energy use between developed countries and developing countries, it also shows fairly large differences among the developed nations, which represent additional opportunities for interpretation. Sample 6.27 on page 158 is a modification of the simple bar chart that uses color effectively. Prepared by the Environmental Protection Agency, it not only depicts total metals discarded in the United States periodically, but indicates the amount of each five-year total that is recovered. By coloring the recovered sections green, the subtle message is that this is the preferred action, while the use of red sends a message of alarm that the rest of the metals are being wasted.

Similarly, pie charts express simple contrasts effectively, particularly when you are trying to show divisions of resources or shareholdings—who gets the largest slice of the pie, as the saying goes.

A twist on this approach is shown in Sample 6.28, on page 159, a pie chart developed by the Reproductive Division of the United States Department of Health and Human Services. In this chart, the largest piece of the pie represents undesirable results (no clinical pregnancies). The smallest portion is separated slightly from the pie to emphasize its importance in displaying positive results and is then subdivided to show additional information—types of pregnancies. In Sample 6.29 on page 160, the Centers for Disease Control takes a similar approach in reporting tuberculosis incidents, but creates a completely new pie representing the portion of interest. The pertinent section of the first pie is then connected to the new pie with dotted lines.

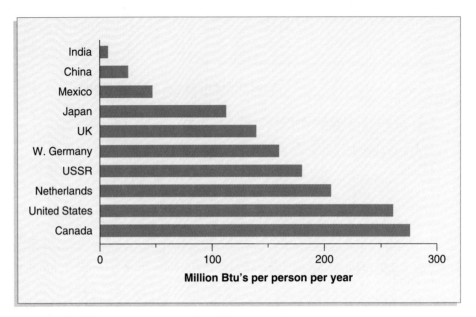

SAMPLE 6.26   Bar chart showing differences in energy use per capita in various countries

Source: Congress of the United States, Office of Technology Assessment, *Changing by Degrees: Steps to Reduce Greenhouse Gases: Summary,* 1991, 33.

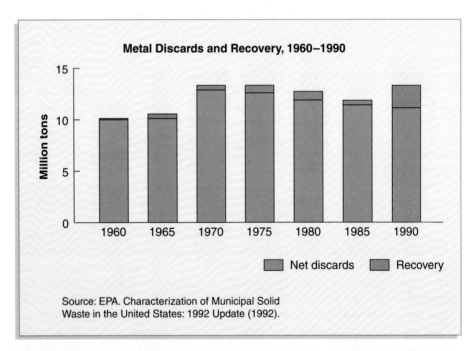

SAMPLE 6.27   Bar graph depicting metal discards and recovery, 1960–1990

Source: Office of Solid Waste, Environmental Protection Agency, Washington, DC [Online]. Revised June, 1996. Metal Discards and Recovery, 1960–1990. Available http://www.epa.gov/ces/ced/ced-2-30.htm. Accessed April 12, 1998.

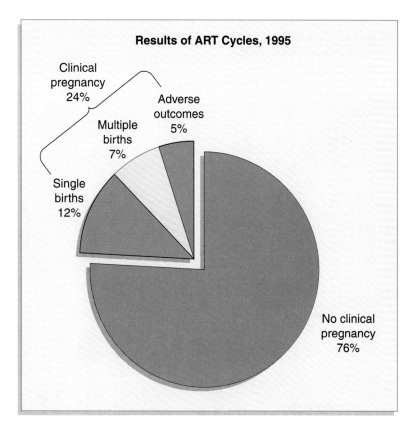

SAMPLE 6.28   Simple pie chart showing 1995 results of Assisted Reproductive Technology, using a separation technique to emphasize the successes

Source: Division of Reproductive Health, Centers for Disease Control and Prevention, U.S. Department of Health and Human Services, Atlanta, GA [Online]. Results of ART Cycles, 1995. Available http://www.cdc.gov/nccdphp/drh/arts/ fig8.htm. Accessed April 12, 1998.

Another kind of low-density graphic that works well for rhetorical emphasis is the *icon* or simple drawing. If done well, a simple image or a combination of images can implant a memorable picture of your information in the user's mind and can help users comprehend a number of points at a glance.

Sample 6.30 on page 161 comes from a student report on alternatives for solid waste management. The reporter is trying to show that traditional methods of waste disposal—such as dumping it into landfills or the sea—are no longer viable. The little drawings were quite effective in the oral report for which they were designed. For a formal written report, however, this graphic may be too informal.

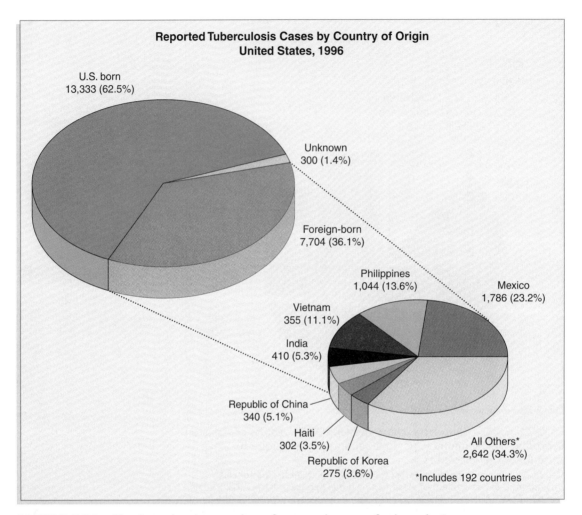

SAMPLE 6.29 Pie chart showing number of reported cases of tuberculosis in the United States in 1996 (by country of origin), with a second pie chart depicting the breakdown of those cases not born in the United States

Source: National Center for HIV, STD, and TB Prevention, Centers for Disease Control and Prevention, U.S. Department of Health and Human Services, Atlanta, GA [Online]. Reported Tuberculosis Cases by Country of Origin, United States, 1996. Updated October 3, 1997. Available http://www.cdc.gov/nchstp/tb/surv/surv96jpg/surv5.jpg. Accessed April 12, 1998.

Cartoons, which are related to icons, can also emphasize key points that you want your audience to pay attention to and remember. They are particularly useful in oral reports. Although some viewers may find that they detract from the seriousness of a presentation, cartoons can take the edge off of potentially threatening or disagreeable topics. Sample 6.31, for instance, was created for a

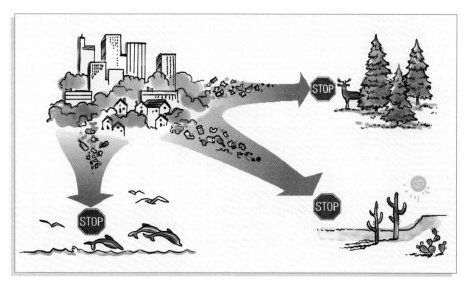

SAMPLE 6.30 Simple drawing showing alternatives that are no longer viable (dumping into the sea and landfills)

SAMPLE 6.31 Motivational cartoon to induce readers to reduce, reuse, recycle, and respond

Source: Office of Solid Waste, Environmental Protection Agency, Washington, DC. *The Consumer's Handbook for Reducing Solid Waste* [Online]. August, 1992. Available http://www.epa.gov/epaoswer/non-hw/reduce/catbook/the4.htm. Accessed April 12, 1998.

consumer handbook to promote alternative waste disposal methods that require more effort and commitment than simply throwing things in a trash can.

The power of low-density graphics to show trends over time and to compare options and alternatives makes them especially attractive for proposals and other documents that argue for change. With such graphics you can effectively make a case for recommendations about future actions. *Line graphs,* for example, work well for arguments based on trends that will probably continue into the future. Sample 6.32 shows a trend that is steadily upward. The amount of land under irrigation has steadily increased since the beginning of the century. Because it did not decrease in any single year, you can safely argue that it will continue to increase.

Area graphs show strong divergences and, like line graphs, are effective in showing changes over time in a dramatic fashion. Sample 6.33 contrasts the projected growth in population in developed countries (where growth is expected to stabilize) with the projected growth in developing countries (where population is expected to explode).

Notice, however, that as you begin to compare more than one variable on a single axis, more interpretation is required, and the contrast loses some of its rhetorical strength. A graphic like the one in Sample 6.33 falls at the mid-

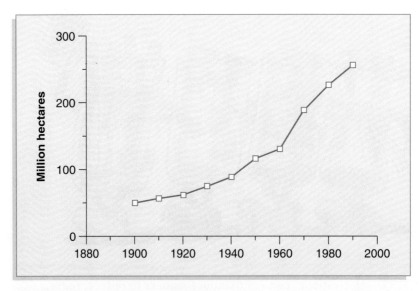

SAMPLE 6.32   Line graph showing world irrigated area, 1900–1985
Source: Reprinted from *State of the World 1989,* 49, edited by Lester Brown et al., with the permission of W. W. Norton & Company, Inc. Copyright © 1989 by Worldwatch Institute.

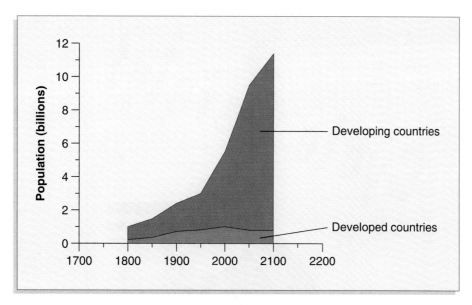

SAMPLE 6.33   Area graph showing historical and projected population growth in developed and developing nations
Source: Congress of the United States, Office of Technology Assessment, *Changing by Degrees: Steps to Reduce Greenhouse Gases: Summary*, 1991, 33.

point on the continuum between low density and high density. The dramatic contrast between the sides of the shaded areas packs the punch of a low-density graphic, but the need for additional interpretation is more typical of high-density graphics.

Use Graphics to Motivate the User

By improving readability or enhancing interest, motivational graphics improve the usability of information in your communication. These graphics typically appear on covers and in introductions, especially in instructional manuals. But you can also use them throughout a recommendation report, a proposal, or a manual, especially if you suspect that the reader's motivation level will be low (as in a health manual on how to lose weight or start an exercise program). In seeking to motivate the reader to act, technical communication borrows heavily from modern advertising's approach to graphics. Indeed, technical sales brochures often share photography (or photographic layouts) with manuals.

Motivational graphics can counter the negative effects of warnings and cautions, which tend to make readers wary of the technology or to make them

SAMPLE 6.34   Motivational graphic
Source: 10 Smart Routes to Bicycle Safety, joint publication of the National Highway Traffic Safety Administration and U.S. Consumer Product Safety Administration.

regard the manual as overly cautious. Sample 6.34, for example, comes from a safety manual on bicycle riding. Typical of motivational graphics, it appeals to the reader's sense of power, freedom, and will to succeed.[6] It suggests that you can still have fun when you obey the rules.

6. See Killingsworth and Gilbertson, *Signs, Genres, and Communities in Technical Communication* (Amityville, NY: Baywood, 1992) 64–67.

A typical motivational strategy in product manuals is to portray attractive models or ideal representations of the prospective user in the act of using the product. Photographs, such as the one in Sample 6.35, are the preferred medium for this kind of graphic, but fine line drawings and paintings can also work.

Because of their evocative power—their ability to bring out emotions and other responses in the reader—photographs are perhaps the strongest motivators you can use in your documents. But like all high-density graphics, detailed photographs can be distracting as well as motivational. Before including any photograph in your document, you need to weigh carefully its possible effects.

From motivational photographs it is only a small step to a very different kind of graphic: the cartoon. Samples 6.36 and 6.37 (on pages 166 and 167) show pages from the U.S. Army's periodical safety manual, *PS: The Preventative Maintenance Monthly.* Though you may think that the use of cartoons diminishes the seriousness of safety instructions, the Army has discovered a big shift in readers' interest levels: Soldiers often ignored the older, more serious-looking safety manuals and directives, putting themselves and others at great risk. But they actually read *PS* magazine.

SAMPLE 6.35   Motivational photographs such as this one send subliminal messages that the product will boost the user's sense of satisfaction on the job
Source: Courtesy of Wolfram Research.

SAMPLE 6.36 Cover from the U.S. Army's publication *PS: The Preventive Maintenance Monthly,* using cartoons to communicate safety information to users with low motivation

Source: PS: The Preventive Maintenance Monthly (issue 467, October 1991) 27.

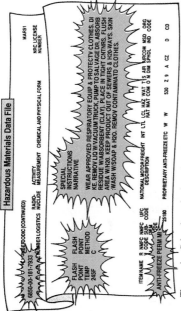

SAMPLE 6.37 Two-page spread from *PS*

Source: PS: The Preventive Maintenance Monthly (issue 467, October 1991) 28–29.

The Army's experience teaches a couple of valuable lessons about illustrating manuals:

- Motivational graphics are good as long as they make the manual more functional (and do not serve merely as slick distractions).
- Despite designers' intuitions about how users will respond to a document, only realistic user testing can provide definitive answers.[7]

A further warning applies especially to graphics designed for international use. Always plan photographs of people, cartoons depicting people, and most other motivational graphics specifically for *localized audiences*—and be very cautious even then. Cultural differences, as discussed in Chapter 2, create a wide variety of responses to seemingly innocent pictures of people.

The young man in Sample 6.35, for instance, may offend users in some countries. While in most parts of the United States pointing is perfectly acceptable, a surprising number of cultures consider pointing to be rude. Working in shirt sleeves without a jacket may also seem offensive, virtually a state of undress, in some cultures. Likewise, the cartoons in the Army manual (Samples 6.36 and 6.37) might work well for a localized audience of American GIs but even among a general readership in the United States, the depiction of the sexy blond in the cartoons would seem exploitive or unnecessarily suggestive to many audiences.

Graphics experts with international experience often recommend a program of high caution, if not strict avoidance, when it comes to including people, body parts, relations between the genders, and even animals in pictures. For example, William Horton advises authors of computer documentation to

- avoid pictures of people altogether, and try to get by with pictures of equipment, computer screens, and functional graphics.
- show hands only in the act of performing operations, and remove all signs of gender (long fingernails, rings, rough knuckles, etc.).
- when you have to show people, use unisex figures, stick figures, or stylized characters without signs of gender or racial identity.
- change graphics for each locale, taking color bias and other cultural factors into consideration with each change.[8]

Sample 6.38 illustrates one trend in how to show bodies in motion without violating taboos pertaining to the human body—the use of stylized "crash test dummies" in place of race- and gender-specific figures. Sample 6.39 goes even further in this direction, resorting to simple stick figures. This sample

7. For other examples and aspects of the Army's commitment to "new-look" publications (including cost considerations), see Benjamin D. Meyer, "The ABCs of New-Look Publications," *Technical Communication* (First Qtr. 1986) 16–20.

8. William Horton, *Illustrating Computer Documentation* (New York: Wiley, 1991) 207–217. See also Nancy L. Hoft, *International Technical Communication* (New York: Wiley, 1995) 263–273.

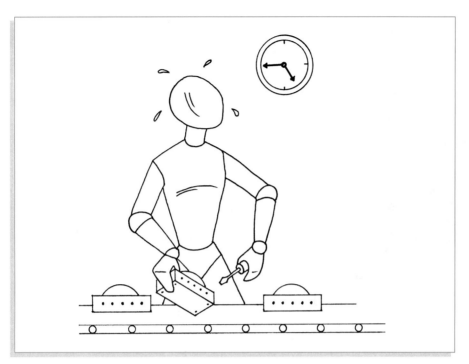

SAMPLE 6.38   Use of stylized figure to depict body in motion without offending cultural norms (Art by Gary Floden)

SAMPLE 6.39   Use of stick figure to depict body in motion without offending cultural norms

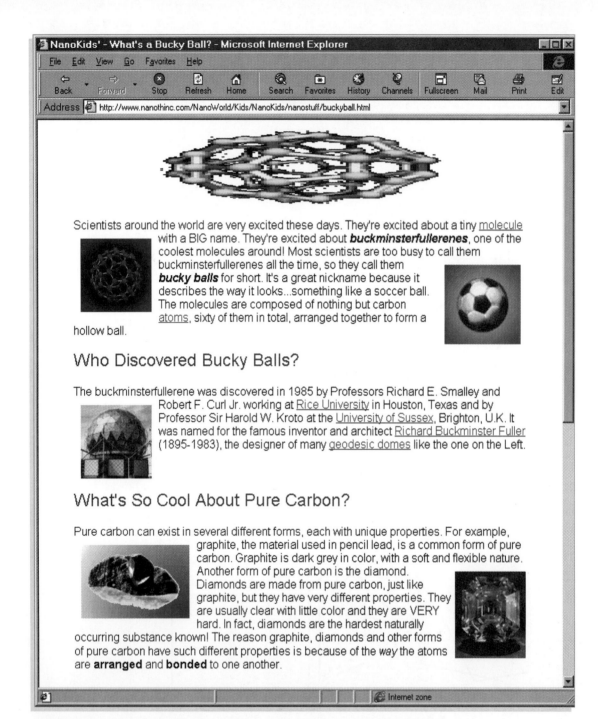

Scientists around the world are very excited these days. They're excited about a tiny molecule with a BIG name. They're excited about **buckminsterfullerenes**, one of the coolest molecules around! Most scientists are too busy to call them buckminsterfullerenes all the time, so they call them **bucky balls** for short. It's a great nickname because it describes the way it looks...something like a soccer ball. The molecules are composed of nothing but carbon atoms, sixty of them in total, arranged together to form a hollow ball.

Who Discovered Bucky Balls?

The buckminsterfullerene was discovered in 1985 by Professors Richard E. Smalley and Robert F. Curl Jr. working at Rice University in Houston, Texas and by Professor Sir Harold W. Kroto at the University of Sussex, Brighton, U.K. It was named for the famous inventor and architect Richard Buckminster Fuller (1895-1983), the designer of many geodesic domes like the one on the Left.

What's So Cool About Pure Carbon?

Pure carbon can exist in several different forms, each with unique properties. For example, graphite, the material used in pencil lead, is a common form of pure carbon. Graphite is dark grey in color, with a soft and flexible nature. Another form of pure carbon is the diamond. Diamonds are made from pure carbon, just like graphite, but they have very different properties. They are usually clear with little color and they are VERY hard. In fact, diamonds are the hardest naturally occurring substance known! The reason graphite, diamonds and other forms of pure carbon have such different properties is because of the *way* the atoms are **arranged** and **bonded** to one another.

SAMPLE 6.40 A website using graphics to provide relief from reading, enhance the visual unity of the page, and increase the reader's curiosity about the text.

Source: http://www.nanothinc.com/NanoWorld/Kids/NanoKids/nanostuff/buckyball.html. Reproduced with the permission of Nanothinc.

also demonstrates the sophisticated abilities of some of the newer software programs; it was created using a graphics program built into one of the more widely-used word processing programs.

The use of human figures to enhance motivation is clearly a risky tactic, though the payoff in localized documents may be high. However, it is certainly not the only strategy. You can greatly improve reader motivation just by constructing tasteful pages that give readers' eyes a break by adding handsome graphics to chunks of text.

Sample 6.40, a World Wide Web page on "Bucky Balls," illustrates the use of pictures and color headings to emphasize key points while increasing curiosity about the text. The similar shapes of the objects, which tie in to the key point of the writing, also give the page a visual unity. Avoiding human figures, the page steers clear of cultural bias.

In Chapter 7, we discuss design issues concerning the relation of text and graphics in more detail.

Use Graphics to Increase Your Credibility

You can use graphics to build confidence in your proposed plans and recommendations by showing structures and details of processes, products, or relations that have not yet been realized or that are hard to explain in words.

An *organizational chart* like the one in Sample 6.41 can be used in a proposal to show a project's management structure. It indicates that you have planned the project in detail.

Engineering drawings, schematics, and other technical depictions of designs and plans also show that you have a strong command of the details involved in your project. The drawings in Sample 6.42 (page 172)—from project descriptions for electronic designs—are typical of such graphics.

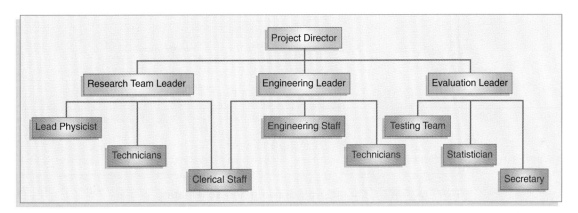

SAMPLE 6.41 Organizational chart

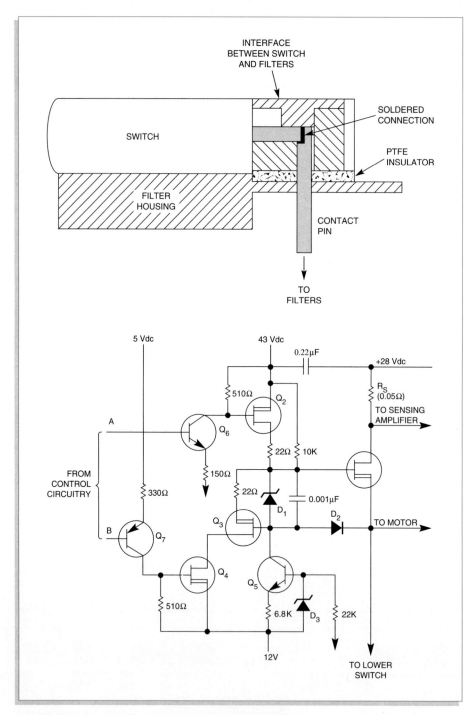

SAMPLE 6.42 Engineering drawing and schematic
Source: NASA TechBriefs 9, no. 3 (Fall 1985) 40–41.

Finally, *blueprints, line drawings, photographs of prototype products, and artist's renderings* can also enhance your proposal's credibility. They allow the audience to develop a clear visual image of your plans.

Sample 6.43, showing an architect's plan for an apartment complex in Beijing, China, offers a good example of the power of well-chosen graphics to communicate across cultures. The artist rendered several views of the buildings to convey a substantial amount of technical information—even to an audience that cannot read Chinese.

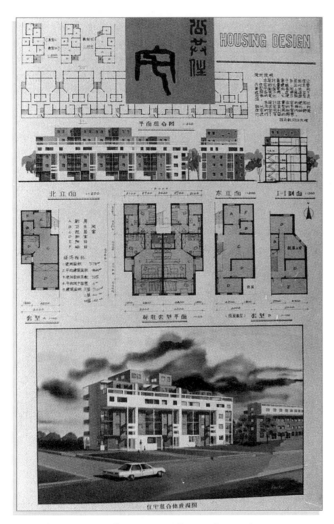

SAMPLE 6.43   Blueprint and artist's renderings

Source: Kai Liu, University of Illinois School of Architecture.

What Graphics Can Do for You and Your Reader: Summary

To help you in planning and designing documents, Guidelines at a Glance 10 summarizes the various functions of graphics discussed in this chapter. The "Form" column is far from exhaustive. You can probably think of other graphics to achieve the same effects in a different way.

Producing Report-Quality Graphics

Once you have sketched your rough ideas for graphics, the degree of professionalism you will need to produce the graphics will depend on the context in which you are working. Different technical writing instructors and different organizations require different degrees of product quality. Some large companies assign artists to project teams immediately, and authors never have to produce more than a rough sketch. Some small companies, on the other hand, require their authors to produce an entire text without additional help. As desktop publishing becomes more accessible and sophisticated, this policy is likely to become more prevalent. If you can produce your own lists, tables, charts, graphs, and drawings, then you can save money for your company and have more control over the quality of the presentation. Expertise of this kind is valuable even in team settings, especially if the team doesn't include a professional graphic artist.

The final sections of this chapter provide a few general guidelines to help you face the challenge of producing your own graphics. You can probably get additional help from the consultants and advisors in your college's computer center or graphics lab. Many copy centers and print shops also provide limited consultation (usually for a fee).

Producing Lists and Tables with Word-Processing Software

The key to making readable lists and tables is a good software package for word processing. All of the sample lists and tables in this chapter were originally produced on a personal computer with a readily available and easy-to-use word-processing program.

The best-selling software for both Apple Macintosh and IBM-compatible computers—packages such as Microsoft Word® and WordPerfect®—have the features you need for this kind of work:

- options for creating bullets (•) and other symbols that align with the text
- formatting capabilities that allow for easy tabbing, tiered indention, and multiple rows and columns
- formatting capabilities that let you insert lines and borders of various sizes and shapes around the text in tables

GUIDELINES AT A GLANCE 10
Functions and forms for technical graphics

Function	Form
Preview and review key points	Outlines, lists, tables, flowcharts
Orient the audience in space and time	Maps, flowcharts, Gantt charts, organizational charts, diagrams with labels, line drawings, photographs
Indicate functions and direct action	Line drawings with directional pointers, photographs indicating motion
Summarize and interpret complex data	Tables, line graphs, and other high-density graphics
Display and emphasize key points	Bar charts, pie charts, area graphs, simple line graphs, simple tables, icons and simple drawings, cartoons, and other low-density graphics
Motivate the user	Icons, drawings, cartoons, and photographs
Increase your credibility	Organizational charts, schematics, blueprints, drawings, photographs, and other displays of specific details

If you have software with these features, learn to use them; in technical communication you will need them often. If you are making a decision about which software to buy, be sure to choose a package with these features.

Producing Graphs and Charts— with and without Computer Software

Many technical authors—such as engineers and architects—are trained to produce high quality graphics either by hand or by computer (or by a combination of methods). If you do not have formal training, you will probably need to study some of the books listed at the end of this chapter and learn to use at least one of the graphics packages that are available for personal computers.

Many of the samples in this chapter were created by using such packages (see, for example, Samples 6.24 and 6.25). To get a sense of the power—and the limitations—of graphics software for personal computers consider briefly the different efforts required to produce a simple line graph by traditional methods and by using a graphics program on the computer. First, Table 6.1 shows the steps for producing a high-quality line graph by hand.

TABLE 6.1

Five stages of graph production.

1. **Calculating and constructing the grid**
   - Horizontal scale (usually time): Check that the data has a full, sequential, and regular set of numbers.
   - Vertical scale (usually values): Check the range of values. If all are high numbers, consider a base value starting with the lowest. If there are sharp drops or rises in the values, consider devising a broad-based diagram.
   - Do a simple sketch of the whole design to get an impression of the patterns created by the statistics.
   - Select the proportions of your grid to fit within the available space.
   - Estimate the area needed for annotation, adding the title, scales, and source information.
   - Construct the grid with sufficient intervals to let you plot the data.
   - Draw the grid with a pencil.
2. **Plotting the data**
   - Plot the data as small dots or crosses. If there is more than one line, use different colored pencils.
   - When complete, go through all the data backward, checking each dot or cross against your source.
3. **Drawing the graph**
   - Ink in the connecting lines between the points using either hand-drawn lines or dry-transfer lines.
4. **Adding the annotations**
   - Ink in the two axes and the subdivisions on the scales.
   - Add annotation to the scales (not all the divisions of which may need to be numbered).
   - Keep the intervals on the final drawing simple and not congested with details.
   - Add labels, titles, sources, and other information by hand or other lettering techniques.
5. **Adding decorative details**
   - Color in important areas. If the artwork is for a printer, put in tones in accordance with his or her advice.
   - Add decorative elements, such as a symbol or picture, to indicate the subject.

Source: Bruce Robertson, *How to Draw Charts and Diagrams* (Cincinnati, OH: North Light Books, 1988) 102–103. Copyright © 1988 by Bruce Robertson. Used with permission of North Light Books.

Next, let's look at the effort required to produce a graph with the help of a graphics program. As you will see, the task is faster and easier, but you give up a great deal of control and some quality. Here are the steps required to do a simple line graph on a typical program marketed for personal computers:[9]

1. Open a new file and be prepared to enter your data in columns, as shown in Figure 6.2.

9. Cricket Graph® for Apple Macintosh.

Untitled Data
1
2
Column 1
Column 2
1

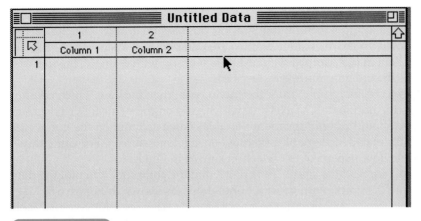

FIGURE 6.2

Opening screen in graphics program.

2. Enter the data into the different columns. In this program, it doesn't matter which data sets you put in which column. Other programs may require you to distinguish from the start between *independent variables,* such as dates, which do not change because of the forces you are measuring, and *dependent variables,* the data that your study actually produces. In Table 6.1, Bruce Robertson refers to the independent variable as "time" and the dependent variables as "values." You might encounter any of these terms in the different software packages you use. In the sample data shown in Figure 6.3, the independent variable is time, and the dependent variable is number of students. (Notice that as you type in the data for column 2, the program adds a third column in case you are planning a more complex chart—a graph with enough lines for two or three dependent variables.)

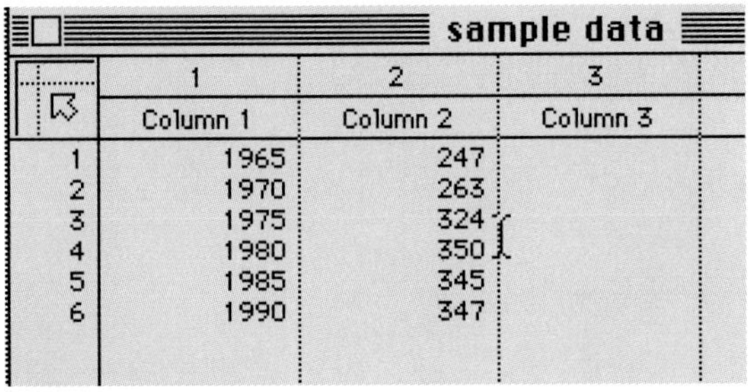

sample data
1
2
3
Column 1
Column 2
Column 3
1 1965 247
2 1970 263
3 1975 324
4 1980 350
5 1985 345
6 1990 347

FIGURE 6.3

Data (independent and dependent variables) entered into columns.

3. Now you can select the kind of graph you want to make from the program's menu. Your choices include line graphs, double line graphs, pie charts, bar charts, column charts, scatter graphs, area graphs, and a few others. For the simple data in our sample, we need only a graph with a single line to show changes in student enrollment over time.

4. If you select "line graph" from the menu, you must indicate which column goes on the horizontal (X) axis and which goes on the vertical (Y) axis. Usually, you should put independent variables (such as time) on the horizontal axis and dependent variables (values) on the vertical axis. For our sample data, therefore, we make the selection shown in Figure 6.4.

5. Now you are ready to select "new plot," and the computer will plot the data and draw the graph with generic annotations, as shown in Figure 6.5.

6. As a last step, you can edit the annotations, taking out the generic labels (such as "Column 1") and putting in your own descriptive labels. For our sample data we really need only a single annotation showing the meaning of the dependent variable ("Number of students"). The meaning of the independent variable (years) is obvious. The result appears in Figure 6.6. Be sure to title the graph.

The final graph shows that you lose some freedom by opting for computer-generated graphics. This program automatically adds the year 2000, for example, and, more significantly, it automatically truncates the graph, beginning with the value of 240 rather than 0. As Darrell Huff pointed out years ago in his now classic book *How to Lie with Statistics*, the truncated graph creates an arti-

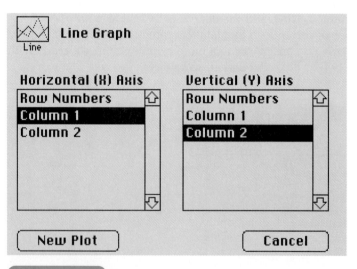

FIGURE 6.4

Indicating the orientation of data on the graph.

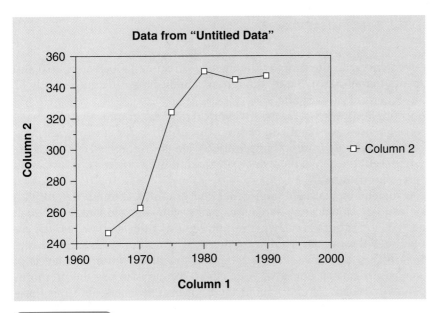

FIGURE 6.5

Computer-developed plot with generic labels ("Column 1" and "Column 2").

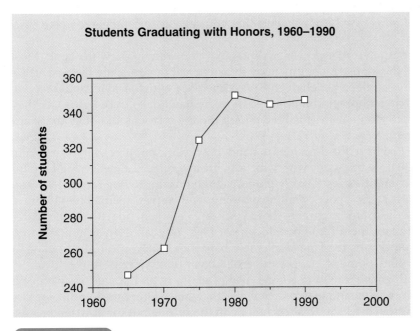

FIGURE 6.6

Final graph with edited annotations.

ficially steep curve. In the sample, the increase in students between 1970 and 1980 appears tremendously large, though in actuality the difference is only 87 people, a 33 percent growth rate.

This example shows that, as in all rhetorical matters, graphical choices have *ethical* consequences. So when you review your graphics for effectiveness, you should also consider their ethical implications. Do your graphics fairly and accurately represent the actual data, or have you unfairly crafted the information to support your argument?

If you remember that computer-generated graphics, through default values, may lead you down some dangerous paths, you can compensate for such deficiencies. Good production always requires good critical review, and the energy saved and the overall results produced by graphics software packages make them highly attractive. Future programs promise even higher levels of quality and user control. For now, your best recourse is to use the programs with open eyes, making sure to avoid the built-in deficiencies and misrepresentations you may encounter.

Producing Flowcharts, Diagrams, and Drawings with Computer Software

For producing simple diagrams and flowcharts, *draw software* works very well. Such programs offer a spectrum of boxes, circles, straight and curved lines, arrows, and other graphical elements that you can easily combine to make basic drawings.

The computer offers fewer advantages if you want to draw freehand. Most users find the mouse more unwieldy than a pencil or pen. Only the best artists I know have attained a high degree of competency with *illustrator* and *paint programs*. Many authors find it easier to draw by hand or to trace figures, or they rely on a professional artist to render high-quality cartoons and other drawings or paintings.

Once you have the drawings in hand, however, computer *scanners* can help you to integrate the work with your art and text files. Scanner technology for personal computers is still fairly crude, though it is improving rapidly. You can already easily scan good, clear line drawings and convert them to files in draw, paint, and illustrator programs. Then you can clean them up with the computer tools and place them into your document files. You can even scan black-and-white photographs to achieve tolerable halftones that you can manipulate in computer files. Some scanners can process color photographs, too.

But all scanners require a great deal of memory and power. They are still slow to work with, and they produce files that take a long time to print. We can expect improvements in this technology in the future, but for now the process of producing professional-quality artwork through scanning remains challenging for technical authors.

Using Clip Art

Ready-to-use packages of images and cartoons, known as *clip art,* are available on computer disks and in CD-ROM files for all kinds of computers. The advantage of clip art is that it is usually formatted in a compact manner and designed to be fast and easy to place into computer text files.

Although clip art is quite handy and often very useful, you should be aware of its shortcomings. First, there's a temptation to use an image simply because you have easy access to it. Be sure to select only pictures that really work well with your text. Second, there's a tendency to overuse clip art. Don't fill your text with distracting images; make sure every picture has a communicative function. (See Chapter 7 on document design.) Third, many readers have a low opinion of clip art. Recognizable clip art cheapens a text for them. Indeed, some clip art is rather cheaply produced. So apply the same standards of quality in selecting clip art that you would apply to work that you commissioned or produced yourself.

Finally, clip art is not internationalized and may be offensive even to many American audiences. Pictures of people and animals abound, with no regard to cultural sensitivity. Symbols and colors in clip art usually reflect a Hollywood or New York City standard. Your best bet is to avoid clip art altogether in international technical communication.[10]

Taking and Selecting Photographs

Although slick decorative photographs are not necessary or appropriate for most technical documents, it is important that any photos you use present your subject matter clearly. If you have taken a photography class or understand the basic principles of camera operation and photographic composition, you can probably produce photographs of a quality that is appropriate for most documents.

The same elements that create an effective prose style—clarity and conciseness—also determine the effectiveness of photographic illustrations. Whether you are taking photographs yourself or selecting them from noncopyrighted sources (such as U.S. government publications), you need to remember the following points:

- The central subject of the photograph should be clear and sharply focused.
- The photograph should have a moderate contrast level. Photographs with too much or too little contrast will not reproduce well and may obscure your central subject.
- The background should be simple and free of distracting images.

10. Hoft, *International Technical Communication,* 272.

As was mentioned earlier in the chapter, the issue of background is critical. Make sure that the central subject of your photograph (the object you want to emphasize in your discussion) is clear and easily distinguished from background. This can be achieved through lighting, selective focus (the subject is sharply focused; the background and foreground are not), tight framing (the central subject fills the image field, eliminating distracting background material), or placing the subject against a solid background such as a white card or posterboard. Additionally, many computer software programs for manipulating and editing photographs offer a variety of tools and methods for highlighting the central subject.

Also try to ensure that all the photographs you select are of a uniform quality; don't mix amateur with professional photographs in the same document, for example.

When your photographs are ready, you can either scan them into your text (if your computer and printer are powerful enough) or have them printed on separate sheets with clear labels and textual references. Decisions about whether to use color or black and white will depend on the production method

Selective focus, a moderate contrast level, and judicious cropping make this photo suitable for use as an illustration.

High-contrast lighting is an effective way to distinguish a subject from background material.

you are using. In general, it is best to print black-and-white photos from an original photo or negative and to print color photos from a slide or transparency. For report-quality graphics you may include the photographs themselves; then such decisions will depend on whether you need color to show certain features of your subject.

Other considerations in whether to use black-and-white or color photos may be cost and the audience, depending on how many copies of the report are to be printed and how "slick" an appearance is required.

The following paragraphs contain information that may be better previewed or reviewed in graphic form. Decide what kind of graphic best suits each problem, and then provide a sketch of the appropriate graphic(s).

EXERCISE
6.1

1. The local health club offers a variety of membership options. If you are a student, you can join at the "Fitness" level for $23 a month or the "Executive" level for $30 a month; the latter includes an upgraded locker room, sauna privileges, and towel service. You also have the option of a "Nonprime Time" membership, which limits the hours you may attend the club. This will save you $7 a month at either level of membership. All of these are also available for the nonstudent at an additional $10 per month. Single-visit passes are also available at both levels for $6 and $10, respectively.

2. Linda, Joe, Helen, and James are brainstorming strategies for their office's recruiting efforts. Helen wants the group to concentrate on getting referrals from their current employees; Linda thinks that the company should invest in the local technical career fair to find new prospects. However, Joe reminds his co-workers that they have to obtain permission from corporate headquarters before taking either of these steps. James thinks the group should continue sending referral forms in the paychecks of employees, but Helen suggests that they research the success rate of that endeavor, since it is costly and hasn't produced noticeable results. Joe recommends altering their current job application form to include a request for the names of prospective job candidates.

3. Having acquired all of the necessary parts for his new fish tank, Dave is ready to set it up. Before he fills it with water, however, he'll have to clean all of the components with a mild detergent, rinse the gravel, and place the undergravel filter trays and gravel in the bottom of the tank. Once the water is in, Dave needs to check the pH; if it needs to be changed, he should do it now. Other water treatments, such as those that help fish adapt to the new tank, should be added next. After the water has been variously treated, Dave should check its temperature; if it's too low, the heater ought to be put in and turned on. Finally, the air pumps should be placed on top of the undergravel filter tubes and turned on, beginning the filtration process. He can add fish in about a week.

EXERCISE
6.2

The following cases highlight situations in which orientational and functional graphics could be very helpful. Plan and make rough sketches for the needed graphics in each situation.

1. Your real estate developing firm's biggest client wants to know how the last six available lots are situated among the twenty-two total lots in your newest subdivision. She is hoping to see something by this afternoon and is especially interested in corner lots.
2. You've decided to rearrange the categories of books in your bookstore. To do the job, your employees will need to know where each category and each shelving unit should be. Categories of books include fiction, biography, travel, cooking, hobbies, self-help, children's, humor and games, automotive, and reference.
3. Your company sells computer parts to private users, and to keep costs low, you encourage customers to install the parts themselves. The customers, who are comfortable enough with their computer knowledge to take their machines apart, need to know how to identify the following basic PC components: the CPU, the RAM slots, expansion slots, and power supply.
4. Some tennis players prefer the parallel theory of doubles play: Instead of keeping one partner at the baseline and the other close to the net, they will stay directly beside one another throughout each point. As a tennis coach, you want to encourage some of your advanced players to experiment with this technique, and you show them how it works in comparison with more traditional doubles play.

EXERCISE
6.3

Part 1: Assessing Visual Needs.   The following situations call for graphical support of various kinds. For each situation, plan and sketch the kind of graphic that would best convey the necessary information to the reader.

1. After surveying 1000 students about their eating habits, you find that 140 of them are very concerned about nutrition, 380 are somewhat concerned, 405 show slight concern, and 75 are not concerned at all.
2. The entering class at a university has an average high school grade-point average of 3.14, compared with 2.98 last year and 2.90 and 2.87 the two preceding years. However, the average score on the university-administered entrance exam has gone from 2.62 three years ago to 6.56, where it stayed for two years, to this year's average of 2.54.
3. You have been asked to give a short presentation about the dangers of tanning to a social organization at your college. Before you report the scary statistics, though, you want to focus your audience's attention on the subject.
4. The annual inventory in your sporting goods store came up with the following numbers. In the high-grade line, the store has 12 gray sweatshirts, 8 navy blue sweatshirts, 9 royal blue sweatshirts, 13 green ones, 18 red sweatshirts, 14 purple sweatshirts, 11 white ones, and 10 black ones. In the medium-

grade line, the store has 4 green sweatshirts, 6 purple ones, 6 royal blue ones, 8 white ones, 5 black ones, 7 navy blue ones, 2 gray ones, 5 green ones, and 8 red ones. In the economy-grade line, the store has 4 royal blue sweatshirts, 6 navy blue, 3 green, 4 black, 7 purple, 8 royal blue, 3 gray, 8 white, and 6 red.

5. Your company's growth has been steady over the past ten years, but lately the growth has been phenomenal. As personnel director, you must present the numbers at the annual corporate banquet. You have researched the number of employees over the past ten years (20, 26, 32, 42, 58, 70, 102, 134, 166, and, this year, 245). To really add impact to your talk, you also researched the company's projected growth from ten years ago, which went from the then-current number of 20 employees to a projected 110 today. In between, management had predicted a gradual increase of ten employees per year. Meanwhile, your major competitor's employee roster has not registered such encouraging growth. Through public records, you have found that the company's numbers have gone from 34 ten years ago to just 108 today (in between, the company recorded 40, 42, 65, 78, 89, 95, 99, and 106 employees).

Part 2: Analyzing your Graphical Choices.   Which of the graphics that you planned for the situations in Part 1 would be considered high-density? How did you determine which situations were appropriate for high-density graphics and which for low-density graphics? Discuss your choices for each situation.

The following situations require graphics that motivate the reader and/or add credibility to the document. Plan and roughly sketch graphics for each situation.

EXER
6.4 CISE

1. Your student booster club has come up with a proposal delineating the student body's suggestions for improvements in the overcrowded conditions in the school's football stadium. The proposal includes results of a student survey along with financial projections for each of the suggested changes.
2. As a volunteer at your local humane society, you have been asked to design a brochure to be distributed to local schools, pet stores, and veterinary clinics. The brochure should encourage pet owners to have their animals spayed or neutered.
3. The citizen's transportation advisory committee in your state is sponsoring a competition to create a new billboard that encourages commuters to carpool.
4. You will be presenting your business plan to a group of prospective investors in a few days. You want to provide a private, safe, taxi-style service that transports children to and from school and after school activities for a small fee. In this case you'll need graphics to both convince readers to invest and show them that you have researched carefully.

EXERCISE
6.5

Part I: Producing Planned Graphics.   Look back at the graphics you planned for Exercise 3.3. Produce three of those graphics in report quality.

Part II: Graphics Planning and Production.   The following data can be rendered visually in several different ways. Try to produce graphical versions in at least three different formats using all or some portion of the information given. Consider dividing the information up by equipment, by percentage of full capacity currently being used, by cost, by improvement from the first week to the second, by improvement still needed to reach capacity, or by any other method that will portray useful information visually.

Production times for the new equipment in our print shop are quite satisfactory, especially in light of the fact that the assembly workers are still adjusting to it in their second week of using it. The cutter, which was manufactured to make twelve cuts per minute, is currently making an average of ten and a half cuts per minute, up from eight cuts per minute on the first day it was used. The cutter cost the company $3000, so we are hoping that we can get up to speed on it very soon. The cutter that we used before this purchase cost the company $1100 (secondhand) and cut at a rate of nine cuts per minute.

The binder was even more expensive, at $4500, but it is already binding at the promised rate of one text every two minutes. In its first week we only averaged one binding every two and a half minutes, but we're up to speed now. Our old binder, which the former shop owner sold to us at $790, bound one book every three minutes and fifteen seconds.

The printing press itself cost us $10,350 and will probably take a longer adjustment period before we are fully accustomed to it. It should eventually print 900 pages per minute, but in the first week we printed at a rate of 400 pages per minute once the installation and training were done. Today, we are slowly approaching the mark with 150 more pages per minute than the first week, but we hope to improve those numbers more quickly. The beastly machine that did this job before this purchase printed just 500 pages per minute, so once we progressed beyond last week's rate, our overall page production increased.

The shrink-wrapping machine's current production rate is twenty-five cases of books per hour, but we are hoping to bring that figure up to thirty cases per hour, and we have already improved from twenty-two cases per hour last week. The old machine could wrap about eighteen cases per hour, so our numbers are already much better, and the $2100 price tag of the machine seems justified.

Recommendations for Further Reading

Arnheim, Rudolf. *Visual Thinking.* Berkeley: University of California Press, 1969.

Harris, Robert L. *Information Graphics: A Comprehensive Illustrated Reference.* Atlanta, GA: Management Graphics, 1996.

Hoft, Nancy. *International Technical Communication: How to Export Information about High Technology.* New York: Wiley, 1995.

Horton, William. *The Icon Book: Visual Symbols for Computer Systems and Documentation.* New York: Wiley, 1994.

———. *Illustrating Computer Documentation: The Art of Presenting Information Graphically on Paper and Online.* New York: Wiley, 1991.

———. "Pictures Please—Presenting Information Visually." *Techniques for Technical Communicators,* ed. Carol Barnum and Saul Carliner. New York: Macmillan, 1992: 187–218.

Huff, Darrell. *How to Lie with Statistics.* New York: Norton, 1954.

Killingsworth, M. Jimmie, and Michael Gilbertson. "How Can Text and Graphics Be Integrated Effectively?" *Solving Problems in Technical Writing,* ed. Lynn Beene and Peter White. New York: Oxford University Press, 1988. 130–149.

Lefferts, Robert. *Elements of Graphics: How to Prepare Charts and Graphs for Effective Reports.* New York: Harper and Row, 1981.

Meilach, Dona Z. *Dynamics of Presentation Graphics.* 2nd Edition. Homewood, IL: Dow Jones–Irwin, 1990.

Meyer, Benjamin D. "The ABCs of New-look Publications." *Technical Communication* (First Qtr. 1986): 16–20.

Nord, Martha Andrews, and Beth Tanner. "Design That Delivers—Formatting Information for Print and Online Documents." *Techniques for Technical Communicators,* ed. Carol Barnum and Saul Carliner. New York: Macmillan, 1992: 219–252.

Norman, Donald A. *The Design of Everyday Things.* New York: Basic Books, 1988.

Robertson, Bruce. *How to Draw Charts and Diagrams.* Cincinnati, OH: North Light Books, 1988.

Tufte, Edward. *The Visual Display of Quantitative Information.* Cheshire, CT: Graphics Press, 1983.

Turnbull, Arthur T., and R. N. Baird. *The Graphics of Communication.* 4th Edition. New York: Holt, Rinehart and Winston, 1980.

White, Jan V. *Using Charts and Graphs: 1000 Ideas for Visual Persuasion.* New York: R. R. Bowker, 1984.

Additional Reading for Advanced Research

Bertin, Jacques. *Graphics and Graphic Information Processing,* trans. W. Berg and P. Scott. New York: Walter de Gruyter, 1981.

Killingsworth, M. Jimmie, and Michael Gilbertson. *Signs, Genres, and Communities in Technical Communication.* Amityville, NY: Baywood, 1992.

Killingsworth, M. Jimmie, and Scott Sanders. "Complementarity and Compensation: Bridging the Gap between Writing and Design." *The Technical Writing Teacher* 17 (1990): 204–221.

Kostelnick, Charles. "Visual Rhetoric: A Reader-Oriented Approach to Graphics and Designs." *The Technical Writing Teacher* 16 (1989): 77–85.

Designing an Effective Document

CHAPTER OUTLINE
▼▼▼▼▼▼▼▼▼▼▼▼▼▼▼▼▼▼▼▼▼▼▼▼▼

CHAPTER OBJECTIVES
▼▼▼▼▼▼▼▼▼▼▼▼▼▼▼▼▼▼▼▼▼▼▼▼▼

After you have worked through this chapter,
you should be able to do the following:

- Use general guidelines to design and
  integrate the elements of a page—graphics,
  written text, and white space
- Recognize key design features in the major
  document genres of technical
  communication—reports, proposals, and
  instructional manuals
- Apply general guidelines for special
  user needs in computer-mediated
  communication
- Review documents for effectiveness

In technical writing, document design represents an attempt to integrate the elements of the page: graphics, prose paragraphs, open space, and hybrid elements such as headings, lists, and tables. In the CORE method, planning for design begins as early as the audience-action analysis. The actual work of integrating the elements in the design comes a little later, as we develop graphics for an oral presentation and draft the written document. In addition, document design is a concern that brings the context of use into decision making about production. So it is an important consideration in reviewing and evaluating a document at both early and late stages of development. As Figure 7.1 shows, the activities involved in planning, evaluating, and executing the design for a document are integral to almost every step in the process of developing technical documents.

Some specialists in technical communication would go further, arguing that design plays a key role in *every* activity in the CORE method, even the ones not

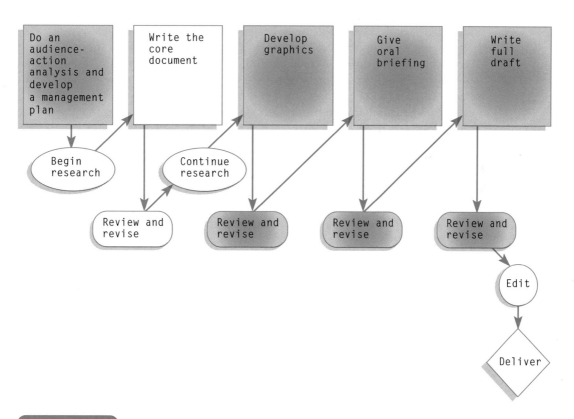

FIGURE 7.1

Opportunities for design decision-making in the CORE method.

shaded in this diagram. But in the early stages of the process, when you are deeply involved in researching and drafting your core document, design considerations should take a back seat. First you must be sure that your document presents solid information. Without this, no design on earth will help it. By the time you reach the other end of the production process—the editing stage—you should have established your main design principles. You have only to ensure that you are applying them consistently.

We use the term *document* in the broadest possible sense. Of course, the term refers to written and printed reports, manuals, and proposals, but it also encompasses brief fact sheets and brochures about businesses and research organizations, resumes for job applications (see Chapter 11), and even warning labels attached to home appliances. In addition, the design principles we recommend, such as the need for chunking information and using color effectively, apply to presentations in different media. With a little modification, you can use these guidelines to create overhead transparencies for oral presentations (discussed in Chapter 8), as well as most online "documents," such as help screens and web sites. After all, most people still use the same terms for electronic communications that they use for written documents; they speak of web "pages" and computer "documents" and "files." Some special considerations do apply to electronic documents, considerations discussed at the end of this chapter, though in modifying principles for electronic documents, you should remember that user expectations and producer conventions are still evolving for computer-mediated communication, so any specific points we make must be tentative.

Other differences to consider in document design include the changes that occur as you move from genre to genre, from designing reports to designing proposals, manuals, scientific papers, resumes, and so on. After presenting guidelines that apply to all documents, this chapter offers some specific advice on designing documents in the main genres of technical communication. In Part Three, we discuss these generic differences in more detail.

Assess Your Design Resources

Before you develop your design, you need to assess the state of affairs in your context of production. At the very least, you should answer the following questions:

- How much time do I have to produce the document? Will there be enough time to test different design features and formats?
- What facilities are available for production? What capabilities do they have? If you think your document needs color to have the right impact, for example, find out whether your organization has a color printer. If not, you need to either rethink your concept or find another source. If you have to compete

with other students or employees for the use of equipment, try to reserve specific time slots. If the equipment is in constant use, also allow for scheduling backups and equipment failures.

- How much money do I have to spend on designing and producing my document? What are the relative costs of design and production options? Professionals always have to work within a budget for a publication, and students have chronically limited sources of funds. When you create an expensive design, be sure you really need it to enhance the communication. Few users (and fewer instructors) are impressed by "glitz." The guidelines below will help you produce appealing, tasteful designs without overspending.

- How much freedom do I have in design? What format requirements does my organization impose? How flexible are these requirements? Companies often reduce document design considerations to format requirements. They do this to save time and money and to use their facilities and personnel in a predictable way. A standard format also allows them to establish a corporate design signature. Perhaps your technical writing teacher will make similar requirements. You may have to use letter-quality printers, place certain parts of your document on certain parts of the page, use a certain style for headings and page numbers, avoid certain kinds of graphics, and always use others. Your freedom as a designer begins after you have fulfilled these format requirements, so it's essential to know them from the start.

Of course, at times it may be appropriate to challenge these requirements. Maybe your company is seeking a new market. Or perhaps user testing has revealed that the company's design signature interferes with usability. When computer companies began to market personal computers for home use in the early 1980s, for example, they had to redesign their manuals for this new market.

To be prepared for such occasions, it's good to know which design features different kinds of users prefer. Research in cognitive psychology and usability testing in industry have contributed a great deal to our understanding in these areas. The guidelines in the following section are based on recent findings.

General Guidelines for Document Design

Once you know the limitations of your production capabilities, you can think realistically about document design from a functional perspective. You can ask, What are the functions of each design feature, and how can I achieve the best effects by matching form and function?

As with style, document design requires that you use good judgment and make good decisions. So don't look on the general guidelines given here as a set of absolute rules and formulas. Instead, look carefully at the needs of your audience and the potential uses for your document before you decide on design features. Then test them on a sample of users before you settle on a final version.

Create Guideposts for the User

Cognitive studies have confirmed that we all need "maps" to sort and process information effectively. In every document you create, each part should relate to every other in a systematic and consistent manner. The organizational scheme you use should be obvious to the reader on every page. You can use three devices to ensure this: a hierarchical structure; informative, consistent headings; and short, digestible chunks of information.

Develop a Hierarchical Structure That Tiers Access to the Information

A hierarchical structure presents different levels of information in descending order, as in a traditional outline:

I. Major Topic
   A. Minor Topic
      1. Subtopic

The topic under roman numeral I is the most general and holds the widest interest for readers. The subtopic under arabic numeral 1 is the most specific and of narrowest interest. As we noted in Chapter 4, using a tiered organizational structure serves two purposes. First, because it is a conventional structure, your readers will know how to use it. They'll look for information at the level best suited to their needs. Second, this structure allows you to broaden your audience. You can accommodate users who have various levels of expertise and commitment to reading.

Use Informative, Consistent Headings Keyed to Your Outline

Your headings should reinforce and reveal your hierarchical organization. They should inform readers about what is in the sections that follow them. Headings that simply mark the beginning of a new section—"Introduction," "Methods," "Results"—reveal nothing about what is to come. Informative headings are more like headlines in a newspaper; they summarize important information in the section. Sample 7.1 shows a set of informative headings used in Chapter 3 of Lester Brown's *Building a Sustainable Society*, entitled "Biological Systems under Pressure."

Such headings show readers where they can find fairly specific information. Even if you are required to use structural headings such as "Introduction" and "Conclusion," you can still use informative headings like these to map the second level of information.

If you are not bound by a company format, you can develop your own plan for placing your headings on the page. You can, for example, center first-level headings over your text, align second-level headings with the left margin, and indent third-level headings like paragraphs. You can reinforce this hierarchical arrangement by using different sizes of type and varying the type style. For

- Deforesting the Earth

- Deep Trouble in Oceanic Fisheries

- Grasslands for Three Billion Ruminants

- Per Capita Consumption Trends

- Future Resource Trends

- Oil: The Safety Valve

SAMPLE 7.1   Informative headings
Source: Lester Brown, *Building a Sustainable Society* (New York: W. W. Norton, 1983).

example, you can use bold 18-point type for first-level headings, bold 14-point for second-level headings, and italic 12-point for third-level headings, as shown in Sample 7.2.

Strategic Blunders in Korea

The first-order headings are centered, bold, 18-point type.

While the U.N. forces enjoyed a number of successes throughout the Korean War, notably the Inchon invasion, they also suffered from a general overzealousness early in the war and an extreme caution that never really paid off late in the war.

Early Zeal

The second-order headings are flush to the left margin, bold, 14-point type.

MacArthur clearly wanted to repeat the success that had brought him fame in World War II. But his zeal for a fast victory blinded him to both the sensitivity of the communist Chinese as he approached their border and the limitations of his own intelligence machine.

The Sensitivity of the Chinese Border

The third-order headings are indented like paragraphs, italic, 12-point type.

As the allied forces approached the Yalu River in the far north of the Korean peninsula . . .

SAMPLE 7.2   Use typographical variation to reinforce the hierarchical arrangement of headings

Divide Information into Short Chunks for Better Usability and Selective Reading   Research has shown that text with frequent breaks—headings, subheadings, paragraph indentations, graphics, and white space—creates what users perceive to be a "friendly" document. You should leave few pages, even double-spaced pages, unbroken by headings. Short sections make it easier for selective readers to find what they need. Say, for example, I'm doing a report on the future availability of forage for cattle. As I open Lester Brown's chapter on "Biological Systems under Pressure," I can look at the headings shown in Sample 7.1, then go right to the section headed "Grasslands for Three Billion Ruminants," and skip the rest.

Use Forecasting Devices and Summaries to Orient the User

Good technical writers prepare their audiences for what to expect. By *forecasting* your main points, you allow readers to skip irrelevant information, and you also prepare their minds to receive information in a structured form. To use a popular metaphor from cognitive science (the mind as computer), you create *empty arrays,* mental slots in the reader's mind for sorting information. For example, if you specify the number of points you plan to make, you create a blank outline that readers can fill in as they read.

In technical writing, good headings do part of the forecasting, but several other techniques are also effective at the beginning of a document:

- *a list of keywords,* indicating what points will be covered
  Example: This presentation on cancer risks will concentrate on three major problems: tobacco, alcohol, and diet.
- *a list of objectives,* stating what the reader should be able to accomplish by reading the document
  Example (from a manual on word-processing software): In the first chapter, you will learn to do the following: 1.) open a new file; 2.) enter texts into the file; 3.) edit the texts; and 4.) save the file for later use.
- *a thesis statement,* conveying the controlling idea or theme of the document Often, this will be a summary of your most important research conclusion.
  Example: "Though research performed in the last decade has suggested that heredity is the chief influence on cholesterol levels, our findings show that diet and exercise can make even more significant contributions."

You can also use these devices at the beginnings of new sections.

In addition to helping your reader look forward, you should also summarize frequently. Remember how helpful it is when a speaker breaks for a moment to remind you of what he or she has just covered? If your mind has wandered, you have a second chance. You can use the same technique in writing to save the reader the trouble of scanning back through long explanations to find key points.

One word of warning: Forecasting devices and frequent summaries can become intrusive. If your document is very short, the reader may feel you are condescending if you repeat your key points three times. Instead, you could begin by saying that you have three main points and then write three clearly demarcated sections, each with an informative heading. Then your readers can easily do their own forecasting and summarizing merely by scanning quickly over the headings.

Also remember that the most powerful summarizing devices are not sentences, but graphics—especially high-density graphics such as tables, flowcharts, organizational charts, and line graphs. You can use them at the beginning or at the end of a section to predict or summarize. Graphics built on an X–Y axis have the power to create structured arrays of information in the reader's mind, much like lists and headings. They also add a comparative dimension missing in a simple list or outline. That is, they can overlay one list on another.

Use Typography to Enhance Legibility

Because word-processing technology is so widespread, design decisions that used to belong exclusively to professional designers now fall to authors seated at their personal computers. When you publish important documents, you will still need to call on publication specialists, but on many occasions you will deliver reports, proposals, and instructional documents printed in-house or on your home computer. Even for some formal publications you will have to deliver camera-ready copy to the printer.

When you prepare camera-ready copy or do your own desktop publishing, follow closely any design guidelines your publisher provides. If no guidelines are available, your chief concern should be legibility.

Because the eye tends to respond favorably to clear contrasts, serif fonts such as Times, Palatino, and New York tend to be more readable than sans serif fonts such as Helvetica and Geneva. Serif fonts, which have more pronounced downstrokes and upstrokes, have more distinct letters. Sans serif fonts are considered more attractive for headings and on posters and viewgraphs because of their clean, crisp appearance; but for text reading, serif type improves efficiency. (See Sample 7.3.)

Your text will also be more legible if you keep font changes to a minimum. Frequent shifts can cause users to pay more attention to the way the information is presented than to the information itself.

Use Highlighting Devices
for Special Emphasis

A speaker can always point to and dwell on a very important item in a list. Writers use highlighting devices to achieve the same effect. By varying the size of

Sans serif fonts like Helvetica, **Chicago,** and Geneva give a clean, crisp appearance in headings, posters, and viewgraphs but should be avoided in long chunks of text because they understate the downstrokes and upstrokes needed for contrast in legibility.

Serif fonts like Times, Palatino, and Bookman increase legibility in longer text chunks.

SAMPLE 7.3   Design typography for legibility

your type, using distinctions such as boldface and italics, or using bullets (•), small icons, or cartoons, you can call attention to particularly important points or give the reader cues about where to focus attention (as in Sample 7.4).

Highlighting devices allow the browsing reader to survey the page quickly, grasping key points and reading more if interest and time allow. They also create a form of tiered access, giving the reader cues about the information's order of importance.

Highlighting devices can also serve as reminders and locators. Headers, for example—the words and page numbers that word-processing programs allow you to place automatically at the top of each page—remind readers where they are in their reading. The most effective headers are brief and specific—a few words that signal the location.

Avoid underlining and words printed in all capital (uppercase) letters, both of which can hamper legibility. These styles are left over from the days of typewriting, when different sizes and fonts of type were not readily available. Studies show that to read efficiently, readers need to be able to see contrasts in letters—the upstrokes and downstrokes of the *d*'s and *p*'s, for example, or contrasts between capital and lowercase letters. Underlining obscures downstrokes, and words in all caps have fewer contrasts for the eye to latch onto, so avoid these devices in your designs.

Use other highlighting devices sparingly. If you boldface every other sentence, for example, the distinction will lose its meaning. You've probably

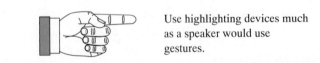

Use highlighting devices much as a speaker would use gestures.

SAMPLE 7.4   Small icons and other highlighting devices help the user to locate key points

known students who, in marking textbooks for study, highlight so many sentences that the *unhighlighted* passages, not the marked ones, stand out most clearly. Take care not to reverse the effect you intended. (See Sample 7.5.)

In addition, page designs that overuse highlighting devices can become so busy that they interfere with fluency. Professional designers strive for consistency and clarity in their layouts.

Use Graphics to Display, Writing to Interpret

Graphics offer a powerful design tool in technical communication. As we saw above and in Chapter 6, you can use pictures, charts, and hybrid elements such as tables and lists with icons to accomplish many design goals:

- To emphasize hierarchical relations among topics
- To summarize and show information at a glance
- To improve your reader's comprehension through visualization and modeling
- To provide impact and emphasis, especially in an age dominated by visually rich electronic media and mass communication
- To break up the task of reading

You cannot afford to ignore these advantages. Neither can you depend on graphics to do all the work. Graphics are powerful devices for display, but they lack the analytical and interpretive power of written prose.

You will have to use written language to complement most of the graphics in your document. Table 7.1 alerts you to the advantages and the dangers of integrating different types of graphics into your design.

Erogonomics is the study of human performance to the *improvement* of the work system consisting of **the person, the job, the tools, the workplace,** and **the environment.**

Putting the Person First

The person is important because it is *people* who stand to benefit from *ergonomics* and because *without people,* business would never get done anyway. When dealing with people in ergonomic design, remember that *training comes only after motivation is established.*

Training works only after the people have been motivated.

SAMPLE 7.5   Overhighlighted or busy text

TABLE 7.1

Advantages and dangers of graphics as design elements.

Graphic	Advantages	Dangers
High-density graphics such as tables and charts	Tremendous summarizing power, accessibility, and ability to show comparative relations	Reader saturation, loss of argumentative and narrative context, loss of coherence
Functional diagrams, flowcharts, and maps	Ability to show total processes at a glance; flexibility in showing complex relationships	Underuse of descriptive and imagistic power of language
Cartoons, photographs, and icons to highlight or illustrate points in text	Additional impact for important points, increased power to control tone (e.g., adding a light touch with a cartoon)	Improper emphasis (text to be illustrated selected because of "visual possibilities" instead of substantive significance)

Use Color for Realism and Contrast

Color printing and reproduction are now readily available (and reasonably priced), so you should consider using color when you need to depict an object realistically. If you were promoting a new ocean floor scanning device that gives color readouts, writing a field manual on how to distinguish various shades of red sandstone in different geological formations, or designing a guide for how to identify different varieties of fall warblers, color illustration would be mandatory.

You can also use color to heighten contrasts that command the reader's attention. If bold text "speaks" loudly, color shouts, especially if it is surrounded primarily by black and white. So be sure that you want strong emphasis wherever you use color. Designers frequently use color for headings. In graphics such as bar charts, color helps fast-scanning readers make careful distinctions. In a flowchart or a line graph, you can use color to emphasize a line that you want the reader's eye to follow with special care.

Just remember that, as with all textual devices, a little color is a good thing. Too much color can create conflicting messages about what's the most important element on the page. Don't use color simply for decoration or to impress the audience. Use it as a powerful rhetorical tool to emphasize the most important information.

Use White Space to Frame Information and Enhance Legibility

The white spaces of a document are important in two ways. First, you can use white space to frame important points for emphasis. That's why it's important to put your most significant points into headings, lists, graphics, captions, and hybrid text elements. White space surrounds these devices, giving them a prominence that words buried in long chunks of text don't have, as Figure 7.2 illustrates.

The second advantage of white space is that it allows the reader's eye to rest and to sort information without confusion. That's one reason why children's books have large print and huge amounts of white space. If each page contains only a single chunk of text and a supporting picture, comprehension is much easier. Of course, white space is expensive; it increases the total number of pages required to cover the same amount of information that crowded texts can cover at a much lower cost. Also, the users of technical information are not children; they are usually quite capable of comprehending information at a very high level.

FIGURE 7.2

White space provides emphasis for key points. The page on the right uses white space to frame important information in the list, the figure, and the caption. Important information would be buried in the long blocks of text that dominate the page on the left, which uses white space only for paragraph indentation.

But ability to read is not the same as willingness to read. Very often, users are unwilling readers. Perhaps they have 200 proposals to review before they can make a decision. Or they may be frustrated by the expensive computer they just bought and are turning to the manual as a last resort. Easy-to-read pages with a generous amount of white space will make such onerous tasks easier to face. For such users it's a good idea to cover no more than 60 percent of each page with unbroken text. For instructional manuals it also makes sense to follow the model of the children's book. Reserve at least a whole page for every discrete task.

With these general design principles in mind, in the next section of this chapter and later in the rest of the book we'll see how various combinations of purposes, users, and documents result in different design demands.

Design in the Genres of Technical Communication

Because of the different relationships among the authors, users, and purposes of the three genres (see Table 5.1 on page 121), each has a different, though overlapping, set of design considerations.

Reports make the heaviest use of interpretive devices. Lists, tables, charts, and graphs interact with complementary chunks of prose to argue for a certain course of future action based on past and present trends. Each part of the report contributes in a tightly unified way to the overall argument.

Proposals are also organically unified. Every element is geared to influence a decision about funding. But there are some differences. Proposal writers often use high-impact graphics, such as photographs and artist's renderings, to add concreteness to descriptions of future plans. These visuals can help to convince decision makers that the author's vision is realistic, and they invite the reader to share the author's vision. If the proposal is divided conventionally into a technical section, a management section, and a cost section, the individual sections are often separate units of design. The technical section relies on detailed figures and drawings such as blueprints and schematics, which are integrated with specialized arguments. The management section includes organizational schemes and promotional materials about personnel and facilities. The cost section is directed to accountants, so tables, graphs, and budget sheets predominate, and text is kept to a minimum.

Manuals and other instructional documents present perhaps the greatest challenge for document designers. Users will want to have the manual open while they are working, so the basic unit of design is the page. The designer must carefully analyze every task and mock up the document in advance to ensure that the end of every page coincides with a logical break in the action. On each page, discrete steps must appear in separate sentences. Many will have a functional graphic along with a warning or caution statement. The manual represents the most radical version of our basic requirement that a technical document should move from simple readability to complex usability.

As you make decisions about genre and design, remember that genres are malleable. They can change or overlap according to the needs of the author, audience, and situation. If necessary, you can create your own subgenres and variations as you confront real communication situations in the working world (or *realistic* situations in your technical communication class). Think of genres not as templates for any document you might create but as flexible models that reflect typical interactions among authors, audiences, situations, and documents.

Special Considerations in Designing Electronic Documents and Web Pages

In the print medium, the document user focuses on *pages.* In electronic texts, the basic units are *screens.* We can apply many of the same design principles to both media if we think of the information *chunk,* or module, as the basic unit of design, and the user's *visual field* (page or screen) as the area within which each chunk must work effectively.

In electronic documents such as online help programs, the visual field tends to be smaller. The temptation is to cram it too full of information, but the best policy for usability is to do the following:

- Be rigorously selective in what to include. Usually the material included is highly functional, with little room for motivational or decorative elements.
- Be strongly hierarchical. Try to get the most important information chunk in the first screen—or better yet, the first sentence or phrase—so that the user only has to change screens or "scroll down" to complete a process or to get secondary information.
- Use only simple graphics, such as icons, that the user can understand with just a glance. See Figure 7.3.

In websites, the users' visual field and graphics capabilities will vary depending on what computer they are using. You can't control exactly what the users will receive, so to keep their frustration to a minimum, concentrate carefully on the things you can control, such as the following:[1]

- Always plan your web site on paper before you begin to construct it and build links. Make a diagram such as the one in Figure 7.4 on page 203.

---

1. For an online manual on designing web pages, see Patrick I. Lynch and Sarah Horton, *Yale C/AIM Web Style Guide,* http://info.med.yale.edu/caim/manual/index.html (1997).

FIGURE 7.3

The "Balloon Help" feature in the Apple Macintosh user interface is a good example of online writing reduced to its shortest form, making use of easily recognized icons. The dialog "balloon," which makes the "trash" feature seem to speak, would be instantly recognized by anyone who had ever read a comic book.

- Design home pages, or first screens, with the idea of making users want to see more and making it easy for the user to act on that desire. Include a series of buttons, ordered in a logical manner, that link users to other parts of your site (taking them to other places in the home file, or linking to other files or "pages"). Create layers of information with tiered access (see Chapter 4), with the most frequently used information no more than one or two links away from the home page.
- Make the titles on buttons informative, as in headings. Users should be able to predict what will appear when they click the button. You may be able to use simple icons as buttons to enhance the predictive effect.
- Be sure to place "navigators"—buttons that allow users to "get home" and find their way back to where they've been—on every page of the site, if not on every potential screenful of information.
- Avoid slow-loading graphics, especially on the first page. Keep graphics small and tightly cropped. Use background color and interesting typography to create the signature effect you're looking for.
- Don't try to make the resolution too high for photographs, detailed graphics, and even colors. High resolution is beautiful, but it can really slow down loading and lead to user frustration.

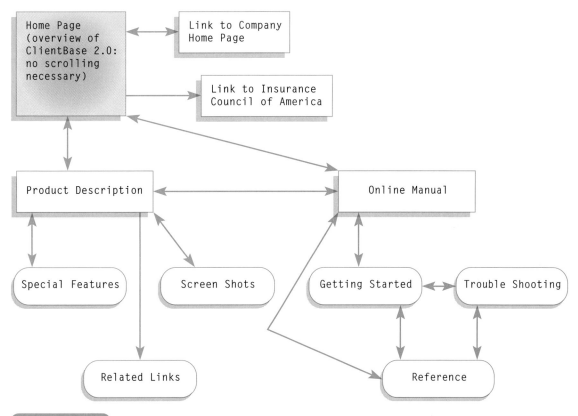

FIGURE 7.4

A "map" for planning a website. The arrows indicate the direction of the links.
A double-headed arrow indicates that a return link exists.

Reviewing Document Designs

With the information in this chapter you should be able to begin the design
process for any document. To complete the process, you'll need to work
through the later chapters on the individual genres.

At this point, you should also be able to review documents with an eye to
improving their design, using the material in this chapter along with the rhetorical, managerial, and ethical skills you began to develop in Chapters 1–6. To
get started on reviewing your own plans or a draft document, use Guidelines
at a Glance 11 to locate trouble spots in documents that require attention and
revision.

GUIDELINES AT A GLANCE 11

Document design

Assessing resources:
- Assess time, money, and technological capabilities available for production.
- List any format requirements that will limit your design.

Organizing and formatting:
- Outline your document, using a hierarchical structure.
- Write informative headings keyed to the outline.
- Make sure the written text is divided into fairly short chunks with few (if any) pages left unbroken by a heading, subheading, or graphic.
- Position headings consistently, using typographical variation and placement on the page to indicate different orders of headings.

Designing for effective emphasis:
- Use effective forecasting and summarizing devices: informative headings, lists of key points, lists of objectives, checklists, and graphical summaries.
- Use highlighting devices such as icons and typographical variation to emphasize key points.
- Use white space to frame important points for emphasis and to allow the reader's eye to rest and sort information.

Designing for visual effectiveness:
- Integrate graphics effectively with written text, using graphics to display and writing to interpret information.
- Use graphics to help readers visualize models of important concepts.
- Use typography and white space to enhance legibility and visual appeal.
- Use color for realism and contrast.

Taking genres into account:
- In reports, consider the entire document as the basic design unit, integrating graphics with textual explanations to create an effective overall argument.
- In proposals, use the section as the basic design unit, employing different strategies for the technical section, the management section, and the cost section. Increase your persuasive power with high-impact, realistic graphics.
- In manuals and other instructional documents, use the page as the basic design unit, assigning a single task to each textual module and placing single steps in each imperative sentence. Integrate functional graphics and appropriate warnings and cautions with corresponding steps.
- If reports, proposals, or manuals are to be posted on the Internet, the basic unit of design is the scrollable screen. Present text in chunks with motivational and functional (even animated) graphics, adding links to other Internet sites as appropriate.

EXERCISE
7.1

Part 1. Assessing resources. With a small group of classmates or on your own, select one of the writing problems listed below. Make a worksheet with three columns. In the left column, list the features—written text, headings, lists, charts, tables, photographs, or anything else—that you think you would need to design an ideal document. In the middle column, write how you would go about producing each feature. For example, would you use your own computer to type the text? Be as specific as you can about which facilities you would need and how you would get access to them in your own context of production. In the right column, estimate the cost in time and/or money to produce the features you would need. Sample 7.6 shows the start of such a worksheet.

If you wish, you can divide the features in the first column among the members in your group. Then go out and get the best estimates you can. Complete your worksheets for the next class, and discuss them with other groups. How realistic do they think you have been in your estimates? Will your context of production allow you to execute the design you envision?

Problem 1.   You are producing a fact sheet on the best undergraduate program in your college's department of electrical engineering. You want to attract the best possible students by stressing the high academic standards and intellectual challenges of the coursework and by detailing research opportunities.

Document feature	Production facility	Costs
10 pages of text	computer in my room; Microsoft Word for Windows	2 hours of typing; do it myself
6 lists	same	extra typing time-- about an hour; do it myself
two orders of headings	same	no extra typing time; do it myself
three line charts--in color	chart program in Bill's room, using his color printer	half hour per chart; get help from Bill (add extra hour just in case)

SAMPLE 7.6   Rough worksheet for document production planning

From a flower-gardening manual:

  You should start off your garden by thinking small. It is best to start a small flower bed than to plant a big one. You can always add to your flower bed later. It is a good idea to draw a diagram of your yard. This way you can roughly sketch in where you want your flower bed to see how it looks. The type of flowers you want to plant will depend on the area you have picked for your flower bed. Some flowers grow well in the shade while others grow well in the sun.

  The next decision you will need to make is what shape you would like your flower bed to be. Most flower beds are rectangular in shape. It is not the most artistic form, but it is easy to mow around. If you want to be artistic, we recommend a curved flower bed. An easy way to see how your flower bed will look would be to take a garden hose and lay it out, outlining the shape of the flower bed you desire. This way you can move the hose around until you are satisfied with the shape you want.

  Once you are satisfied with how the hose outlines your flower bed, you are ready to start preparation for digging. The best way to proceed is to drive a lot of small stakes along the curve of the hose, about two feet apart. This will serve as a guide for spading the flower bed. You can remove the hose after all the stakes are in place. Another tip is to tie a string from stake to stake. This keeps you within your outline while digging.

  Grass and flowers do not do well together, so you need to dig up the grass. Take a spade and follow the line of your string. Place the blade of the spade in line with the string and dig straight down. To make the removal of the grass easier, cut the grass into squares about four to six inches deep. Then you will need to pile the turf out of the way. If you have any bare spots on your lawn, you can use the turf for that. You will need to make sure you clear out any large rocks, weeds, and other debris. Rake the dirt smooth after you finish all of the digging and clearing.

SAMPLE 7.7   Chunk of text in need of document design

Problem 2.   You have designed a smoke alarm for hearing-impaired people. A funding agency has tentatively agreed to fund further research and development but has requested a five-page description of your work to show to possible corporate contributors.

Problem 3.   You worked all summer as an intern social worker in a project designed to help teenage mothers return to high school. You're impressed with the

From a report on business conditions in Russia/Commonwealth of Independent States:

Another prominent concern of many foreign businessmen seeking to conduct business in Russia is the increasing risk to their personal safety. As one US corporate executive revealed to the *Wall Street Journal* after having been confronted by mobsters demanding that he relinquish ten percent of the two million dollar communications firm he managed or forfeit his life, "We always tried to keep our heads down to avoid attracting the interests of Russia's criminals . . . but Moscow is getting to be a more brutal place every month." In fact, Russia's Interior Ministry has reported a seventy percent increase in crimes specifically targeted against foreigners in the past year. In the first seven months of 1993, there were 956 reports of violence against foreigners, compared with just 664 in the entire previous year. Furthermore, nearly one hundred foreign registered automobiles are stolen daily in Moscow, with less than two percent ever recovered. Maxwell Asgari, an American executive managing a Swiss industrial cooperative in Moscow, firmly contends that criminal activity, for the most part, has gone unabated by government officials and the Moscow Organized Crime Unit. "It is not that the current level of crime is so bad, but it is growing rapidly, and I don't see any vehicle or resource that the government has to prevent it," he said.

SAMPLE 7.8   Chunk of text in need of document design

success rate of the project and have promised to help write a funding proposal to a charitable foundation that has shown interest. Now the time has come to make good on your promise.

Problem 4.   You and several other students have created a hypertext guide that provides online help for students using the library's new computer system. Two hundred librarians from across the state have contacted you to purchase the program. They also want an installation and maintenance manual to go with it. You have agreed to write it for a fee of $500.

Problem 5.   You have an idea for reorganizing the principal design team in your company, a toy manufacturer that prides itself on employee satisfaction. You've been asked to present your plan to the board of directors in a brief report.

Part 2. Designing a Plain Page of Text.   In your group or on your own, examine the chunks of text in Samples 7.7 (page 206) and 7.8 (above). What design features do these texts need? Plan a new design, and produce a rough version of one of the samples by the next class. If you cannot produce some of the features the text needs, prepare a mock-up of the feature instead. For example, if you can't supply a photograph, insert a box that contains a brief description of the photograph in the correct position on the page. Compare designs with other groups. Has anyone overdesigned? Has anyone underdesigned? What features does nearly every design have in common? What are the differences? Explore with one another the rationales for the different approaches.

EXERCISE
7.2

Read carefully over one of the following documents:

- the "Bucky Balls" website in Sample 6.40 on page 170 of Chapter 6
- the "Hemodialyzer Reuse" paper in Sample 7.9 at the end of this chapter (pages 209–218)
- "A Guide to Planning a Successful Workshop" in Sample 7.10 at the end of this chapter (pages 219–226)

In a group of your classmates or on your own, use the guidelines for writing and review developed in this chapter to evaluate the potential effectiveness of the document. How could you enhance its strengths and eliminate its weaknesses? What kinds of production resources would you need to make these improvements?

Recommendations for Further Reading

Bernhardt, Stephen. "Seeing the Text." *College Composition and Communication* 37 (1986): 66–78.

Redish, Janice C., R. M. Battison, and E. S. Gold. "Making Information Accessible to Readers." *Writing in Nonacademic Settings.* Ed. L. Odell and D. Goswami. New York: Guilford, 1986: 129–153.

Horton, William. *Secrets of User-Seductive Documents.* 2nd Edition. Arlington, VA: STC Press, 1997.

Schriver, Karen A. *Dynamics in Document Design.* New York: Wiley, 1997.

White, Jan V. "Color: The Newest Tool for Technical Communicators." *Technical Communication* 38, no. 3 (1992): 346–351.

White, Jan V. *Graphic Design for the Electronic Age: The Manual for Traditional and Desktop Publishing.* New York: Watson-Guptill, 1988.

Additional Reading for Advanced Research

Killingsworth, M. Jimmie, and Scott P. Sanders. "Compensation and Complementarity in the Design of Technical Documents." *The Technical Writing Teacher* 17 (1990): 204–221.

Kostelnick, Charles. "Visual Rhetoric: A Reader-Oriented Approach to Graphics and Designs." *The Technical Writing Teacher* 16 (1989): 77–85.

Martin, M. "The Semiology of Documents." *IEEE Transactions on Professional Communication* 32.3 (1989): 171–177.

Rubens, Phillip M. "A Reader's View of Text and Graphics: Implications for Transactional Text." *Journal of Technical Writing and Communication* 16 (1986): 73–86.

Documents for Review

Hemodialyzer Reuse

An Analysis and Recommendations for the
Reuse of Hemodialyzer Cartridges

Submitted to the Association for the Advancement of Medical
Instrumentation
by
Trent Elliot
October 23, 1994

Abstract

With the increasing cost of health care and increasing public awareness on
health issues, the kidney dialysis industry has begun reusing hemodialyzer
cartridges that are labeled "For Single Use Only." This thought strikes fear in many
health conscious Americans. Because of the AIDS scare and various other
diseases, the word "reuse" tends to have a negative connotation when referring to
medical devices. Is the reuse of hemodialyzer cartridges a health risk? Are dialysis
clinics and hospitals reusing hemodialyzer cartridges for economic purposes while
at the same time putting dialysis patients at risk? These are some of the questions
facing the kidney dialysis industry. The purpose of this report is to provide
substantial evidence on why the reuse of hemodialyzer cartridges is advantageous
and to present recommendations to the Association for the Advancement of
Medical Instruments (AMMI) and the directors of hospitals and dialysis clinics.
These recommendations are being made with the intent of encouraging the
multiple reuse of hemodialyzer cartridges. The health considerations, reprocessing
procedures, and economic benefits of reuse will be analyzed with some
comparisons to the single reuse of hemodialyzer cartridges.

SAMPLE 7.9 *Hemodialyzer Reuse:* A recommendation report
Source: From *Journal of the American Medical Association,* 260:14, 2073–2077.

Hemodialyzer Reuse

What is hemodialysis and reuse?

Thirty to fifty thousand Americans die each year from kidney disease. More importantly, ten to fifteen thousand of these people would be suitable candidates for artificial kidney treatment—hemodialysis. Hemodialysis is the procedure most commonly used to remove excess water, metabolic wastes, and excess electrolytes from the bodies of patients suffering from end-stage renal failure.

This process of cleansing the blood in the absence of proper kidney function is carried out three times a week per patient and lasts approximately three hours per session. Access to the patient's blood pool is normally done by placing permanent indwelling tubes inside the cephalic vein and the brachial artery of the patient's arm and connecting them with a shunt when not in use as shown in *Figure I*. A typical hemodialysis arrangement, including the monitoring equipment needed for safe operation, can be seen in *Figure II*.

Hemodialyzer reuse can be described as the practice of reusing the same dialyzer for multiple sessions on the same patient without replacement of filter membranes or other surfaces that come in contact with the blood during the cleansing process. This is accomplished by utilizing reprocessing techniques which include cleaning the blood surfaces followed by disinfection and sterilization.

Current Status

Presently, there are more than 8,000 patients receiving hemodialysis treatment, and 72% of all dialysis procedures carried out in the United States utilize recycled or reused hemodialyzer cartridges with or without the patient's knowledge (Baris and MacGregor, 1993). Recent evidence has shown that properly reused hemodialyzer cartridges are equally effective and safe as well as being more cost-effective than new cartridges. Many of the procedures involved in the reprocessing of hemodialyzers have been automated, which has increased the efficacy of reuse;

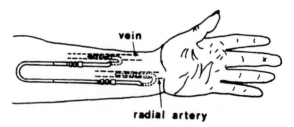

vein

radial artery

Figure I   Arteriovenous shunt and typical cannula system
(Source: Cooney, 1976)

SAMPLE 7.9 continued

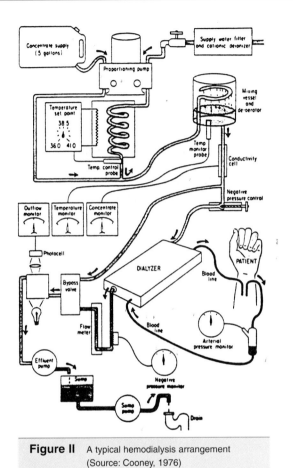

Figure II   A typical hemodialysis arrangement
(Source: Cooney, 1976)

however, the dialysis industry lacks requirements for the use of this automated process. Three of the main types of hemodialyzer cartridges on the market can be found in *Figure III*.

The flat plate dialyzer cartridge is the most commonly used hemodialyzer cartridge in the kidney dialysis industry. The decision to reuse hemodialyzer cartridges relies on three principal issues:

- Do the health benefits associated with reuse outweigh those benefits associated with single use?
- Are the cartridge reprocessing procedures safe?
- What are the potential economic gains associated with reuse?

SAMPLE 7.9 continued

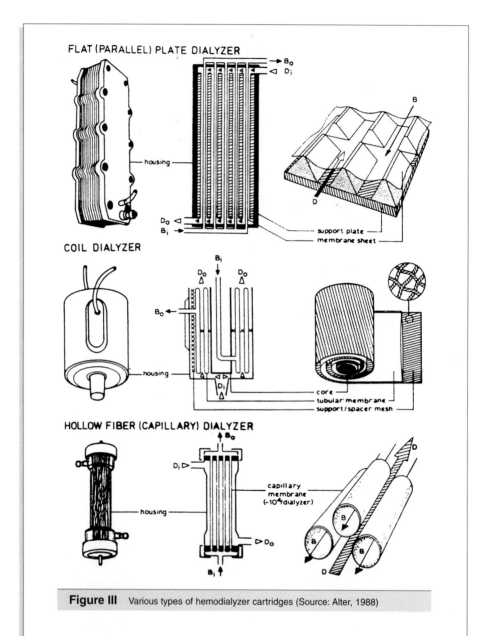

Figure III   Various types of hemodialyzer cartridges (Source: Alter, 1988)

Health Considerations

Concern over the safety of hemodialyzer reuse for patients with chronic renal failure has led to investigations of rates of death, the frequency and duration of

SAMPLE 7.9 continued

hospital admissions, the risk of infection, the frequency of pyrogenic reactions, and the health risks of exposure to disinfectants, particularly formaldehyde.

Rates of death and hospital admission

Investigations of hemodialysis centers and hospitals have not found increased rates of deaths due to dialyzer cartridge reuse. In fact, some investigations have shown that centers that reuse have significantly lower rates of death as a result of hemodialysis complications. In addition, rates of hospital admission due to hemodialysis complications was lower for patients reusing cartridges (Baris, 1993). Additional support for reuse can be found in a study published by K. R. Jones in *Health Service,* which stated that reuse was not among the factors associated with risk of hospital admission for patients with renal disease.

Risk of infection

In addition to the normal reprocessing, the cartridges must be rinsed with a water solution before and after use. This rinsing procedure has been a source of concern with hemodialyzer reuse because of the risk of microbacterial infection from the water supply. Many opponents of reuse say this is reason enough not to reuse, but what they overlook is that all cartridges must be rinsed before use. Therefore, new dialyzers are just as prone to infection as reused dialyzers because they are rinsed using the same water supply. The easiest solution to this problem is to control the water supply levels of microbacteria. The U.S. Centers for Disease Control (CDC) has recommended that a four percent formaldehyde solution be used during reprocessing in order to kill the microbacteria and there must be at least forty-eight hours of disinfection (Baris, 1993). Another possible source for concern due to reuse is the risk of HIV or Hepatitis B infection. Although the same cartridges are reused on the same patients, some rare instances have occurred in which cartridges were given to the wrong patient. With the use of automated reprocessing techniques and adequate training of medical personnel, the spread of infectious diseases is highly improbable (Baris, 1993). Therefore, the risk of transferring infections from one patient to another is not reason enough to not reuse cartridges.

Pyrogenic reactions and first use syndrome

Pyrogenic reactions and first use syndrome are two health considerations that must be taken into account when considering reuse. Pyrogenic reactions include fever, shivering, nausea and hypertension. These reactions result from endotoxins of microbacteria entering the bloodstream from the water supply. As previously mentioned, this water supply is used for rinsing the new as well as the reused cartridges; therefore, pyrogenic reactions are not inherent in just single or multiple use.

First use syndrome symptoms include chest pain, back pain, respiratory problems and muscle cramps. Since 1980, first use syndrome has been reported

SAMPLE 7.9 continued

by patients whose dialysis included a new cartridge. There are two types of first use syndrome. Type A reactions are uncommon but do exist and require medical attention and cessation of dialysis. Type B reactions occur after five percent of dialysis with a new cartridge but do not require complete discontinuation of dialysis (Baris, 1993). Both pyrogenic reactions and first use syndrome can occur in single use; however, only pyrogenic reactions occur in multiple use. Therefore, reuse has the lower probability of reaction. In addition, it has been proven that reused cartridges are more biocompatible than new ones (Alter, 1988). Biocompatibility is extremely important in that the more biocompatible the cartridge is the less likely it will produce an adverse reaction.

Exposure to formaldehyde

Formaldehyde is one of the most popular disinfectants used in hemodialysis. The problem with formaldehyde results from improper elimination of residuals. The accidental introduction of large doses of formaldehyde into the bloodstream can cause toxic effects. The risk of disinfecting with formaldehyde can be attributed to new or reused cartridges. The risk for both can be avoided by compliance with measures of quality assurance and control. With the concentration of formaldehyde below 10 ppm, the formation of toxins can be prevented. In addition, if the concentration of formaldehyde in the rinsing solution remains below 5 ppm, then the maximum amount of formaldehyde to reach a patient would be negligible (Baris, 1993). In order to monitor the formaldehyde levels, anti-N-like antibodies, antibodies that increase in number due to formaldehyde, should be carefully monitored. Any increase in the number of these antibodies directly shows an increase in formaldehyde (Ringoir and Vanholder, 1987).

Safety

Issues of safety affect not only the patient but also the personnel that operate the hemodialyzer equipment. The personnel that work directly with disconnecting, cleansing or disposing of the cartridges are at risk of HIV or hepatitis B infection. These risks are inherent in both multiple and single use. *Table IV* shows the incidence rate of Hepatitis B in personnel and patients for single and multiple use.

Since the introduction of automated reprocessing machines, the risk has been greatly reduced. Surveys conducted by the CDC have shown that with the use of protective clothing and less manual reprocessing there was not an increase in infection regardless of reuse. Personnel are also at risk of toxic formaldehyde vapors; however, the U.S. National Institute for Occupational Safety and Health has recommended that the air concentration of formaldehyde not exceed 1 ppm for a thirty minute period and concentration levels must be checked regularly (Baris, 1976). With the automated reprocessing machines and standards for control of formaldehyde, the personnel are not at risk with reuse of cartridges.

SAMPLE 7.9 continued

	Hemodialyzer; incidence rate, %			
	Not reused		Reused	
Year	Personnel	Patients	Personnel	Patients
1976	2.3	2.6	2.5	2.7
1980	0.7	1.0	0.7	0.8
1982	0.5	0.2	0.4	0.4
1983	0.2	0.2	0.2	0.2
1986	0.1	—	0.1	—

Table IV   Incidence rate of hepatitis B (Source: Baris and McGregor, 1993)
Source: Reprinted from, by permission of the publisher, CMAJ 1993; 148 (2).

Discussion

Safety, reduced rates of death and hospital admission, decreased risk of infection, pyrogenic reactions, first use syndrome, and exposure to formaldehyde all alter a patient's health differently according to multiple or single use. Safety, on the whole, is a factor in both multiple and single use because personnel have to handle the cartridges whether or not they are reprocessing or disposing of them. Therefore, safety should not be an issue in deciding whether or not to reuse. Rates of death and hospital admission both decreased with reuse. Risk of infection can be somewhat of a factor if not controlled properly. Pyrogenic reactions occur regardless of multiple or single use; therefore, pyrogenic reactions can also be ruled out as a deciding factor of whether to reuse or not. First use syndrome only occurs in single use; therefore, it is a strong indicator for reuse. Exposure to formaldehyde can result from both multiple and single use. In all, safety, pyrogenic reactions, and exposure to formaldehyde exist regardless of multiple or single use. Reduced rates of death and hospital admission, decreased risk of infection and first use syndrome all strongly support reuse. With the evaluation of health benefits, performing multiple use over single use is seemingly more advantageous.

Reprocessing Procedures

In order to reuse cartridges, a sound reprocessing procedure must be in place. The reprocessing procedure includes cleaning of the blood surfaces within the cartridge followed by a disinfection and sterilization process. The cartridges are then stored and reused by the same patient up to twenty times. Current studies show that cartridges reused five to twenty times are actually more biocompatible than new cartridges (Baris and McGregor, 1993). Today, many hemodialysis centers and hospitals have fully automated reprocessing machines that provide standardization and reliable testing of clearance and pressure. Clearance and pressure are tests performed to check the efficacy of the cartridge and to ensure that it is still performing properly (Hemodialyzer, 1989). The advent of automated

SAMPLE 7.9 continued

reprocessing equipment has further reduced health risks favoring reuse. Studies indicate that using automated reprocessing does not change the dialyzer function or biocompatibility over prolonged use (Gagnon, 1985). The majority of the problems associated with the reprocessing are not due to the procedure itself but instead are due to the lack of guidelines and regulations concerning reprocessing. Currently, no regulations exist ordering centers and hospitals to use automated reprocessing techniques. Also, no regulations exist to control the licensing of equipment or to train the medical personnel who operate the equipment. This lack of uniformity and lack of regulations cause dialysis centers and hospitals to set up their own system of guidelines and regulations which results in confusion.

Economic Benefits

Financial considerations are very important in the current health care system. Hemodialyzer cartridges on the average cost thirty to thirty-five dollars each. Although they are relatively inexpensive, cartridges represent substantial medical costs when considering the number of people receiving dialysis treatment and the frequency of the sessions. On the assumption that a patient receiving long term dialysis undergoes 156 sessions yearly, the total savings resulting from just five reuses would be approximately $3217 per patient per year (Baris and McGregor, 1993). Most centers and hospitals reuse cartridges up to twenty times before discarding. So for an average dialysis center or hospital, the reuse of hemodialyzer cartridges for each patient would constitute savings anywhere from 1 to 1.5 million dollars. A cost analysis of the benefits involved with reuse can be seen in *Table V (a)*. *Table V (b)* further illustrates the economic benefits of reuse by plotting the annual cost per patient versus the number of reuses. The savings are drastic for the first few reuses; however, they begin to level off after the fifth reuse.

Savings of this magnitude should not be ignored but have been in recent years. The Federal Government provides funding for one cartridge per patient per session regardless of whether the dialysis center or hospital reuses or not. This leaves many unallocated funds which could be used to provide better facilities or redistributed to other areas of the health care industry.

Conclusion and Recommendations

The health considerations, advancements in reprocessing procedures and economic benefits associated with the reuse of hemodialyzer cartridges are sufficient to justify that reuse is more beneficial and advantageous than single use. The following recommendations are being made to AAMI and the directors of hemodialysis centers and hospitals:

• Standardize the reuse of hemodialyzer cartridges on a national basis
• Emphasize further improvements and quality control in the area of reprocessing

SAMPLE 7.9 continued

Hemodialyzer: cost per session, $

Component†	Not reused			Reused		
	S1	S2	S3	S1	S2	S3
System	1.01	0.76	0.61	4.92	3.69	2.96
Personnel	1.50	1.50	1.50	3.00	3.00	3.00
Operation	0.68	0.68	0.68	1.19	1.19	1.19
Hemodialyzer	35.00	30.00	25.00	–	–	–
Total	38.19	32.94	27.79	9.11	7.88	7.15
Total for 5 sessions‡	190.95	164.70	138.95	74.63	64.46	56.39
Total for 10 sessions‡	381.90	329.40	277.90	120.18	103.86	92.14

*Situation (S) 1 = 30 patients per reconditioning system and a cost of $35 per hemodialyzer, S2 = 40 patients per system and $30 per device, and S3 = 50 patients per system and $25 per device.
†See the assumptions listed in Appendix 1.
‡For the reused hemodialyzers the figures include the cost of the first use plus the cost of subsequent sessions.

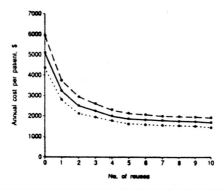

Table V   Cost analysis showing the economic benefits of reuse
(Source: Baris and McGregor, 1993)

Source: Reprinted by permission of the publisher, CMAJ 1993;
148 (2).

- Establish standards for reprocessing and the training of medical personnel
- Reallocate the funds involved with single use to improve the hemodialysis facilities and other areas of the health care industry

With the implementation of these recommendations, I feel that the prolonged reuse of hemodialyzer cartridges in the treatment of end-stage renal failure constitutes the most stable form of renal replacement therapy provided that adequate cartridge reprocessing is applied.

SAMPLE 7.9 continued

References

Alter, M. J., et al. (1988). Reuse of hemodialyzers. *Journal of the American Medical Association* 260–14, 2073–77.

Baris, E., and McGregor, M. (1993). The reuse of hemodialyzers: an assessment of safety and potential savings. *Canadian Medical Association 148,* 175–81.

Cooney, D. O. (1976). *Biomedical engineering principles: an introduction to fluid, heat, and mass transport processes.* New York: Marcel Dekker, Inc.

Gagnon, R. F., and Kaye, M. (1985). Dialyzer performance over prolonged reuse. *Clinical Nephrology 24–1,* 21–7.

Hemodialyzer reuse (1989). *Quality Assurance Guidelines for Hemodialysis Devices,* 10: 1–5.

Ringoir, S., and Vanholder, R. (1987). Influence of reuse and of reuse sterilants on the first use syndrome. *Artificial Organs 11–2,* 137–9.

SAMPLE 7.9 continued

A Guide to Planning a Successful Workshop

Davon Taylor

SAMPLE 7.10 *A Guide to Planning a Successful Workshop:* An instructional manual

"What? I'm presenting a workshop?" These words shouldn't send you into a panic. People are often asked to present information or training in a workshop setting by their employers or organizational affiliations. For many novice presenters the idea of developing a program from scratch and then presenting it (without putting everyone to sleep) is terrifying. It shouldn't be. This manual will lead you through a step-by-step process of developing content, designing materials, organizing the logistics, and practicing your presentation to make it a smashing success. An example workshop is developed after each phase in the boxes.

You should realize that every topic and situation will vary, so feel free to deviate from this order. You may or may not use every step. Let your creative juices flow, and let's get started!

Phase 1: Develop the content.

The most important part of any workshop is the content. You can be the smoothest presenter in the world, but talking without actually saying anything is a waste of everyone's time.

Step 1: Define the objectives.

It is important to have a clear picture of *why* you are conducting the program. Determine if the goal of your workshop is to:

✓Entertain
✓Disseminate new information
✓Train others to perform a task

Your answer will affect the development of your content. For example, disseminating information will focus on communicating in an interesting and memorable manner. Training others for a task will focus on making sure they understand all the techniques involved.

Step 2: Identify the audience type and size.

Your presentation style and complexity will depend very much on the audience's

✓Age
✓Education
✓Experience with the topic
✓Interest in the topic

Be sure to consider this in every stage of development. For example, be sure that the speech you use will be on the audience's level. (Don't use big words for ten-year-olds.) If the audience is not there of their own free will, they may not be  especially interested in the material. Design your material to catch their interest and relate the subject to something they care about.

SAMPLE 7.10 continued

The size of the group will determine the feasibility of activities. A group of 200 won't be able to introduce themselves individually, but a small group of 15 will work much better together if they *do* get to know one another individually.

Step 3: Form an outline.

Write an outline of the *main points* you wish to make. If it is a training program, outline the steps in order. Be sure to include an *introduction* that relates the topic to your audience and previews what will happen. Also include a *conclusion* that ties it all together and summarizes the main points.

Step 4: Consider the time frame.

The best way to ruin a great presentation is to run overtime. The participants have scheduled the allotted amount of time for your workshop—don't lose their respect for you and your information by overdoing it. Estimate the time needed for *each section* of your presentation, and practice each section to determine the accuracy of your estimate. Although you may want to have extra material prepared just in case you finish very early, plan to communicate all necessary information in the time provided.

Step 5: Fill in the details.

Now decide exactly what you want to say. Make your outline more specific by providing examples and illustrations of your main points. Add in facts and figures you wish to use. This is the meat of the presentation—where the points are explained in the time provided.

Phase 2: Design the supporting materials.

People learn differently, so using both audio and visual senses is important. You will probably wish to have visual aids such as slides, transparencies, posters, videos, or handouts to supplement your oral presentation.

Step 1: Make them relevant and helpful.

Visual aids are supposed to *help* the audience understand your presentation. If they are slides, overheads, or posters, they should reiterate what you are saying in a concise manner. If they are handouts the participants will keep, they may be designed to provide greater detail or data to be kept as a source.

Step 2: Make them clear and usable.

It is sometimes tempting to become so caught up in making visual aids captivating that they don't fulfill their purpose. Although brightness and graphics are helpful, don't sacrifice clarity for cuteness.

SAMPLE 7.10 continued

EXAMPLE: Phase 1

1) I want to teach the participants how to set goals.
2) My audience will be high schoolers, so I'll speak simply and use examples that relate to students.
3) I.   Introduction
   II.   Prioritize
   III.   Visualize
   IV.   Verbalize
   V.   Conclusion
4) 1 hour total time period; 10 minute introduction; 45 minutes on 3 steps; 5 minute conclusion.
5) *Introduction (10 min.)*
   Stress importance of goal setting; dream vacation game; introduce 3 steps
   Prioritize (20 min.)
   Write down goals in five categories of family, friends, school, physical, & spiritual; choose one in each category that can be accomplished within one year and ten years; balloon game
   Visualize (10 min.)
   Free throw game; draw picture of one goal actually happening; share with partner
   Verbalize (15 min.)
   Write down for each goal a deadline, rewards, obstacles, help sources, and daily tasks; write letter outlining plan of action and what help you will need; exchange letter and address with a partner and vow to check up on each other's progress
   Conclusion (5 min.)
   GOAL—Go Out And Lead

Step 3: Plan sequence and order logically.

Be sure that your materials follow the source of your presentation and that they provide for a smooth flow.

Step 4: Make the materials interactive if possible.

An old adage says, "Tell me and I forget; teach me and I remember; involve me and I learn." Ask questions or have fill-in-the-blank handouts and require the audience members to be a part of the program. Games are a wonderful way of illustrating a point. Small group discussions let the participants interact. Try to follow the 80/20 rule as much as possible; it requires participants to do 80% of the activity and the presenter to do 20%. Use any means you can dream up to get them *involved*.

SAMPLE 7.10 continued

> **EXAMPLE: Phase 2**
>
> Workbook to be handed out will have: a) exciting graphical representations of prioritize, visualize, and verbalize; b) pages to write down goals in all five categories with instructions to highlight top priorities; c) blank page to draw visualization picture; d) chart to verbalize the components of each goal; e) place for address and goal of partner to keep in touch with.
>
> I will have overheads that match the workbook and markers to fill in blanks with audience's examples. I will have posters and tape for dream vacation game and Nerf basketball and hoop for free throw game. I'll also need balloons and string for balloon game.

Note: If you plan activities that require splitting into groups, have a definite method of separation. Some possibilities are birthday months, interest groups, color-coded nametags, numbering off, or first letter of the name.

Phase 3: Organize the logistics and facilities.

The surroundings and atmosphere can make or break a presentation, so don't ignore them.

Step 1: Obtain the needed equipment.

Determine what equipment you will need, i.e., overhead projector, slide projector, microphone, etc. Find out if it will be provided or if you are responsible for obtaining it, and act accordingly.

Step 2: Plan the seating arrangement.

The shape of the room will determine the best arrangement of chairs. Make sure that all the participants will be able to see you and your visual aids clearly. Having the seating block's width greater than its length is preferable because they feel closer to you (Figure 1). If group action is to take place, you may prefer round tables instead of rows. If participants will be required to move around into different groups, ensure that they have room and it won't be chaotic. Don't forget to consider columns, windows, and other distractions in your decisions.

SAMPLE 7.10 continued

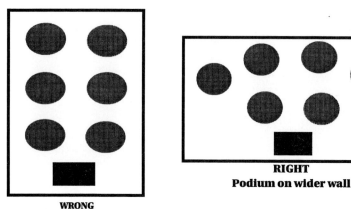

WRONG
Podium on narrower wall

RIGHT
Podium on wider wall

Figure 1. Optimal room orientation.

Step 3: Control the temperature.

Find out where the thermostat is and who controls it. Nothing is more miserable than being cold or hot during a workshop. Know how to keep the room at a comfortable temperature. Find out ahead of time—you don't want to have to interrupt your workshop to chase down a maintenance worker!

EXAMPLE: Phase 3

1) Facility provides overhead projector and cordless microphone. I bring all copied materials, game materials, and supplies.

2) Because the participants will need to move to their dream vacation spot, I need 4 clusters of 12 chairs each and 4 clusters of 13 chairs each. This equals exactly 100 chairs. I want to avoid having extra chairs to prevent participants who are reluctant to get involved from spreading out and isolating themselves. Each cluster should consist of two rows, both facing the front. There should be spaces between each cluster.

3) The thermostat on the south wall controls the room. Set it on 70 degrees at least one hour before starting.

SAMPLE 7.10 continued

Phase 4: Practice the actual presentation.

You've heard it a million times—practice makes perfect! Don't waste wonderful content and materials by neglecting your actual presentation skills. While running through the actual presentation, remain aware of these five aspects.

Step 1: Plan your appearance.

Of course you should always be neat and well-groomed, but you should also dress according to your subject and audience. If your workshop is a painting demonstration, you'll be prepared with a painter's apron. If it is a computer presentation to business professionals, you'll dress in business-like attire. Furthermore, it is very important to be comfortable in whatever you wear to calm your nerves and give yourself a boost of self-confidence.

Step 2: Watch your speed.

Make sure you are speaking slowly enough to be understood but not dragging at an artificial pace. You are not effective if not understood. Double-check that all of your planned material will fit in the allotted time. Be sure to allow for pauses to rest and to emphasize important points. A few quiet seconds after a pertinent statement lets the information sink into the participants' heads.

Step 3: Check your volume.

Make sure that you are speaking loudly enough that even the back row can hear you. If you are using a microphone in the presentation, practice with it ahead of time. Be sure to know how to adjust its volume and how to hold it without squealing. Volume variation is an excellent way to hold the audience's attention and to emphasize certain points.

Step 4: Practice using the materials.

Even the simplest tasks of advancing slides or transparencies can be difficult when you are nervous. Practice pointing to the screen or referring to the handouts. If you are demonstrating a task, have a dress rehearsal of the entire process.

Step 5: Be prepared to adapt as needed.

No matter how many times you practice, you can't control other people or situations. Be prepared to adapt as necessary to the audience's mood, time changes, location moves, equipment failure, etc. Think through in your head what you could easily delete from your program without ruining its message. Consider how you can make the points without your props. Most importantly, have your mind set to *be flexible!*

Presenting an interesting, informative, and effective workshop requires preparation in four areas: developing the content, designing the materials, organizing logistics, and practicing the presentation. Follow the simple steps and suggestions outlined above and you will soon be the most demanded workshop facilitator around!

EXAMPLE: Phase 4

1)   I want to look professional, but also be able to move easily with the speed of teenagers. I'll wear nice slacks and blouse with a blazer.

2)   I plan to keep things moving during the games and activities and then slow the pace down a bit when making strong points.

3)   Because this age group is prone to whisper and be easily distracted, I will use a microphone to be heard clearly. The microphone also helps to establish an environment of professionalism and authority that affects their behavior.

4)   PRACTICE! I will store my overheads in a three ring binder in the right sequence to minimize confusion.

5)   I'll think about what I will do if: a) someone is unruly or disrespectful, b) a game or activity flops. or c) I have an odd number of people for partner activities.

WORKSHOP PREPARATION CHECKLIST

Did you:

☐ Develop the content considering your objectives and the audience's characteristics?

☐ Make sure everything fits into the allotted time?

☐ Include illustrations, examples, facts, and figures to support your main points?

☐ Design supporting materials to be helpful and easy to understand?

☐ Include interactive activities to keep the audience involved?

☐ Provide for all necessary equipment and become familiar with the facilities?

☐ Practice your presentation considering your attire, speed, and volume?

☐ Perform a dress rehearsal using all materials?

☐ Prepare yourself mentally to be flexible, to relax, and to do a wonderful job?

SAMPLE 7.10 continued

Planning and Delivering Oral Presentations

CHAPTER OBJECTIVES
▼▼▼▼▼▼▼▼▼▼▼▼▼▼▼▼▼▼▼▼▼▼▼▼▼▼▼

After you have worked through this chapter,
you should be able to do the following:

- Plan and prepare oral presentations using
  a graphics-oriented method of developing
  visuals
- Deliver an oral report based upon your
  understanding of rhetorical standards
  widely accepted among technical
  professionals

E ffective oral communication skills are essential for technical professionals at every level. On formal occasions you'll be called upon to present the results of months of research and planning before an audience that may well determine the future of your career. More frequently, you'll give short presentations and briefings to your colleagues and departmental managers. Your mastery of oral delivery techniques as well as your understanding of the different demands of these occasions will determine the effect you have on your audience. In addition, you will need good oral communication skills to produce effective written documents, especially if you are collaborating with others. For this reason the CORE method—this textbook's basic approach to the process of developing technical documents—treats oral presentation as an integrated element, as Figure 8.1 suggests.

This chapter will provide specific guidance for developing your oral communication skills and preparing an effective presentation. Building on what

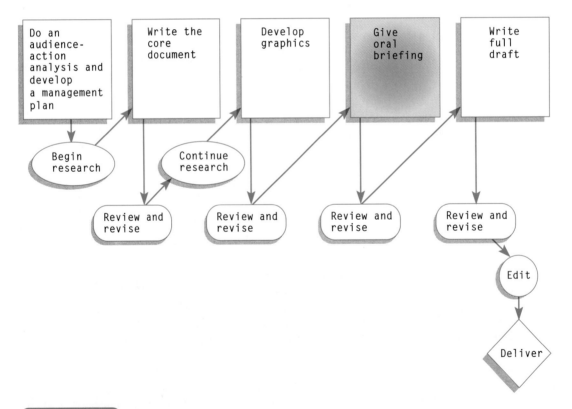

FIGURE 8.1

The place of oral presentation in the CORE method of document development.

we learned in Chapters 6 and 7, we will begin with a graphically oriented, audience-sensitive approach to planning oral presentations. From the start, it is important to realize that audiences of technical professionals prefer visually rich presentations that are tailored specifically to meet their information needs. In addition, the process of preparing visuals is a planning tool for your presentation as well as your final document. It will sharpen your sense of focus and bring concreteness to your ideas.

Planning Oral Presentations: The Mixed Media Approach

The standards for what makes a good speech have changed drastically over the last hundred years. Before the introduction of television and radio—in the golden age of speech communication in the United States—people would sit for hours listening to skilled orators delivering long, complex addresses. The Lincoln–Douglas debates, which lasted for several days, were the norm; the incredibly brief Gettysburg Address was the exception. The pattern of long speeches and patient audiences continued well into the twentieth century. But with the introduction of electronic media—which engage us in fast-paced, multisensory experiences—there was a shift in users' expectations. Now, after a half-century of television, we should assume that even the most sophisticated technical audiences prefer presentations that do the following:

- involve as many of the audience's senses as possible, with special emphasis on visual appeal
- address their particular needs (like the ever-multiplying cable television channels, which cater to a wider and wider range of special interests and tastes)
- give them instant satisfaction by presenting information in small, readily digestible bits (like the block between television commercials: eight to ten minutes)

To meet these new expectations, technical communicators have developed a mixed media approach for both oral and written communication. The best speeches and documents in industry, business, and government now mingle the positive qualities of spoken, written, and electronic communication and try to minimize the negative qualities of each medium. Table 8.1 briefly summarizes the strengths of each medium.

The new mixed media presentations strive to reproduce the immediacy and "friendliness" of interpersonal communication, the precision of good writing, and the appeal to a variety of senses and reading styles inherent in electronic media. These presentations are directed less to the patient listener and more to the busy, selective *user* who is now the typical consumer of technical information.

TABLE 8.1

Strengths of communication media.

Medium	Strengths
oral communication	The actual presence of the speaker allows for communication through gesture, body language, questions, and responses, resulting in a high level of interactivity.
written communication	Writing increases the range of dissemination along with the number of possible users. It makes fewer demands on the user's memory because it creates a record that can be consulted again and again. It also increases the user's independence.
electronic communication	Technologies like the telephone, television, and the new computer technology encourage interactivity and multisensory engagement (unlike writing); they can reach a broad range and number of users, and (like writing) can produce a recordable reference base.

Adding Visual Power to Oral Presentation

To take full advantage of the potential of the mixed media approach for your oral presentation, you'll need to design effective visuals (photographic slides, computer screen projections, or overhead projections) that concentrate your main points into a few well-chosen phrases and sentences and, wherever possible, connect these words to visual images. Like television news programs, educational programs, and computer software, your presentation should display your most important points visually. Since your audience is likely to be performing an action or making a decision based on the information you present, you want to make the presentation *memorable*. Research suggests that the most effective way to do this is by combining oral and visual communication. Table 8.2, based on a study by the Rand Corporation, illustrates this important finding.

TABLE 8.2

Briefings that blend oral and visual communication are most memorable.

Method of Presentation	Recall 3 Hours Later	Recall 3 Days Later
oral only	25%	10%
visual only	72%	20%
blend of oral and visual	85%	65%

Source: Peter White, "How Can Technical Writers Give Effective Oral Presentations?" *Solving Problems in Technical Writing,* ed. L. Beene and P. White. Copyright © 1988 by Oxford University Press, Inc. Reprinted by permission.

Visual communication not only is an aid to memory; it also caters to the expectations of typical users in an age in which frequent visual stimulation is the norm. Moreover, the development of visuals provides a powerful method for planning your report. It forces you to distill your overall content to the most essential information and then seek ways to make that information vivid and memorable.

Creating Visuals

The visuals for your oral presentation should help the audience follow the structure, as well as understand the content, of the presentation. Using as few words as possible, they should enable your audience to learn information independently of your speech. Knowing that your audience's attention will stray at various times in the speech, follow the lead of politicians, newscasters, and others who use the media for speeches: *Tell them what you're going to say, say it, and then tell them what you said,* constantly providing opportunities for the wandering mind of the busy, usually preoccupied listener to reenter your discourse. Ideally, your audience should be able to get the gist of your presentation from reading your visuals, even if they don't hear a word you say.

Arrange your information hierarchically as you prepare your visuals. Some of your information will naturally have less immediate importance to your reader. Background material or historical reviews, for example, are less important than items that require immediate action. You may want to present the least important information orally without much visual support—or eliminate it altogether. But all of your key points should always be repeated systematically and should be supported with strong visual images, perhaps even multiple images. Points that require action—recommendations or instructions, for example—should receive the greatest emphasis.

Say, for example, that a public health official is giving a presentation on cancer risks to a group of safety officers from major corporations. Their interest focuses on how her information can improve their clients' behavior both in and out of the workplace. So her primary purpose is to show what research says about avoidable risks. Her main point could be that by recognizing sources of cancer risk, employees can develop good health and safety habits. They can avoid the sources of risk where possible and minimize necessary risks. In support of this point, she will discuss six sources of risk: tobacco, alcohol, diet, radiation, workplace hazards, and pesticides. She begins her preparation by using her computer to compose an initial visual that lists the six main areas of concern (Sample 8.1).

She makes the characters large enough to read, using 24-point type. She also uses a clear typeface or "font"—Helvetica in this case. She makes her list grammatically parallel; every item is represented by a single noun. When she speaks, this will enable her to form good sentences as she moves through her presentation.

SAMPLE 8.1   Overhead slide with keywords corresponding to each major part of the presentation

To increase the visual power and embed the items in her viewers' memories, she could even add small icons or cartoons to the list, as in Sample 8.2.

This basic list can serve both as a general outline and as the basis for an overview of the report (*telling the audience what you're going to say*). The list reveals the overall structure of the talk as well as a condensed form of the content. The speaker plans to use this early in the presentation, perhaps first. As she speaks about the first topic, tobacco, she will state her main point in a memorable way, perhaps recalling a version of a familiar sentence: "As we all know, the Surgeon General has determined that smoking is hazardous to your health and is likely to cause cancer." The icon of the burning cigarette would reinforce her point. Then she will go on to elaborate on each of the six areas of concern, presenting an overhead slide for each one, with most of the information in visual form—statistics and charts or graphs showing incidences of cancer deaths among heavy smokers, for example.

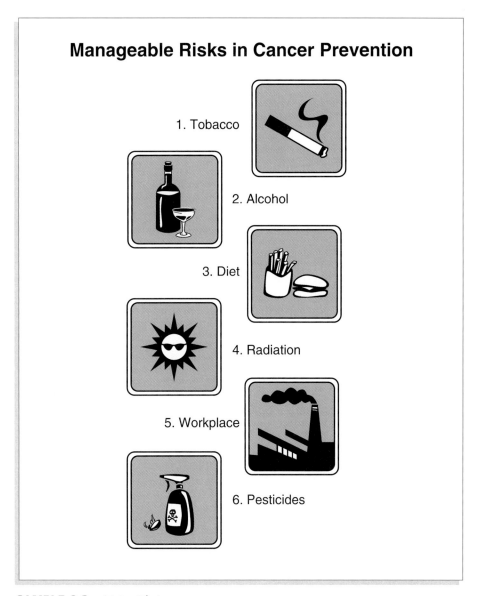

SAMPLE 8.2   List with icons

She must be careful, though, not to overburden her listeners with too many details either in the narrative or in the graphics. Each section of the speech and each accompanying visual should strive to make one point clearly. As we saw in the discussion of high- and low-density graphics in Chapter 6, the audience for an oral presentation cannot absorb as many visual details as can the reader of a technical document. So the health officer planning her oral report must select only the most telling statistics for graphs that will have a strong visual effect. For example, she might select trends that show definite peaks and valleys, as in Sample 8.3.

The type of graphics you use will be determined partly by the content of your presentation. If you are presenting an engineering design, for example, you will need to show some schematic drawings and perhaps artistic renderings of your ideas. If you are discussing landscape or environmental design, you will need photographs, sketches, and blueprints. Management plans and business proposals need tables, graphs, and charts. (Review Chapter 6 for ways to use each graphic most effectively.) But whichever type you use, oral presentations demand that graphics have a strong focus; they should communicate their point clearly without too much additional explanation (either on the slide itself or in your speech).

As you prepare your visuals, keep in mind that your audience will probably only remember as much information as you can get on the visuals—if they can remember even that much. If you want your listeners to retain fine details or complex arguments, you can either prepare handouts for them to carry home and read later or refer them to published writings. In this way you can use the

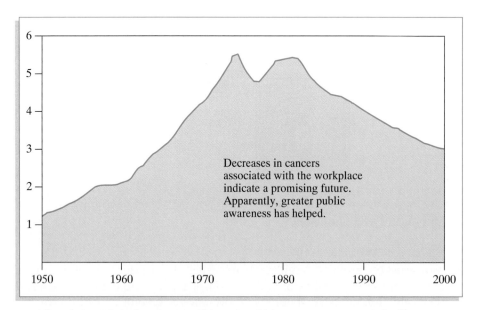

SAMPLE 8.3   High-density graphics should have a strong visual effect (fictional data for illustration only)

An initial overhead slide listing all the points you plan to cover
will help your audience follow the structure and remember the
content of your presentation.

strengths of written communication to compensate for the weaknesses of oral
communication and high-impact visuals.

　　If you have a choice about the equipment you can use to project your visuals, think about what will make you comfortable as well as how to achieve the
effect you want. The overhead projector is an extremely useful instrument. It allows you to glance down and read the words on your transparencies while you
project them onto the screen for the audience. The quality of the projection is
not as high as the quality you can achieve with photographic slides, but the use
of the overhead allows you more contact with the audience. You can leave the

GUIDELINES AT A GLANCE 12

Planning oral presentations and preparing visuals

1. First, define the audience for the presentation. What special interest does this audience have in your research or your technical expertise? What kind of information will satisfy audience members' need to act effectively or make good decisions? Form a mental image of this audience and what your listeners will do with the information you provide.

2. With an image of your audience in mind, write down the main point you wish to make in your presentation. The point should have a direct bearing on your listeners' future actions or decisions.

3. Divide the main point into a series of subordinate points; then compose a list or simple outline of these points. (See Chapter 6 for instruction on developing lists, outlines, and tables.)

4. For each point, create a memorable graphic or series of graphics. (Again, review Chapter 6 on the functions of different types of graphics.)

5. Organize your talk and plan your visuals so that you have at least two minutes to discuss each viewgraph. Keep the words on your visuals to a minimum.

6. For the final version of your visuals, use large type (14- to 24-point) and clear fonts. Don't use more than two different fonts (such as Times and Helvetica) or more than two different styles (such as italic and bold) on any single viewgraph. Otherwise, the viewgraph will appear "busy." (For more on typography and design, see Chapter 7.)

lights on and move back and forth between the projector and the screen, pointing to portions of the image if necessary. (Remember, however, to face your audience when you speak. Don't address the screen!)

You can also combine your methods of projection. The health officer in our example, for instance, could project photographs of sunbathers onto one corner of the screen that is already projecting her overhead transparency on types of radiation risks to avoid. Remember, though, that when you use equipment, especially in such elaborate combinations, you should rehearse with the equipment that you will use in the presentation to ensure that you are comfortable with its operation. Also, it's a good idea to check the microphone and other equipment in the room that you will use before the actual presentation. Many times, you will not have been able to use that equipment in rehearsal.

Guidelines at a Glance 12 describes how to plan oral presentations keyed to the preparation of your visuals.

Producing Your Visual Aids

How you physically produce your visual aids depends on your resources (money, time, and equipment), your technical expertise (almost anyone can make an overhead transparency, but some special skill is required for photographic slides), and the facilities in which you will be making your presentation. If you have access to good computer resources both in production and at the point of presentation, one of the most impressive kinds of presentation uses computer-generated overhead projections, which can accommodate photographic slides as well as simulate old-style overhead projections. You produce your visuals with the help of a computer program such as Microsoft's Powerpoint®.

Such programs have established the standard for oral presentations these days. But they do have drawbacks. For one thing, presenters tend to rely on the default categories of presentation graphics, with the result that all presentations start to look alike. Presentational formats take on the sameness we associate with clip art. And familiarity breeds contempt. Another problem is that you have to use not only an overhead projection device but also a computer to generate your images. You have two devices that can possibly malfunction, so the chances of trouble greatly increase. We recommend that you always print out your Powerpoint® slides and create backup overhead projections or handouts in case of mechanical failure.

Producing transparencies for an overhead projector is a fairly simple matter. The best way is simply to create paper versions of your overheads, then copy these onto 8½″ × 11″ transparent films on a photocopy machine. (Make sure that the machine you use is capable of accepting transparencies. Not all of them are.)

Be sure to rehearse with the visuals. Work with a partner or team member to make sure in advance they are legible from an appropriate distance.

Rehearsing and Delivering Your Presentation

Your oral narrative should fill in information and examples that support the points made in the minimal notes and graphics on your visuals. But you should use no notes during the presentation other than those that you share with the audience. This is an absolute must if the room is darkened for slides.

To achieve the best results in this kind of speaking, you do not need to write out your narrative word for word. In fact, you may run into serious problems if you try to memorize a long, prewritten text; you will almost certainly panic if you forget your lines. But you do need to write out an outline at some point, practice speaking from it enough times that you can deliver your presentation without faltering, and then rehearse while you are manipulating your support equipment. Nothing, but nothing, can kill a presentation like a miscued slide or a faulty sound system.

Rehearsing Effectively

As you begin practicing for your presentation, keep in mind that when you project graphic forms onto a bright screen in a darkened room or when you lead a user through a hands-on training experience, you draw attention away from yourself. But you should not neglect the need to develop a credible and engaging persona. "Persona" literally means "mask." The word implies the role we take on whenever we give a performance, and an oral presentation is definitely a performance.[1] You will be judged not just by what you say but also by the impression you make. By experimenting a bit with your persona, you can establish a special rapport with your live audience.

Our agriculture students, for example, often adopt a "farmer-scientist" persona, positioning themselves between the worlds of university research and practical food production. While presenting the latest scientific information, they use the accent and some of the mannerisms of country folk, which most of them know intimately from their childhoods on a farm or ranch. Likewise, education students may adopt a teacher persona as they present the latest research from education labs. Petroleum engineers project their experience in the oil field through their language and mannerisms. If you can handle a persona well, without insulting the audience by slipping into caricature, then you can engage the audience at its own level even as you impress them with your expertise.

To get the best sense of the image you project, try videotaping your rehearsal. Nothing is better for helping you see whether your persona is credible and authoritative or whether you need to make adjustments in your style. The camera will reproduce unforgivingly all of the places that need work.

Guidelines at a Glance 13 will also help you during your rehearsals.

Performing with Confidence

When the day of your presentation arrives, you will of course be nervous, and there is no foolproof cure for stage anxiety. Certainly, a full command of your subject matter and plenty of rehearsal will help to alleviate nervousness, and there's no substitute for experience. But even well-prepared, experienced speakers get nervous sometimes. Guidelines at a Glance 14 gives five techniques that speech communication specialists recommend to put your nervous energy to work for you.[2]

In addition, you can relax by thinking of your presentation as an interactive learning session rather than a solo performance. Encourage your listeners to

1. See Peter White, "How Can Technical Writers Give Effective Oral Presentations?" *Solving Problems in Technical Writing,* ed. Lynn Beene and Peter White. (New York: Oxford University Press, 1988) 191–204.

2. For many of these techniques we are indebted to Michael and Suzanne Osborn. See their widely used book, *Public Speaking,* 3rd Edition (Boston: Houghton Mifflin, 1994) 48–50.

GUIDELINES AT A GLANCE 13

Rehearsing your oral presentation

- *Make your purpose and position clear from the start.* Especially strive to make one point clearly.
- *Use variations in voice pitch, loudness, and tone to emphasize your most important points.* A monotone will put people to sleep, and varying your voice at inappropriate times will confuse your listeners.
- *Avoid pitfalls of oral presentation that send the wrong message.* Long pauses and breakdowns in fluency suggest a lack of familiarity with the topic. Monotone and failure to make eye contact suggest a lack of enthusiasm. Slow or awkward transitions or fumbling with your visuals suggests a lack of planning. In fact, all of these problems may arise if you don't rehearse enough. Don't risk giving the wrong impression. Rehearse!
- *Stay within your allotted time at all costs.* If listeners want to hear more from you, they will ask questions or request a more complete report in writing or at a later date. Most people experience difficulty staying awake and concentrating on a presentation after 20 minutes, especially in a darkened room. If you must go beyond this amount of time, plan to take a break and turn on the lights.
- *Beware of jokes.* Even innocent-seeming humor may offend or bore your audience.

GUIDELINES AT A GLANCE 14

Five techniques for dealing with nervousness

1. Before your presentation, practice *cognitive restructuring.* Instead of thinking, "Nobody is interested in what I have to say; I'm not really an expert on this topic," force yourself to think over and over again, "My audience really needs this information; they won't get it unless I bring it to them."
2. Before your presentation, practice *positive visualization.* Form a mental picture of yourself standing before an engaged and happy audience. See yourself succeeding.
3. Before your presentation, develop a *positive attitude toward your audience.* Think of them, for example, as your valued clients, the people who will help you pay the bills and advance your career.
4. Just before and during your presentation, remember to *breathe slowly and regularly.* Take some deep breaths during breaks in your speech, using your diaphragm so that you feel your stomach expand with each breath. Nervous people tend to hold their breath, which only increases the body's anxiety response.
5. During your presentation, *act confident,* even if you don't feel that way. You may even convince yourself.

mull over data with you. Have them develop their own examples of your points. Tell them to imagine themselves doing the actions you describe and to ask questions or offer comments based on their own experiences. Answer their questions with confidence and honesty. Don't bluff when you don't know an answer; simply say that you don't know and can provide that information later.

GUIDELINES AT A GLANCE

Don't get into arguments with aggressive questioners, but don't back down either. Try to maintain a professional demeanor and reply with the best information you can. Have additional supporting information at your fingertips (or better yet, in your head) so that you can cover questions about details with confidence.

Remember that your audience will think of you as the leader. If you are forced to speak with little time to prepare—say your boss asks you to find out everything you can about the competition in a particular market and bring a report to the board of directors by next week—then try to become as much of an expert as you can in the time available. Don't try to bluff your way through. Simply discover as much as you can and present with a clear head the best talk you can muster.

Confidence is the key to effective oral presentation. You need to project a tone of calm and knowledgeable assurance. Only through careful planning, knowledgeable design, a full control of your subject matter, and effective rehearsal can you achieve the necessary confidence and project it to the audience.

Special Strategies for Training Sessions

If your purpose is actually to instruct your audience in the use of an instrument or in the performance of a process, you will need to find more ways to involve them in your presentation. Ideally, workshop and training sessions should be interactive, hands-on experiences. You should spend no more than 20 percent of your allotted time telling people what to do.

If you're giving a short course on how to use a new computer program, for example, it's best to have each participant seated near a terminal and, if possible, to use either an overhead projector or (better yet) an interactive network to guide the users through clear exercises. Long lectures at the beginning of training sessions may have little or no effect later. There is something psychologically deadening (and logically inconsistent) about hearing imperative instructions ("Do this, do that") while being forced in reality to do nothing but sit and listen. When you say, "Start the machine," your listener's body will gear up for action. Your audience will grow weary if it has to remain passively seated. And weariness can lead to frustration and hostility, attitudes that will ruin the opportunity for effective action later in the workshop.

Of course, you can't simply invite users to start working without any instruction. Through some combination of lecture, demonstration, and graphics you will need to prepare them for participation and stimulate effective action. You have at least five ways to do this.

1. *You can give a simple overview.* Even hands-on training sessions usually begin with a quick look at the whole procedure being taught. Limit this overview to ten or fifteen minutes. Engage listeners as fully as possible by appealing

to the imagination, creating strong visual impressions that help them see themselves in positions of action. The graphics you use can even substitute for action if necessary.

For example, imagine that you are giving instructions on how to build a composting bin. Unless you do it in an industrial shop, it will be hard to involve the audience directly. Even a demonstration would be dull and awkward, since much of the work is repetitious hammering, nailing, and wiring. Instead, you develop a good set of clear graphics, which you can project and distribute as handouts after the presentation. (If you give them out during the presentation, users may read rather than listen and participate.)

To compensate for a lack of participation and active demonstration, your graphics and narrative should have a strong motivational element. You could show how composting helps the environment (by improving the soil and saving landfill space) and the pocketbook (by saving money on garbage pickups and fertilizer). The idea is to leave the audience with a burning desire to go out and perform the tasks you describe as soon as they can.

2. *You can give a simple demonstration.* In situations in which you can perform the task fairly quickly and with little possibility of failure, demonstrations can be very effective. If it would be dangerous or expensive for the audience to participate—as with chemical experiments—demonstration is definitely the way to go. Use these guidelines:
   - Observe safety measures carefully.
   - Be sure to arrange the seats and perform the actions so that everyone can clearly see what you're doing.
   - If certain actions or objects are hard to see, supplement the demonstration with overhead visuals. (You may need a helper or team member for this.)
   - Give complete overviews and memorable summations so that the users can perform the same activities effectively after they leave the demonstration.
   - Do everything possible to avoid failure; test props in advance and have backups ready.

   This last point is worth some emphasis. In addition to backup equipment, you should have a contingency plan in case something goes wrong. For example, you can prepare a quick overview of the procedure that requires no demonstration or participation but depends instead on visuals.

3. *You can mix demonstration and participation.* Computer trainers realized some time ago that the best approach combines an overhead projector with an active computer terminal. The trainees sit (either alone or in small groups) in front of the terminal in a classroom or lab equipped with a podium for demonstration and several user stations. The trainer demonstrates a single task, then invites the audience to attempt the same thing at one of the stations. The trainer moves from station to station to answer questions and give specific directions.

4. *You can use a presentation team to choreograph narration, visual displays, and demonstration.* While a narrator describes the process, using overheads or

slides, another presenter (or group of presenters) can demonstrate the process. This method is particularly effective if the trainees need to understand the whole procedure before they begin to work on the parts. But do not keep the audience inactive too long. Limit your presentation to ten or fifteen minutes.

One group of students in our class used this approach to demonstrate cardiopulmonary resuscitation (CPR). One of the group members went fairly quickly through a step-by-step narration of the procedure while other group members worked to resuscitate a life-sized rubber dummy they had borrowed for the occasion. Because the live demonstration was inherently visual, the narrators used overhead graphics only for preview and review.

5. *You can make a training video of your demonstration.* If you have access to good videotaping and editing facilities, you can move beyond the limitations of the lecture hall, training center, or conference room. Before you decide to tape your presentation, however, consider carefully the advantages and disadvantages of this medium. Videotaping allows users to watch training films at their own work sites or at home, on their own schedule. Users don't have to show up for an expert's presentation; they can take the expert with them. But the presenter cannot establish a rapport with the audience and cannot know whether listeners are following the procedure correctly. Again, we see that every medium, even the most advanced, has inherent limits.

EXER**CISE**
8.1

Evaluate the six transparencies in Sample 8.4. Determine whether they are helpful, attractive, informative, and audience-appropriate. Does the author appear to have followed Guidelines at a Glance 12? Go over the transparencies point by point, writing down the strengths and weaknesses as if a coauthor had asked you to improve the work. Once you have written down your evaluation of these visuals, plan and sketch out better ones, solving the problems you noted for the sample set. Share your notes and sketches with your discussion group or with the class as a whole.

EXER**CISE**
8.2

Plan a short presentation (between seven and ten minutes) based on one of the following situations. Fill in any details you need to make your presentation effective. Be sure to consider how you should organize your presentation, how your audience will react, and how you can support your speech visually. Sketch out a rough set of visuals. Discuss your plan in your discussion group or with the whole class.

1. As a junior partner in a real estate development firm, you were given the task of assessing a piece of land for development into an apartment complex. The piece of land seems ideal: It is a gravel parking lot just a block from a major university. The current owner is eager to sell, but she cannot

Preventing Lower Back Injuries
A Program in Ergonomics
by Marcus Duggins

Ergonomics - applying the study of human performance to the improvement of the work system consisting of the person, the job, the tools, the workplace, and environment

1. The Problem

2. The Program
   • Abdominal Supports
   • Worker Training
   • Management Training
   • Continuous Ergonimic Evaluation and Research

3. The Benefits

continued

SAMPLE 8.4   Transparencies for an oral presentation:
"Preventing Lower Back Injuries: A Problem in Ergonomics"

The Problem

1. Back injury is the single most costly type of occupational health disorder.

2. Insurance pays $11 billion annually for back injury claims.

3. The average case costs $7400.

4. Back injury is second only to the common cold as the leading cause of absence from work.

5. The average injury causes a 14-day absence.

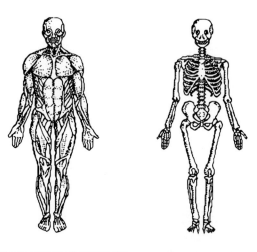

SAMPLE 8.4 continued

lower the price past $1.75 million for the land. Several of your partners are eager to move forward with the project, but your analysis shows that the venture would not break even for eleven years. Convince your colleagues that the project is ill-advised, and fill in enough information to help them discourage investors.

Step 1. Abdominal supports aren't enough.

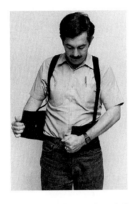

1. Pressure is applied to the abdominals.

2. Lower back is supported.

The pressure keeps the vertebrae from moving and prevents separation during lifts that can lead to
- severe muscle strain
- pinched nerves
- herniated disks

Problems with using the support belt:
1. The effectiveness is still largely undocumented.
2. Use of the belt may merely shift the location of likely injury.
3. Workers may not wear the belts properly.

SAMPLE 8.4 continued

2. Spurred into action by a report about the mistreatment of many rejected racing dogs, you have been volunteering at the local greyhound adoption agency for a year. The greyhound adoption agency offers the dogs to the public, after a reference check, for a small fee, and provides shots, check-ups, and spaying or neutering. The director has appointed you to the committee

Step 2. Enhance awareness through worker training.

Education programs *for workers* usually focus on simple changes in behavior, such as

1. Proper technique in heavy lifting

2. Good posturing and work space design

3. Avoidance of unnecessary reaching, twisting jarring, or repetitive lifting

SAMPLE 8.4 continued

that is responsible for the agency's expansion into the next town. To obtain the necessary permits for operation, the committee must make a short presentation at the next city council meeting that explains the operations of the agency.

Step 3. Create a better workplace through management training and ergonomic research.

Educational programs *for managers* usually focus on simple changes in behavior, such as

1. Safety in task design – e.g., reduce repetitive lifting.

2. Better worker selection.
   • Regular physical exams to determine suitability for tasks assigned.
   • Assignment of task according to experience and conditioning.

3. Continuous research and design.
   • Determine high-risk jobs.
   • Evaluate mechanical aids, and redesign if necessary.
   • Evaluate all operations involving lifting.
   • Monitor progress continually.
   • Include ergonomics in all designs.

SAMPLE 8.4 continued

3. You have been working at the your company's computer help desk for long enough to have become frustrated by some of the careless errors that keep recurring, too often, with the same employees. Your supervisor has given you permission to offer all employees a half-day seminar in which you and

The Benefits of Ergonomic Planning and Evaluation

1. Reduced absences from work.

2. Reduced worker's compensation and insurance costs.

3. Increased productivity.

4. Increased confidence and willingness of workers to perform the job.

SAMPLE 8.4 continued

your co-workers will demonstrate and explain the company's computer system. To recruit employees for the seminar, you have to give a short presentation at next month's staff meeting.

4. An acquaintance of yours who teaches at a local high school has asked you to be a guest speaker at the high school's Career Day. She suggests that you

demonstrate one part of your job, discuss its various challenges, or describe a typical day at your job.

5. After ten months of research, your project is finally complete, and you are quite satisfied with the depth of your findings. Although your advisor is enthusiastic about what you've done, you must present your research at a department meeting to receive credit for the project.

Recommendations for Further Reading

D'Arcy, Jan. *Technically Speaking: Proven Ways to Make Your Next Presentation a Success.* New York: American Management Association, 1992.

White, Peter. "How Can Technical Writers Give Effective Oral Presentations?" *Solving Problems in Technical Writing.* ed. Lynn Beene and Peter White. New York: Oxford University Press, 1988: 191–204.

Additional Reading for Advanced Research

Killingsworth, M. Jimmie, and Michael K. Gilbertson. *Signs, Genres, and Communities in Technical Communication.* Amityville, NY: Baywood, 1992.

McLuhan, Marshall. *The Gutenberg Galaxy.* Toronto: University of Toronto Press, 1962.

———. *Understanding Media: The Extensions of Man.* New York: McGraw-Hill, 1964.

Ong, Walter J. *Orality and Literacy: The Technologizing of the Word.* London: Methuen, 1982.

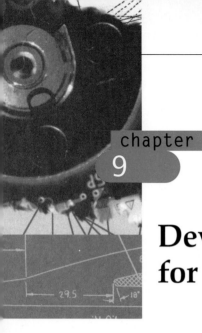

Developing Your Writing Style for Technical Documents

CHAPTER OUTLINE
▼▼▼▼▼▼▼▼▼▼▼▼▼▼▼▼▼▼▼▼▼▼▼▼▼▼▼

Strive for an Open and Active Technical Writing Style

Opening Access: The Rhetoric and Ethics of Technical Style

CHAPTER OBJECTIVES
▼▼▼▼▼▼▼▼▼▼▼▼▼▼▼▼▼▼▼▼▼▼▼▼▼▼▼

After you have worked through this chapter, you should be able to do the following:

- Recognize the features of an active and open style of technical writing
- Use general stylistic guidelines to write and revise sentences and paragraphs that are effective for technical documents
- Recognize the key rhetorical and ethical issues associated with an active and open style

In Chapters 6, 7, and 8, we saw how technical writers use graphics, oral communication, and combinations of media elements to arrive at good communication designs. In this chapter, we focus more specifically on the task of writing, as we turn our attention to prose style.

Defining style can be very difficult because it has multiple meanings, as Dan Jones points out in an excellent introductory chapter on the topic.[1] For our purposes, we can define style simply as the choices you make about which words and what kinds of sentences or paragraphs to use in your writing. While document design deals with "macro" issues of writing—decisions that affect the whole document or large sections of the document—style deals with "micro" issues: decisions that work below the level of the paragraph.

Another meaning of style has to do with "documentation style," such as the APA "style" of citing sources. That is not our concern in this chapter. Here we are interested in stylistic principles that affect your writing at a more fundamental level.

Like all rhetorical concepts, style brings into play principles that affect both production and review of information products. We must attend to style whenever we draft or revise technical documents. As Figure 9.1 indicates, the CORE method of producing documents involves several rounds of writing, reviewing, and revising. Style is important at each of these stages.

To achieve an effective style, you need to become a critical reader of your own writing, putting yourself in the place of the information user. This requires both the right attitude and the right skill. The right attitude is a willingness to look hard at what you've done and change it if necessary. This may be more painful than you think. We tend to value (and sometimes overvalue) anything we get on the page; after all, we think, every word gets us that much closer to finishing the writing task. But if we are to create quality products, then dedication to good craftsmanship must take precedence over the drive to simply "get it done."

In addition to patience (and pride), stylistic craftsmanship requires the ability to use your knowledge of language to make and execute decisions about your writing. You need to know what options are available to you in English prose and which of those will best suit your audience. This chapter discusses some key options as well as the preferences of typical technical users. It also provides opportunities for you to practice applying these options in reviewing and revising.

Strive for an Open and Active Technical Writing Style

When you write for typical technical readers—busy, selective information users—you must accept that no one is going to read every word you write and

1. Dan Jones, *Technical Writing Style* (Boston: Allyn and Bacon, 1998), Chapter 1.

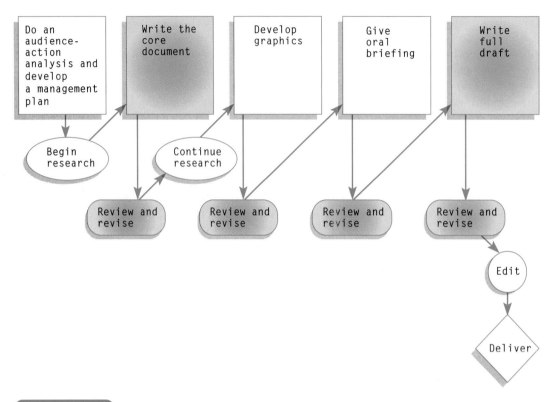

FIGURE 9.1

Writing, reviewing, and revising activities in the CORE method.

no one is going to have much patience with prose that makes great demands on their energy and time. Your typical audience is composed of readers whose main goal is fast access to information that will determine further actions and communications.

Researchers and practitioners in technical communication have learned from years of user testing and work with trial audiences that certain stylistic qualities tend to be more effective than others. Together, these qualities contribute to a general style that is *open*, in the sense that it creates the widest possible access to information, and *active*, in the sense that it facilitates the translation of language into action.

Following are a number of guidelines for achieving an open and active style in your own work. Think of them as tools of the trade. As you work through the guidelines, however, keep in mind one proviso: Style is never a matter of absolute formulas and unalterable laws. Stylistic preferences change from audience to audience, from place to place, from culture to culture, from time to time, and even from person to person. What works well in a computer manual will not work well in an academic article. What Brazilian readers consider eloquent may seem pompous to American audiences. The forthrightness encouraged in the United States may be considered rude in Japan or in many European countries. What once passed as the ordinary language of educated people now seems irritatingly verbose in most technological cultures. User testing is probably your best bet for determining whether the style you've cultivated really works. Even then, your test group may not adequately represent your final users.

The key to stylistic success is flexibility. The same attitude that keeps you willing to revise and improve any individual sentence should also keep you on the lookout for shifts in your typical audience's preferences and expectations. These shifts may lead you to reconsider some of the stylistic guidelines on which you have come to depend as a writer. Guidelines for style are never hard and fast rules. They are merely summations of current knowledge about typical audiences, and that knowledge will change as audiences change.

Cultivate a Fluent Style

In her book on manual writing, Susan Grimm says, "Write the way you talk, but more carefully."[2] What could this mean?

The phrase "more carefully" relates to the inherent limits of written communication. As we saw in Chapters 3 and 4, a writer does not have the advantage of a partner in spoken conversation, who is present to correct oversights, omissions, or misrepresented facts; the writer has to get it right the first time. So to be effective, you have to write more precisely than you talk.

What good writing should have in common with good speech is *fluency.* In speaking and writing, fluency affects the audience's response to the communication. A fluent speaker is one who uses language in such a way that the listener's attention never focuses on the language itself. The words flow without uncomfortable pauses, mispronunciations, malapropisms, or strange idioms and figures of speech. Listeners can focus on the content without distraction.

Fluent writing is similar. If the text flows smoothly from beginning to end, readers can get information with minimal resistance without having to turn back or pause over difficult and confusing passages. A reader can pay attention to *what* you have to say instead of *how* you say it. Such a style is critical for nonspecialist readers, including experts reading outside their own fields of specialization.

In technical writing, five particular stylistic issues will have a strong impact on your fluency: the "voice" in which you address your audience, your emphasis on agents and actions in sentence structure, your use of technical terms

---

2. Susan Grimm, *How to Write Computer Manuals for Users* (Belmont, CA: Lifetime Learning Publications, 1982).

or "jargon," your ability to avoid language that is inflated or disrespectful to your audience, and your construction of internally consistent and parallel lists and sentences.

Choose the Right "Voice"   In conversation we feel perfectly comfortable including ourselves or other people directly in our speech. We say "I" or "we" when we refer to ourselves and "you" when we refer to the person or people we are addressing. In writing, however, teachers and students find these three little words a source of endless worry and dispute. "Can we use first-person and second-person pronouns?" is one of the most frequently asked questions in technical writing classes.

The problem is that many years back, a cadre of so-called authorities on writing pronounced that the pronouns *I, we,* and *you* were the hallmarks of "informal" writing. This kind of writing, they added, was inappropriate for writing assignments at the college level or beyond. But today these strictures are loosening up. Authorities acknowledge that distinctions between "formal" and "informal" writing are often hazy and can be counterproductive. Research has shown that when sentences have subjects such as "the user" and "one," readers must go through a kind of translation process: " 'The user,' " they stop and think. "That's me." But when instructions begin with words such as "You should attach" or simply "Attach," they can skip the translation step and go right to the action.[3]

This direct form of address also has an ethical advantage. Unless we tell readers that "you should never mix hydrochloric acid and ammonia," we may not convince them of the responsibility they should feel in a lab. If we are to meet the goal of talking to our users through our writing, then instead of avoiding the conversational *I* and *you,* we should use these direct forms of address whenever we can get away with it.

In instructional writing, for example, the aim is to get from information into action as quickly as possible, so the most fluent language allows the user to get more easily from the printed page into the workplace. Direct address ("You do this" or simply "Do this") is clearly the best voice for instructional writing.

But not all technical writing is instructional. In decision-making documents such as reports and proposals, your audience may take offense if you give them imperatives. You may be overstepping your bounds as an advisor and taking on the role of dictator if you write, "You should fund the project at the rate proposed by Group A." Even "I recommend funding the project" may be too direct because it puts the focus on you rather than on the decision at hand. The best course might be to say something like, "Funding the project will result in at

---

3. See Linda Flower, John R. Hayes, and Heidi Swarts, "Revising Functional Documents: The Scenario Principle," *New Essays in Technical and Scientific Communication: Research, Theory, Practice* (Amityville, NY: Baywood, 1983) 51–68.

least three benefits for our production team," and then list the benefits. In this instance you would avoid using *I* and *you* not because they are too informal or because of some rule about their usage in technical writing. Instead, you would want to avoid a directness inappropriate for an advisor. Herein lies an important lesson: Style communicates a message of its own—usually a message about the relationship between the reader and the writer.

As a new technical writer, also be aware that old traditions die slowly. Be sensitive to what *most* people do in your profession or workplace. Many scientific writers still think it is bad form to say, "I checked the titration every hour." Instead, they use a passive construction: "The titration was checked," because they think it reflects the impersonal objectivity needed for scientific inquiry.

Under the pressure of strong evidence that overuse of the passive voice hampers readability,[4] though, some scientific authors have begun to rebel. *I* now appears much more frequently in formal scientific papers. But many writers prefer the plural *we* to the singular *I*. The use of *we* provides an agent for an active verb ("*We* checked the titration") without drawing the same negative attention from readers that the use of *I* still seems to provoke. *We* is also appropriate because most scientific papers have several coauthors.

Connect Agents with Actions in Your Sentences   Part of the problem with the passive voice is that it obscures who is doing what. For example, if a computer manual tells the user, "The data are now entered," the meaning is unclear. Has the computer automatically entered the data? Should the user enter the data at this point? Has some operation that the user performed caused the data to be entered? It would be better to use the active version of the sentence, clearly specifying the agent and the action involved. Say either "The computer will enter the data automatically" or "Now you can enter your data."

In scientific writing, similar confusions can result from the use of the passive voice. In the following passage, for example, notice how hard it is to distinguish between findings reported in the literature and the authors' own findings, even with the parenthetical references to the names of researchers:

> The existence of a singularity phenomenon (SP) has been verified many times (Smith, 1984; Jones, 1988). But the exact nature of the event remains unclear. The possibility that it actually comprises two or three distinct phenomena was raised by research using electron-optical methods (Jones, 1989; Rama, 1990). With the introduction of laser-based methodology (Rama et al., 1992), the hypothesis could be reconsidered and laid to rest once and for all.

---

4. See Thomas Warren, "The Passive Voice: An Annotated Bibliography," *Journal of Technical Writing and Communication* 11 (1981): 271–286, 373–389. More on the passive voice below.

The following revision uses the active voice and connects agents directly to the actions they perform, thereby clarifying who is responsible for each action:

> The research of Smith (1984) and Jones (1988) verified the existence of a singularity phenomenon (SP) but did not go very far in specifying its exact nature. Later researchers used electron-optical methods (Jones, 1989; Rama, 1990) to show that the events observed in the first experiments may actually comprise two or three distinct phenomena. Using the laser-based methodology introduced by Rama et al. (1992), we were able to establish that the SP is in fact a single, though complex, phenomenon.

In this case, then, the price of using style to identify with the scientific research community is simply too high. It causes confusion over who can take responsibility for which discoveries and innovations—a confusion that the use of *we* and the active rather than the passive voice corrects.

The obscuring of responsibility can lead to ethical problems. When it is important to know who should take credit or blame for an action, the use of the passive voice can cause problems much worse than a lack of fluency. Consider the following example:

> The problem with accounting has now been thoroughly analyzed. It was found that the missing funds were misappropriated. Categories of spending were mixed. People were paid for work that was never done. The omissions were never reported.

In this case the use of the passive voice amounts to a refusal to name names and point to the agents responsible for unethical actions. To make credit and blame clear, you would have to restore the active voice, as follows:

> We have now completed our investigation into the problem with accounting in Department A. We find that the departmental management misappropriated funds, mixed categories of spending, and paid the contractors Smith and Jones for work they did not do. The first auditors never reported this misconduct and never cited the contractors' failure to complete the work.

You may think that the context of the report would have made it clear who did what to whom in this case, so such an explicit assignment of agents and actions would be unnecessary. But remember that busy readers are selective and may want to skip quickly to the conclusion. They may not have absorbed enough about the context to understand the relation of agents to actions by the time they read this crucial paragraph.

To achieve an active and open style and to improve your fluency, you would do well to increase your use of strong active verbs, name agents explicitly (especially human agents), and connect agents with their proper actions.

Avoid Misuse of Technical Jargon   Using the technical terms of a particular research field is one way to show your identity with a specialized com-

munity. Economists, education specialists, anthropologists, and physicists all have their own technical jargon. As a kind of shorthand, jargon terms are useful. They save space and time. Biologists pack more than a hundred years of evolutionary theory into the term "natural selection." Computer scientists have coopted a term "interface" from the biological sciences to describe the combined techniques and technologies that users employ to make a computer work.

But jargon also has a self-limiting effect on your audience. Members of your in-group will find jargon-filled prose quite fluent; "natural selection," for example, would give no pause to a trained biologist. For others, however, this term might be an obstruction. Even people who are highly educated in another field might have difficulty because they do not often use the term.

Before you use a technical term, then, consider the match between the context of use and the context of production. If you are writing up experiments from your research lab for users from other research labs, for example (a typical scenario for a scientific paper), you can probably use a great deal of jargon and still remain fluent. But when the context of use is dominated by people from another discipline or community of research, you'll need to use a less technical vocabulary. An engineer explaining a new design to a city council or a company's board of directors could not use the technical vocabulary of machine specifications, for example. Nor could a physician use the jargon written on a patient's chart (intended for nurses and other doctors) in a conversation with the patient. When you do use technical terms, you should define them carefully and explain their importance. The doctor, for example, might use a word such as "angina" to indicate to the patient that doctors understand the condition well enough to have a routine terminology for it. Such a use of jargon, if handled well, could even be reassuring. The key is to be aware of the possible effects of language upon your audience, then select, define, and explain words accordingly.

Avoid Pompous and Disrespectful Language   Using unexplained technical jargon outside the field in which people understand it shows disrespect for your audience. The tone you project suggests pomposity and arrogance.

Worse yet is the use of merely inflated language—obscure or rarely used words and complex, convoluted sentences that you could easily replace with simpler expressions. Why say "rubbish receptacle" when you mean "trash can," for example? Or why write, "The power mechanism should be activated at this point in time" when you could say, "Now switch on the power"? The only reason to use inflated expressions like these is to impress your audience with your education or erudition. However, people are rarely impressed by such language. They are more likely to be irritated by having to pause over dense passages and translate them into everyday terms.

TABLE 9.1

Alternatives to sexist language

Instead of this . . .	Write this:
mankind	humanity
manmade	manufactured, synthetic
businessman	businessperson, executive, manager
man hours	work hours
congressman	member of Congress, representative
career girl	professional woman

Source: Donna Gorrell, *A Writer's Handbook from A to Z* (Boston: Allyn and Bacon, 1994) 149–150. This is an excellent brief handbook that also contains other examples.

Another form of disrespect is the use of *sexist language.* Some will argue that using "he or she" instead of the old generic "he" to refer to unidentified persons is a matter of academic fashion or "political correctness." But most style manuals disagree; they present nonsexist usage as their standard, warning against the generic "he" as well as such gender-biased terms as those listed in Table 9.1.

You should likewise avoid expressions that reinforce cultural or ethnic stereotypes. Don't imply, for example, that women are more emotional than men or that some ethnic group—Korean, Hispanic, Italian, or Anglo Protestant—is lazier or harder-working than some other.

As you revise your work, try to regard every potential reader as a unique individual. If your language offends the reader, drawing attention to his or her difference from you, you may lose that reader's good will, and you will certainly interrupt the flow of the reading experience. The population of the workforce is changing rapidly, adding more women and becoming more ethnically diverse. In all their diversity, your readers are your clients; without their good grace, your job may be on the line.

Construct Internally Consistent Lists   A less ethically sensitive but no less important factor in maintaining a fluent style is the construction of internally consistent lists. In technical documents, lists often interrupt the flow of text material, so they should not cause additional disruption by being difficult to read and interpret. When you create a list, make sure that every element in the list is grammatically equivalent, whether each element is a single noun, as in the following list:

- *bucks*
- *does*
- *fawns*

or a phrase, as in the following:

- *bucks that have inferior antler structures*
- *does with evidence of disease*
- *fawns born with obvious genetic defects*

or even a whole sentence:

- *Bucks should not have inferior racks.*
- *Does should not exhibit evidence of disease.*
- *Fawns should exhibit no obvious genetic defects.*

Do not mix single words with phrases or phrases with sentences. Keep the list items parallel.

   If you are finishing off a thought with a series of options, keeping the list items parallel may be a challenge. Here's a list with nonparallel options:

If the project funding is insufficient, then we may do the following:

- *seek a suitable alternative*
- *the project may die*
- *we will shelve it and reconsider it three months from now.*

Here's the same list with parallel items:

- *seek a suitable alternative*
- *let the project die*
- *shelve the project and reconsider it three months from now.*

   The introduction to a list should always be a full sentence. Write "The items are as follows:. . . ." not "The items:. . . ."

   If your list has a second level, indent the secondary items under the primary ones. Use a consistent numbering system, and be sure to keep the structure parallel at all levels, as in the following:

A vacuum cleaner includes the following parts:

1. *A hose assembly*
   - *flexible tubing*
   - *rigid tubing*
   - *elbow connectors*
2. *The cleaner engine . . . etc.*

Construct Parallel Sentences   In addition to keeping your list items parallel, make sure your sentence constructions are consistent and parallel as well. Poor constructions can dam up the flow of your prose and bring the reader to a standstill. In sentences with items in a series connected by a conjunction such as *and, or,* or *but,* double-check the structure: *The range manager should be on the lookout for bucks with inferior racks, does with evidence of disease, and fawns with obvious genetic defects.*

In sentences with correlative conjunctions such as *both . . . and, either . . . or,* and *not only . . . but also,* make sure the words and phrases after each segment of the conjunction match up structurally:

- *Both* **the acid** *and* **the base** need to be checked every five minutes.
- You should *either* **run the virus check** *or* **erase the file.**
- We need *not only* **to hire more accountants** *but also* **to get a new spreadsheet.**

It's all too easy to lose track of where you are when you're writing such sentences and end up with something like "We need not only to hire more accountants but also a new spreadsheet." So it's a good idea to check your work whenever you encounter these possible trouble spots.

Make Your Information Action Oriented

In the sections above, we saw how addressing readers directly, connecting agents with actions, and providing users with language that suits their context and shows them respect can ease the way toward action by creating a fluent reading experience. Now we will examine ways to present information that encourage action.

Use Effective Modifiers to Interpret and Recommend   Although one of the goals of technical communication is to present information in a fair and balanced way, another is to give other people the advantage of your expertise. So when you are reporting, don't just provide neutral information; tell your users what effect it could have on their actions. Connect facts with previous actions or with future possibilities (or both). Don't just say, "Our profits fell by 5% last year." Say, "Because of mismanagement in our shipping operation, our profits dropped 5% last year." Or "Because our profits fell by 5% last year, we may be forced to lay off a number of workers in the next quarter."

Notice that these are not simple sentences. Modifying clauses such as "because our profits fell by 5% last year" are necessary for effective explanation and interpretation. These dependent clauses—groups of words that, like sentences, have subjects and verbs but cannot stand on their own—serve as important qualifiers by setting conditions, describing, defining, and explaining. Evaluative adjectives such as "effective" and "correct" and adverbs such as "adequately" and "poorly" can serve a similar interpretive function.

Notice how in the following sentences the adjectives, adverbs, and dependent clauses in italics help the author to explain or interpret facts and events. If these modifiers were removed, how much less would the sentences communicate?

1. The proposed dump site cannot *adequately* meet the *best* standards of waste disposal *because it lacks the recommended clay base.*
2. *Whereas Department A functioned at a profitable margin,* Department B failed to achieve its minimal goals.
3. *If you neglect the calibration in the initial heating,* the process will fail to produce *good* results.
4. The blue crystals, *which are mined in Albania,* are *very difficult* and *expensive* to replace.
5. We must be sure not to alienate the Albanians, *who are crucial to the mission as proposed.*

Two cautions on the use of modifiers are in order: First, make sure that you use the main clause of a sentence, not its dependent clauses, to make the point you want to emphasize. In Sentence 2 above, for example, the emphasis falls on Department B's failure, whereas the relative success of Department A is used only as a point of comparison. Second, when you use an evaluative adjective or adverb, read your sentence over carefully to make sure that you are treating your topic fairly and in the manner you intend. These words can easily reflect an unconscious bias on your part.

Weave Your Sentences to Achieve Cohesion   We have seen how features such as personal pronouns and active voice can increase the fluency of your writing by giving it a conversational flow and flavor. Fluency also makes demands at the paragraph level. To create fluent paragraphs, you need to write sentences that fit together in a cohesive structure. Cohesion results from overlapping old and new information in each sentence, paragraph, and section of a document.[5]

A sentence, as you probably know, is composed of a subject and a predicate. The subject presents a *topic,* and the predicate (the part with the verb) makes a *comment* about the topic. In cohesive prose the topic relates to information that has already been presented, whereas the comment carries the writing forward with new information. An overlapping, woven pattern develops, each subject taking up the information presented in the previous predicate. If we represent each bit of information as a number, then the pattern will look something like this (S = subject, P = predicate): S1»P2. S2»P3. S3»P4.

Consider a set of simple sentences that fail to follow this pattern. Because each subject introduces new information, the paragraph lacks cohesion and fluency.

> *Example (original):* The sales team (S1) left the St. Louis meeting with many new ideas for promoting the new product (P2). Wilcox (S3) liked the direct

5. See Joseph Williams, *Style: Toward Clarity and Grace* (Chicago: University of Chicago Press, 1990), Chapter 3; see also Martha Kolln, *Rhetorical Grammar* (New York: Macmillan, 1991), Chapter 1.

mail approach (P4). Customers (S5) can be located by mailing lists purchased from professional organizations (P6). Then we (S7) can send them a brochure on the product (P8).

When every sentence starts with new information, the chance of saturating the reader and destroying the fluency increases. It's far better to make each sentence half old and half new, as in the following revised version. The subjects and predicates are interwoven.

> *Example (revised):* The sales team (S1) left the St. Louis meeting with many new ideas for promoting our product (P2). Our favorite (S2) was the direct mail approach (P3). Direct mailing (S3) is best for companies like ours that have a good idea of who will buy the product (P4). Additional customers (S4) can be located by mailing lists purchased from professional organizations (P5). The old customers and the new ones (S5) will receive a product brochure at their business addresses (P6).

For a variation on the pattern, you can put the old information in an introductory phrase instead of in the subject itself. The important thing is to begin with the old and lead readers toward the new.

The same is true for transitions between paragraphs, sections, chapters, and even whole documents. For example, a literature review at the beginning of a report shows how previous research flows into the information presented in the current document.

Many matters of style are less of a concern during the writing of first drafts and more of a concern in the revision and editing stages of the writing process. But sentence weaving is a useful technique to incorporate right from the start. It will keep you on task, reminding you where you have been and where you are going.

Position Information for Emphasis and Coherence   Another means of increasing the readability and action orientation of your writing is to employ the classical technique of rhetorical emphasis: Put the most general and most important information at the beginnings and endings of paragraphs and chapters, where hurried readers will look first.

Think of your own experience. When you're late for class or cramming for an exam, which parts of your assigned text do you read? If you are a typical busy reader, you will focus on the following (in the order that time usually allows):

1. The title
2. The abstract or summary at the beginning and list of conclusions at the end (if available)
3. Headings, figure and table titles, and captions for figures and tables

4. The first and last page of every chapter
5. The first and (less frequently) the last sentence in each paragraph

To appeal to readers who are as busy as you are, use these positions for your most important information.

In writing a paragraph, your aim should be to create coherence in addition to cohesion. A paragraph is cohesive if it flows nicely from one sentence to the next. It is coherent if it seems to be unified around a single topic. Joseph Williams, a University of Chicago linguist and the leading authority on teaching style to students of writing, says, "A reader will feel that a paragraph is coherent if she can read a sentence that articulates its point."[6] This *point sentence* can, by itself, answer the question, What is the point of this paragraph?

Williams goes on to say that the point sentence must appear in a position of emphasis. Usually, it should be the first or second sentence or the very last one in the paragraph. In technical documents the point sentence should usually come first. However, if the first sentence is a transition from the previous paragraph or section, the point sentence should come second. In an introductory paragraph, which must spend some time setting the stage, the point sentence may come even later.

There is one obvious advantage to putting point sentences first; if your readers read only the first sentence of every paragraph, they will get the gist of the information. Sample 9.1, for instance, shows a set of point sentences from a section on ecological economics in Lester Brown's book *Building a Sustainable Society.*[7] The explicit audience, in addition to the general public, consists of busy policy-makers, particularly people in government—good representatives of the typical readership of technical reports. As you can see, the first short paragraph sets up the main idea in three sentences. From then on, the first sentences of each paragraph (2–11) does most of the work.

Another strategy that technical writers often use to catch the attention of fast readers is to break paragraphs fairly frequently, creating more eye-catching positions of emphasis. But if the breaks are too abrupt or illogical, or if the reader runs out of time or patience reading all the extra first sentences, then this strategy may backfire.

Reduce Density and Weed Out Wordiness   As you draft sentences and paragraphs and begin looking back over what you have written, keep thinking

---

6. Joseph Williams, *Style: Toward Clarity and Grace* (Chicago: University of Chicago Press, 1990) 97.

7. New York: Norton, 1983. For a full rhetorical analysis of this fine example of technical writing for the general public, see M. Jimmie Killingsworth and Jacqueline S. Palmer, *Ecospeak: Rhetoric and Environmental Politics in America* (Carbondale: Southern Illinois University Press, 1992), Chapter 7.

¶ 1 The Law of Diminishing Returns was first articulated by David Ricardo, the nineteenth-century English economist. He reasoned that at some point additional food could be produced only by extending cultivation onto less fertile land or by applying ever more labor and capital to land. In either case, returns would diminish. (p. 117)

¶ 2 Initially based on calculations for wheat in the United Kingdom, Ricardo's formulation has a compelling logic. (p. 117)

¶ 3 As the eighties begin, . . . interest in Ricardo's analysis is reviving. (p. 118)

¶ 4 While Ricardo's concern with the diminishing quality of new land was initially unfounded, it is now being borne out. (p. 118)

¶ 5 Just as the quality of new cultivable land has declined, so too efforts to raise land productivity do not pay off as handsomely as they once did. (p. 118)

¶ 6 Efforts to expand the fish catch represent another clearcut case of diminishing returns. (p. 119)

¶ 7 With energy as with food, efforts to expand supplies eventually meet with diminishing returns. (p. 120)

¶ 8 Diminishing returns also govern the mining industry. (p. 120)

¶ 9 The capacity of the earth's ecosystem to absorb waste also brings diminishing returns. (p. 121)

¶10 Investment in scientific research—long the answer when productivity lagged—may itself be experiencing diminishing returns. (p. 121)

¶11 In retrospect, Ricardo appears to have been ahead of his time. (p. 121)

SAMPLE 9.1   Set of point sentences

Source: Excerpted from *Building a Sustainable Society* by Lester R. Brown, with the permission of W. W. Norton & Company, Inc. Copyright © 1981 by Worldwatch Institute.

about "talking" your writing. It's even helpful to read out loud, reminding yourself as you stumble over complex passages, "Don't write a mouthful!"

Be on the lookout for high concentrations of technical terminology, abstract concepts, and long, complex, and indirect sentences. Consider the following paragraph, for example, from a specialized journal on environmental toxicology. In this paragraph the sentence construction is so indirect and the terminology so dense that the effect is almost like reading a foreign language.

Example (original): Nonpoint source pollution can be defined as the diffuse input of pollutants that occurs in addition to inputs from undeveloped land of similar genesis. Agricultural nonpoint sources of pollution significantly altered water quality in 68% of the drainage basins in the United States, and in nearly 90% of the drainage basins in the north central region of the United States. Agricultural sources are probably the major contributors of suspended and dissolved solids, nitrogen, phosphorus and associated biochemical oxygen demand loadings in U.S. waters.[8]

For readers trained in toxicology this is fairly fluent prose, but nonspecialists would find it rough going. To make it speakable, you would need to spread out the information over a series of shorter sentences, define terms, use fewer abstractions, and add more concrete nouns:

Example (revised): Nonpoint source pollution has become a major problem in the United States. It occurs when pollutants flow into waterways along with the usual runoff of soil and minerals. Over two thirds of the drainage basins in the country suffer from nonpoint source pollution. In the north central United States, where agriculture plays a major role, nearly 90% of the drainage basins have been affected. Agricultural uses of pesticides and fertilizers account for the heaviest concentrations of these suspended and dissolved solids, which include nitrogen, phosphorus, and other chemicals that make heavy demands on the oxygen supply in the water.

Now each sentence contains no more than one abstract concept, and there are fewer abstractions, many of which were unnecessary (such as "diffuse input" and "genesis"). There are more concrete nouns ("soil" and "minerals" instead of "inputs from undeveloped land"), and the movement of the paragraph is from general point to general region ("the country") to more specific region ("the north central United States"). Also, every sentence after the first one has an active verb ("occurs," "suffer," "plays," "account").

Warning: When we alter sentences in this way, we may change their meaning. To some extent, the medium *is* the message, so it is always important to weigh the potential costs of revision against the potential gains in reading ease.

Another prose problem you'll want to avoid usually goes by the name *wordiness,* though this can mean many things. It does not refer simply to long passages or ones that have a lot of words; often you need many words to get the job done. Wordiness refers to unnecessary words—or unnecessarily long and pompous words—verbiage that slows reading without enhancing meaning. You can use Guidelines at a Glance 15 as a checklist for revising your work if wordiness is a problem in your writing.[9]

---

8. Stephen J. Klaine, et al., "Characterization of Agricultural Nonpoint Pollution: Pesticide Migration in a West Tennessee Watershed," *Environmental Toxicology and Chemistry* 7 (1988) 609.

9. These guidelines are based on Joseph Williams's discussion of simple and complex wordiness. For a fuller discussion, see his *Style: Toward Clarity and Grace,* 116–133.

GUIDELINES AT A GLANCE 15
Common problems with wordiness

Problems	Examples
Using redundant pairs of words	*basic and fundamental, questions and problems, hopes and desires*
Using redundant modifiers	*true facts, important essentials, final outcome*
Using redundant categories	*period of time, red in color, slick in appearance, slimy to the touch*
Using meaningless modifiers	excessive use of words like *basic, basically, really, definitely, certainly, individual, certain,* and *particular,* as in the sentence "Our basic problem is fundamentally one of a certain concern with an individual accountant in a particular department—Department A, that is."
Using pompous words where simpler ones will do	*endeavor* rather than *try* *utilization* rather than *use* *termination* rather than *end* *initiation* rather than *start* *implement* rather than *start, carry out,* or *begin* *prior to* rather than *before* *subsequent to* rather than *after*
Providing excessive detail	describing the internal workings of a computer in a manual for users who will never need to know how the thing works
Using a phrase where a word will do	*the reason for, for the reason that, due to the fact that,* or *in light of the fact that* instead of *because, since,* or *why; on the occasion of* or *in a situation in which* instead of *when; it is crucial that* or *it is necessary that* instead of *must* or *should; in order to* instead of *to*
Repeating hedges and/or pronouncements	*in my opinion, it is our studied opinion that,* or repeated use of *usually, often, sometimes, almost, virtually, possibly, apparently, seemingly, in some ways, to a certain extent, more or less*

Once you've located these trouble spots and learned to avoid the kinds of constructions in the "problem" column of the table, you will probably have picked up a few general strategies, such as those listed in Guidelines at a Glance 16 on reducing wordiness.[10]

10. These points receive a lively treatment in Jack Rawlins's *The Writer's Way,* 2nd Edition (Boston: Houghton Mifflin, 1992).

GUIDELINES AT A GLANCE 16
Reducing wordiness

- Cut out pretentious language. You can do without impressive-sounding, pseudo-academic phrases such as *fundamentally it is important to note that . . . , the fact is that . . . ,* or *it is a commonly held opinion that. . . .*

- Reduce redundancy. Planned repetition may be highly effective in oral presentations, but it is quickly overdone in writing. Depend on forecasting devices such as headings, point sentences, and summary graphics, and let it go at that. (See Chapter 7.)

- Look for the news in every phrase, sentence, and paragraph, and put the rest of the words to the test of necessity. You may find that in a sentence such as "The project is profitable and scientifically valid," the only important words are the modifiers "profitable" and "valid." Can't you cut the sentence and import the adjectives into another, more substantial sentence?

- Reduce transitional words and phrases. As long as your paragraphs and sections are well structured, you can often eliminate "glue" words and phrases such as *moreover, furthermore, thus,* and *now that we have seen how X works, let us turn to Y.*

Opening Access: The Rhetoric and Ethics of Technical Style

These instructions for achieving an open and active style in technical writing bring us back to the complex issue of access. Although your audience may be highly specialized, it is still sound ethical practice to write for the widest possible audience.

For example, if you are writing about a high-level theoretical question in nuclear physics, few people will be conversant with the mathematics needed for even a basic understanding of your subject matter. But you could try to include graduate students as well as leading professors in the field by explaining your concepts with clarity and an energetic, active style. Turgid, densely abstract writing impresses few really important readers. So always write to communicate with your readers, never merely to impress them.

Perhaps the best strategy for addressing multiple audiences is to use tiered access (see Chapter 4). General readers should be able follow your presentation far enough to get the information they need, and technical readers should be able to go further, to pore over every word and figure. Technical writers often use executive summaries for this purpose. These short documents, placed at the beginning of a report, summarize for nontechnical managers the key results, conclusions, and recommendations of the technical project detailed in the full report.

In addition, you can organize your documents internally for tiered access. Your information should flow as follows:

- From simple to complex
- From information pitched for a broad readership to information pitched for special audiences or experts
- From information that builds a general awareness and guides broad-based decision making to information that informs specific technical actions

In addition, the recommendations we have made in this chapter urge you to take responsibility for the information you write. As we noted, writers who use the passive voice may be trying to slip free of demands for accountability and responsibility. A worried accountant may write, "An error was made in account 35642G in the last fiscal year." (Who made the error?) A military officer could write, "The village was bombed." (By whom?) Or, to protect an expert group's power of decision, a writer may adopt a deliberately incomprehensible scientific style for a public document like an environmental impact statement or a recommendation report on housing. If people cannot understand the issues, they are not likely to cause trouble.[11]

We urge you to hold fast to the value of open access to information. Fluent, active writing and effective display and interpretation of information will create "openings" in your document, doors of access for busy problem solvers. Through these doors will come readers whom you may have missed in your audience analysis. You should realize, however, that one day in your professional life you may have trouble following our advice. Someone may urge you to be less than clear and open, to narrow your audience rather than widen it. In this case you must rely on your conscience.

EXERCISE
9.1

Part 1. Revising in class.   The following sentences could be improved with some attention to voice and to the relation of agents to actions. Revise them to make them clearer. You may have to supply some fictional names or agents to make the revisions work. Discuss your revisions with a small group of your classmates or with the whole class. How does revision change each sentence? What did you have to add to each sentence to make it active and clear? What do the additions tell you about how passive voice and indirect expressions obscure information?

1. After tubes C and D are connected, the user should insert parts A and B.
2. Once the possibility of nonpoint source pollution was established (Craig and Watterford, 1991), the question of toxicity could easily be confirmed.

---

11. See M. Jimmie Killingsworth and Dean Steffens, "Effectiveness in the Environmental Impact Statement: A Study in Public Rhetoric," *Written Communication* 6 (1989): 155–180.

3. If the work order had been clearer, the omissions in the January job would not have been led to such an unacceptable fulfillment of the contract.

4. One must always turn off the electricity before opening the back of the machine.

5. The cause of such infections has been located in the improper handling of bandages during the earliest stages of dressing the wound.

6. If the user is concerned about security, the PROTECT option can be activated from a network mode.

7. Following activation of the PROTECT option, operation can proceed as usual.

8. It was found that the report had been misfiled, and the neglect to pay the contractors was therefore identified.

9. The structure of the yeast cell has been known for many years. The structure was verified by applying the usual dye test. Then the dissection was begun.

10. The user should beware of overusing the test format. Overuse can lead to errors in data delivery. If such errors should appear, the networking capability is to be disabled.

Part 2. Revising in the library.   Go to the library and photocopy an article from a major journal in your field. Examine the article for examples of pronoun usage. Also check how frequently the author uses active and passive voice. Highlight any places where fluency or clarity may suffer because of indirect style. In the margin or on a separate sheet of paper, try revising a few of the trouble spots. Bring the article and your revisions to class for discussion.

Part 1. Paraphrasing in class.   One way to explain concepts from your own field of specialization is to paraphrase passages of specialized prose, using terms that lay readers can understand. In groups of four or five students, try paraphrasing the following passages for an audience that is not trained in the area of specialization represented by the prose. Draw on the knowledge of your group members, and use the dictionary if necessary. If no one in the group is familiar with the discipline from which the jargon comes, make a list of all the words and phrases that your group considers to be jargon. Share your lists and your paraphrases with the rest of the class.

EXERCISE
9.2

1. Continued operation of the Space Shuttle over the next 20 or more years leads to a high probability of the occurrence of one or more instances in which an automatic landing capability will be needed to minimize landing risk. At least two basic situations might result in the need for an automatic landing. The first would involve the inability of the crew to see the landing runway due to factors such as deteriorating weather in the landing site after the deorbit burn, a partially or fully obscured windshield, or smoke in the cockpit. The second would involve the inability of the crew to perform a safe landing due to subtle or obvious incapacitation. The requirements for an

automatic landing system to meet these situations must encompass hardware, software, and flight rules that are appropriate in terms of functional capabilities and reliability for those flight conditions or scenarios deemed by analysis and risk management decisions to require automatic landings.

Source: Committee on Science, Space, and Technology, *Hearings before the Subcommittee on Science, Space, and Technology* (Washington, DC: Government Printing Office, 1993).

2. The Government cannot simply adopt standard commercial documentation and only order commercially available data. It needs to consider all elements of the acquisition equation from intended use and deployment to the maintenance/support concept, including warranties. The Government and contractor may need to agree that documentation will be made available to the Government if the contractor stops making the product or goes completely out of business. It is important to note that commercial buyers and users of large systems, such as worldwide airplane fleets, seem able to operate with significantly fewer data deliverables than the Government.

Source: U.S. Defense Systems Management College, *Commercial Practices for Defense Acquisition Guidebook* (Washington, DC: Government Printing Office, 1992).

3. Differences in coal accessibility and recoverability vary among regions. The accessibility/recovery factors for underground resources are low compared to the same factors for surface resources. Therefore, supply regions where underground resources dominate will have a low overall recovery proportion, and regions where surface resources dominate will have a high overall recovery proportion. Examples of the former are supply region 9 (Pikeville, KY) and supply region 21 (Charleston, WV). Examples of the latter are supply region 95 (Powder River basin, Montana) and supply region 96 (Powder River basin, Wyoming).

Source: U.S. Department of the Interior, *Economic Effects of Western Federal Land-Use Restrictions on U.S. Coal Markets* (Washington, DC: Government Printing Office, 1990).

4. The DST noted that there were few dissemination problems in the north. This was likely due to the extensive coordination among offices and with the user community even before the event began. **Offices in the northeast were especially adept at relating the magnitude of this storm by wording products in such a way that the targeted audience understood exactly what was being forecast. This should not be underestimated as a causal factor for the low mortality rates in areas severely impacted by the Superstorm.** In fact, Pennsylvania Emergency Management Agency (PEMA) reported 52 total deaths associated with the event. Of these, 48 were post-storm deaths due to shoveling the heavy snow accumulations. This trend of higher post-event deaths was evident in other northeastern states as well, although most states did not report these deaths as definitively as Pennsylvania.

Source: National Oceanic and Atmospheric Administration, *Superstorm of March 1993, March 12–14, 1993* (Washington, DC: Government Printing Office, 1993).

5. Business services industries will also generate many jobs. Employment is expected to grow from 5.2 million in 1990 to 7.6 million in 2005. Personnel

supply services, made up primarily of temporary help agencies, is the largest sector in the group and will continue to add many jobs. However, due to the slowdown in labor force participation by young women, and the proliferation of personnel supply firms in recent years, this industry will grow more slowly than during the 1975–90 period, although faster than the average for all industries. Business services also includes one of the fastest growing industries in the economy—computer and data processing services. This industry's rapid growth stems from advances in technology, worldwide trends toward office and factory automation, and increases in demand from business firms, government agencies, and individuals.

Source: U.S. Bureau of Labor Statistics, *Tomorrow's Jobs* (Washington, DC: Government Printing Office, 1993).

Part 2. Paraphrasing in the library.   As a follow-up to this exercise, go to the library and find a specialized article in your own field. (You may use the one you copied for Exercise 5.1.) Copy a paragraph verbatim, and then write as accurate a paraphrase as you can manage. Bring your work to class, and discuss it with your group or with the class as a whole. What difficulties did you face in the exercise? How did you handle them?

Revise the following sentences to eliminate unnecessarily elaborate, pompous, disrespectful, or offensive language.

EXERCISE
9.3

1. Having now completed the task of the annual review, we are in a strong position to recommend an impressive salary increment of 3.1% for every employee if he has served the company for a period of one year or more.
2. No one should complain about the makeup of the committee. We've got the required black, Hispanic, and female members.
3. Our group interfaced with group C and got sufficient input to realize our objective of cooperating on certain selected projects to be determined at a later date.
4. The user should always enter his name into the identification slot provided in the opening screen of the user interface.
5. Our group worked well together, thanks to the housekeeping talents of Ms. Berry and the passionate commitment of our Italian colleague, Mr. Santini.
6. It is now apparent that sufficient attention has been paid to the problem of employee tardiness to have concluded that lateness due to family illness and emergencies far exceeds the usual unpromptness occasioned by such unplanned exigencies.
7. We should soon be able to open the new Asian branch of the research lab and, thanks to the Korean work ethic, we ought to be able to commit more funds to research than to overhead.
8. Every man in this operation should now see the merit in our plan, which reduces the number of man hours required for each job and adjusts the profit margins to a higher than expected figure.

EXER **9.4** CISE

Rewrite the following sentences to improve their fluency, using all the strategies suggested so far. Change passive to active constructions. Use a direct voice and human agents whenever possible. Also translate technical terms, make structures parallel to enhance readability, and use good modification to point the way toward action. When you have completed your work, discuss your revisions in a small group or in the class as a whole.

1. Cooperation is essential among our partners. It is necessary for them to be warned of the consequences of choosing independence over company welfare.

2. There are four problems new users often encounter:
   - difficulty reading the user interface,
   - they are unable to recover from errors,
   - impatience with the machine, and
   - failing to read the instructions.

3. The exerciser should, upon completion of the first step, move quickly to the second one.

4. The meal is usually ingested before feeding the young. That is a sign of the age-structured hierarchy among the mountain lions.

5. Our findings suggest that either runoff from agricultural lands is a bigger problem than heretofore thought, or it is only one of many unidentified sources of river pollution, some of which have not been identified.

6. Cell division occurs just at the point when the lipocytes are expanding. They are expanding due to the internal heat increases.

7. Users don't have to sit around and wait for the computer to finish its calculations. By choosing the "quit" command, the user can leave the station. The computer will keep working if it is not actually switched off.

8. The replacement of the workers in Cell C must be accomplished if the managers are going to achieve their goals. Productivity is lacking in that cell.

9. A cursor will appear on the screen when the calculation is completed. At that point, the data is entered.

10. I recommend that you use the resources developed in the 1994 fiscal year. That way, expenses for the project can be covered, and pressure on resources will be relieved consequently.

11. The use of systemic insecticides is not necessary to get the best results in cotton crops. For most pests in this region, they have limited effectiveness, and there is a human factor of danger.

12. The burner was lit just before the explosion occurred. We think it may have happened because:
    - the acid was too close to the flame;
    - electrical charges can trigger explosions;
    - we have not yet ruled out the possibility of spontaneous combustion.

13. The junior partners will either have to take up the new client burden or senior management will be stuck with it.

Rewrite the following paragraphs to increase cohesion and to emphasize key points. Be sure to "weave" the sentences and place good point sentences early in each paragraph.

EXER CISE
9.5

1. All too often, companies in today's market are neglecting children when it comes to their marketing efforts. This technique worked in the past but will have trouble in the future. If a company wants to be successful in the 21st century, strong relationships must be formed between the firms and their customers, especially children. Even if these firms possess no child-related products, they must think of the future. Gaining the friendship and trust of a youngster now could plant the seed that will bring him or her back in the future. Many firms are divided on this topic. Companies believe it to be an all or nothing issue. They are either targeting children in full force or not at all. Those who are avoiding the issue will soon realize that the children of today are the growth potential of the future. Not going after this market segment could greatly jeopardize business expansion. Kid marketing is the way of the future and cannot be overlooked unless we want to be left behind.

2. The automatic irrigation system can also do the job of humans better and more efficiently. The system can be programmed to water in the middle of the night or any irregular hour that is just not feasible to arrange with an employee. It can water the container plants evenly and with less water wasted, because the sprinkler heads disburse the water evenly over the field and do not commit the human error of putting too much water in one pot and not enough in the next. Since the conservation of our natural resources is a major topic of discussion, our nursery will use its well water more efficiently with the installation of an automatic irrigation system.

3. Most of the games in the paintball category are versions of "capture the flag." But paintball has a serious twist: Your opponents are not just trying to tag you, they are trying to shoot you with a .68 caliber paint ball! All of the players are armed with an airgun which shoots the paint projectile out at about 300 feet per second. The paint balls are thin, rigid plastic spheres filled with a water-soluble "paint." The guns are powered by either small carbon dioxide cartridges or refillable compressed air canisters. There are pump and automatic versions available. Typically, two teams take the field and attempt to capture the other team's flag and bring it back to their home base without getting shot. Players who get shot are considered "dead" and must walk quietly off the field with their guns raised above their heads. Players dress in black or camouflage clothing and are equipped with goggles and a face mask for protection. New innovations include silencers, sniper paint balls, paint and smoke grenades, and body armor. New products are being developed all the time to keep the game fresh.

4. Uncle Sam's Pecan Company would reap many benefits from exporting. We can add sales volume and potentially quadruple our market by exporting. Exporting will help Uncle Sam's decrease production costs through economies of scale. Through working with other countries and people who utilize new and different technology, we may be able to apply cutting-edge technology

to our domestic operations. Uncle Sam's may be able to increase its line of products through information gained by working with our international counterparts. An example of this may be the exchange of breeds of pecan trees from Mexico or Israel to keep our orchards fresh and producing to their highest capacity. Exporting will also decrease the risk of doing business by selling to more diversified markets. It will be easier to ride out business cycles by exporting our pecans. Now is a perfect time to export because the dollar is competitive on the world market. Finally, exporting is a U.S. government priority, which affords us easier access to information and financing.

EXERCISE 9.6

Either on your own or in a group, rewrite the following paragraphs. Try to improve cohesion by weaving the sentences to get the proper flow from old to new information. Improve coherence by placing a clear point sentence at the beginning. Break up logjams of overly abstract and convoluted prose to increase fluency. Reduce wordiness if possible. When you have completed your work, discuss each paragraph with the whole class.

1. Intensive properties are independent of the mass of material in the system. Extensive properties depend on the mass of material in the system. All thermodynamic properties basically have to be either intensive or extensive. Extensive properties can be transformed into intensive properties by dividing the total system mass. Mass and volume are the particular extensive properties. Pressure, temperature, density, volume per unit mass (specific volume), and mass and mole fractions are intensive properties.

2. Much land suitable for landfills cannot be used for dumping garbage and waste due to the fact that the land is already really being used for other indispensable human initiatives and endeavors, such as agricultural development projects. More frequently, however, attempts to locate a suitable individual landfill site fail because of social and political pressures from landowners or residents. They strongly object to having a landfill nearby. Whatever the cause, locating sites for new landfills is becoming increasingly more and more difficult.

3. The editing function is among the most useful human interface elements available on the new system. To activate it, you have to push the button marked command on the keyboard pad. At the same time, you push the "e" key. The same button will activate the basic print function if you push the "p" key while holding it down. And then there's the function you get when you press the "q" key while holding down the command button and the shift key at the same time. So be aware of that command button. It can really help you.

4. Deer management is an old and ancient art. Farmers and ranchers can make a great deal of money and get pleasure, too, from managing for trophy bucks. Managing the deer herd, a practice that comes down to us from the old European royal families, has only become a science since the development of the high-technology genetics-related practices known as culling and

artificial insemination. Culling is easy if you like hunting, but artificial in-semination takes more time and money. The manager has to capture the does and milk the bucks, then inject the semen into the does and watch them carefully. With culling, you just shoot and destroy any bucks that don't have the right kind of rack or size. Though animal rights activists don't like you to shoot animals, it is my opinion that this practice is good for the deer over-all. It's better than letting them starve on overgrazed land due to the fact that their natural predators have been eliminated.

5. For years, industrial practices favored the SMR method of machine hy-bridization. The engineer inserts the P and Q stems of the interactor into the jacks marked "input" on the receiving amplifier. The L and M inputs are left open. The S, M, and R transmitters are activated. Another engineer or tech-nician has to be at the receiver at this time to simultaneously record the sig-nals. So long as hyperfluentization is maintained on all transmitters and so long as the L and M inputs don't give off too much feedback, the SMR method has been shown to be a powerful code transmission effecting method.

Recommendations for Further Reading

Flower, Linda, John R. Hayes, and Heidi Swarts, "Revising Functional Documents: The Scenario Principle." *New Essays in Technical and Scientific Communication: Research, Theory, and Practice.* ed. P. V. Anderson, R. J. Brockmann, and C. R. Miller. Amityville, NY: Baywood, 1983: 51–68.

Gorrell, Donna. *A Writer's Handbook from A to Z.* Boston: Allyn and Bacon, 1994.

Jones, Dan. *Technical Writing Style.* Boston: Allyn and Bacon, 1998.

Kolln, Martha. *Rhetorical Grammar: Grammatical Choices, Rhetorical Effects.* New York: Macmillan, 1991.

Lanham, Richard A. *Revising Business Prose.* New York: Scribner's, 1981.

Rawlins, Jack. *The Writer's Way.* 2nd Edition. Boston: Houghton Mifflin, 1992.

Williams, Joseph M. *Style: Toward Clarity and Grace.* Chicago: University of Chicago Press, 1990.

Additional Reading for Advanced Research

Killingsworth, M. Jimmie, and Michael K. Gilbertson. *Signs, Genres, and Communities in Technical Communication.* Amityville, NY: Baywood, 1992.

Killingsworth, M. Jimmie, and Jacqueline. S. Palmer. *Ecospeak: Rhetoric and Environmen-tal Politics in America.* Carbondale: Southern Illinois University Press, 1992.

Lanham, Richard A. *Analyzing Prose.* New York: Scribner's, 1983.

Selzer, Jack. "What Constitutes a Readable Technical Style?" *New Essays in Technical and Scientific Communication: Research, Theory, and Practice.* ed. P. V. Anderson, R. J. Brockmann, and C. R. Miller. Amityville, NY: Baywood, 1983: 43–50.

Editing and Polishing

CHAPTER OUTLINE
▼▼▼▼▼▼▼▼▼▼▼▼▼▼▼▼▼▼▼▼▼▼▼▼▼▼▼

The Editing Process
Level 1: Substantive Edit
Level 2: Language and Mechanical Style Edit
Level 3: Format and Copy Clarification Edit
Level 4: Screening and Integrity Edit
Level 5: Policy and Coordination Edit

CHAPTER OBJECTIVES
▼▼▼▼▼▼▼▼▼▼▼▼▼▼▼▼▼▼▼▼▼▼▼▼▼▼▼

After you have worked through this chapter,
you should be able to do the following:

- Recognize the need to edit your documents
  responsibly and, when necessary, to
  cooperate fully and productively with
  professional editors
- Recognize the different types and levels of
  editing
- Follow a systematic process for thoroughly
  editing documents

I n this chapter we come to the last stage of document production: editing, the process of making final revisions and preparing your document for delivery. Figure 10.1 shows the place of editing in the CORE method.

As a technical professional, you will sometimes do all your own editing and sometimes have help from colleagues, professional editors, and clerical staff. Whatever your situation, you should always follow the *editorial imperative:* Assume full responsibility for the quality of your document, even if you have others to help you achieve your communication goals. Be on guard against the kind of mental laziness that leads writers to think, "Well, the secretary will catch my spelling errors, the technical editor will fix the punctuation and equations, and managers can worry about whether the publication suits company policy. So I'll just attend to the weightier problems of technical content."

If you rely too heavily on others, they may let you down, or they may simply return the document to you, unable to help. In either case you will have to rewrite later—after you have "finished" your part of the work. To prevent such

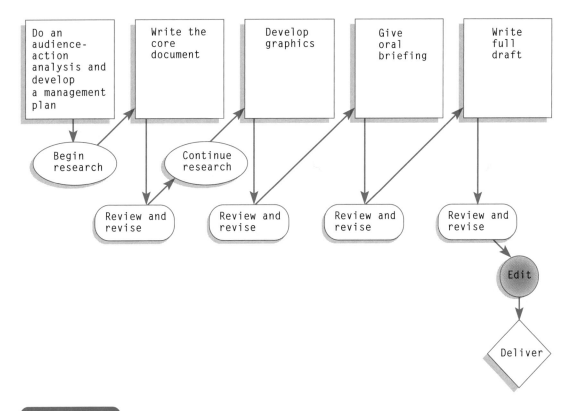

FIGURE 10.1

Editing as the last step of document preparation in the CORE method.

headaches, get it right before you send it out. The guidelines in this chapter will help you meet that goal.

The Editing Process

If you have followed the CORE method in developing your reports, proposals, and manuals, you will have been revising your work as you go. With the information user in mind, you will have written and rewritten at every stage of your work. So what's left to do?

Editing versus Revision

The revision that an author does throughout the writing process differs somewhat from the editing done toward the end of the process. "Revision" literally means "seeing it again." It involves reconceptualizing the document—its overall organization, its tone, its approach to interpretation. By the time your paper is ready for editing, most of these big-picture issues should be resolved. Now the time has come to apply the finishing touches, to attend to the details, to correct mechanical problems, to smooth the flow of sentences and connect creaky paragraphs, and to make sure that the approach is consistent throughout the document.

Revision and editing may well be overlapping activities. The difference is largely one of attitude. In revising, you should remain open to making major changes in the document. In editing, you should push toward finalizing the written product, although editing does sometimes reveal the need for a more substantial revision.

The Editorial Imperative

In addition to the different attitudes that distinguish revision from editing, there may also be a division of labor. In many companies, technical authors must turn their documents over to technical editors toward the end of the writing process. In others, editors may work with the author early on. Eventually, after editing, the document passes into the hands of managers, who give (or withhold) final approval on matters of policy and quality.

Even as you strive to follow the editorial imperative and to become as independent as possible, you should realize that editors and managers are not merely obstacles to completing the job. They are people who can save you much embarrassment and also help you communicate better with your audience. Often, editors and managers have more direct contact with end users than do authors; they serve as an effective intermediate link in the process of converting information into action.

Here are some guidelines you can follow to help editors and managers improve the communicative value of your work:

- *Create the highest quality document you can before you pass it on to the next person.*
- *Keep an open mind about changes that your editors suggest.* Be inclined to accept rather than reject their advice. They may have a better understanding of the context of use or of company policy than you do.
- *Answer all queries and complete all revising tasks that the editor or manager assigns to you as carefully, completely, and quickly as you can.* The editor will usually make every change he or she is capable of making in the text but will sometimes solicit your help with an *author's query,* a question in the margin of the document, usually involving some matter of content or technical language that only you can resolve. Consider this an opportunity to make one more contribution to the excellence of the document. Don't give in to the temptation to "get it over with" too quickly, thereby sacrificing the quality of your communication. You will probably regret your haste when the document appears in print. And your reader will almost certainly regret it!

Types and Levels of Edit

In an effort to systematize relations between authors, editors, and managers, technical editors Robert Van Buren and Mary Fran Buehler at the Jet Propulsion Laboratory (JPL) in Pasadena, California, developed a system for editorial management called *The Levels of Edit.* It has won international approval as a standard procedure in the field of technical communication and has been adopted by a number of organizations in government and industry across the United States.

The system was designed to prevent the kind of troubles that most editors face every day. Because authors present their manuscripts in all stages of completeness, editors often find it extremely difficult to say how long it will take (or how much it will cost) to do the editing. Any particular document may need several kinds of editing—ranging from a simple repair of grammar, punctuation, and spelling errors to a policy review or further research.

To deal with this variety, Van Buren and Buehler's system covers five levels and nine types of edit. The system is summarized in Table 10.1. An X beside each type of edit indicates whether it is included in each of the five levels.[1]

---

1. We have revised the original system slightly to make it more directly applicable to the process of editing student work in technical writing courses. In effect, we have turned the table upside down and reversed the numbers of the levels. Our apologies go to Van Buren and Buehler for taking these liberties with their excellent system. We would also like to acknowledge the generosity of Mary Fran Buehler in helping us to define the goals of this chapter in the few months before her untimely death. She is greatly missed in our profession.

TABLE 10.1

Types and levels of edit.

Type of Edit	Level of Edit				
	1	2	3	4	5
substantive	X				
language	X	X			
mechanical style	X	X			
format	X	X	X		
copy clarification	X	X	X		
screening	X	X	X	X	
integrity	X	X	X	X	
policy	X	X	X	X	X
coordination	X	X	X	X	X

Source: Robert Van Buren and Mary Fran Buehler, *The Levels of Edit,* 2nd Edition (Pasadena, CA: Jet Propulsion Laboratory, 1980) 5.

As a student, you will probably have to edit your technical papers on your own, although you may be able to exchange papers with a classmate in some cases. If you are working independently, we suggest that, because of its wide acceptance in industry and its sheer ingenuity, you use the levels of edit system as a guide while editing your work.[2] Think of the levels of edit as steps in a process for editing your documents. With each new step, you add new procedures while reviewing the changes you made in the previous step. In Step 1, for example, you perform a substantive edit on your paper, surveying the document for complete and correct information. In Step 2 you edit for language and mechanical style, reviewing the changes you made in Step 1. The effect is cumulative.

Figure 10.2 gives a visual representation of the method. The rest of this chapter will provide guidelines to help you put this step-by-step procedure into practice.

The levels of edit system, as adapted here, requires a special kind of concentration. At each level, you should try to keep your attention focused on the particular editorial problem at hand (mechanical style, format, and so forth), giving each one its due. You may catch a comma error or a misspelled word while you are doing your substantive edit, and certainly you should go ahead

---

2. The idea of using the levels of edit as a key to the process of writing, revising, and editing was suggested to us by Charles Campbell, a professor at the New Mexico Institute of Mining and Technology. Professor Campbell, an experienced technical writer and editor, uses the levels of edit to teach freshman composition as well as technical writing and editing.

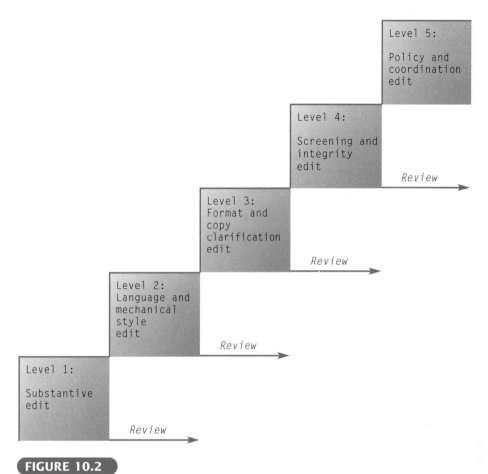

FIGURE 10.2
The levels of edit as steps in a process.

and correct these errors. But try not to be distracted by other problems as you deal with the area of focus at each level.

Level 1: Substantive Edit

Substantive editing covers questions about the sufficiency of the information you present.[3] More than any other, then, this level overlaps with revision. Return one last time to potential problems with the quantity and quality of the

---

3. As Van Buren and Buehler say, "The Substantive Edit deals with the meaningful content of the publication" (*The Levels of Edit*, 23).

researched information you presented in your document. Ask the following questions:

1. *Have I provided enough well-directed information?* Can my readers carry out the actions or make the decisions I expect from them on the basis of the information I provide in each section of the document? Are all the sections clearly directed toward that action or decision? Do the sections support one another in a coherent and reasonable sequence?

2. *Have I provided too much information?* Am I running the risk of oversaturating the reader? Does the information I present seem logical because of my position in the context of production but make no sense in the context of use? Have I cut everything possible so that I do not waste the reader's time?

Remember that the aim of technical communication is to provide information that is both necessary and sufficient for action and decision making—no more and no less.

Level 2: Language and Mechanical Style Edit

This level requires a big step for many students in technical writing classes. Your teacher is likely to assume that you have already mastered most of the nuances of English grammar. But few of us can keep in mind all the rules of grammar, syntax, and usage.

Good reference books can help. In your professional work, you will regularly consult in-house manuals on style provided by your company. Near your workstation, you should also have a good dictionary (such as those by American Heritage, Random House, or Merriam-Webster) as well as a professional style manual and a general handbook of English grammar and usage.

You should own a copy of the professional style manual associated with your field of specialization. Biologists use the *CBE Style Manual*, for example, and researchers in the social sciences and education frequently rely on the *Publication Manual of the American Psychological Association*. The Society for Petroleum Engineering, the American Society for Mechanical Engineering, and other such organizations publish similar style guides.

Since these references mainly serve as publication guides for specialized journals in your field, you should always have on hand a general handbook on English grammar and style to help you in your effort to reach broader audiences. Strunk and White's *Elements of Style* is a perennial favorite; Donna Gorrell's *A Writer's Handbook from A to Z* is a more up-to-date choice for professionals in all fields. Both of these books are short and relatively inexpensive. Philip Rubens' *Technical Writing: A Manual of Style* gives the usual information found in good English handbooks and adds more specialized advice for technical authors.

GUIDELINES AT A GLANCE 17

Editing sections and paragraphs

- *Chunking: Use frequent breaks.* Start a new section or subsection about every 3–5 paragraphs.
- *Keep paragraphs fairly short.* For technical reports and proposals, 100 to 200 words is a good length for paragraphs (about 1/3 to 2/3 of a double-spaced typed page per paragraph). For manuals, paragraphs could be even shorter.
- *Place point sentences first (or early) in every paragraph.* Supporting sentences should follow. (See Chapter 9 for more on point sentences.)
- *Use clear logic to structure paragraphs.* For example, move consistently from general to specific (deductive order), from past to present to future (chronological order), from cause to effect (causal order), from whole to parts (divisional order), or from simple to complex information (pedagogical order).
- *Use first and last sentences to "weave" your paragraphs and sections together.* Move from old information to new information.

Space does not permit us to reiterate all the good advice contained in these reference books, nor can we anticipate all the problems you are likely to encounter as an editor of your own writing. At best, we can provide some general guidelines on the kinds of grammatical errors and stylistic indiscretions that frequently crop up during the editing of technical documents.

Guidelines at a Glance 17–21 will get you started on your language edit, which deals with problems involving paragraphs, sentences, and words. Some of these guidelines review issues we took up in Chapter 5 in our discussions of style. On matters of basic grammar and mechanics, what we offer here is merely a summary of problems that technical writers and editors frequently encounter. For more detailed information on these basic problems, consult your general English handbook and references on style.

Spelling should also be a major consideration for technical writers. Though bad spelling rarely changes the meaning of a document, it very often spoils the professional impression the author is trying to create. Bad spelling, though often dismissed as the province of nitpickers and English teachers, bothers more people than you may realize.

If you habitually misspell words, you should try to reform. Keep a list of words that you commonly misspell, and notice that these usually fall into patterns and categories. The words may end in *-ence,* for example, or have double consonants in the middle. Know your weaknesses, and look these words up in the dictionary until you master their spelling.

Also use the spell-checker that comes with your word processor. Spell-checkers do not catch all misspelled words. They cannot distinguish between *there, they're,* and *their,* for example, or between *two, too,* and *to.* Such distinctions depend on context, and the check for that cannot (yet) be computerized. But even with these limitations, spell-checkers usually catch over 90 percent of

GUIDELINES AT A GLANCE 18
Checking sentence grammar

Rule	Example
Subject–verb agreement: Make sure subjects and verbs agree in number. A singular subject takes a singular verb. Plural subjects take plural verbs. Subjects compounded with *and* take a plural verb. For subjects compounded with *or,* the number is determined by the part of the subject closest to the verb.	■ The **result appears** encouraging. ■ The **results appear** encouraging. ■ **Women** now **compete** with men in the workforce. ■ The **data are** now complete. ■ The first experiment **and** the second one **confirm** our findings. ■ The first experiment **or** the second one **contains** an error. ■ Either the chemicals **or** the equipment **has been** contaminated. ■ Either the equipment **or** the chemicals **have been** contaminated.
Pronoun–antecedent agreement: Make sure pronouns agree in number and gender with their antecedents.	■ **John** presented **his** findings, and **Mary** presented **hers.** ■ **John and Mary** presented **their** findings. ■ **Everyone** brought in **his or her** data to compare. ■ **All the researchers** submitted **their** data.
Verb tense shifts: Verbs change their forms or endings to show changes in tense.	■ Professor Jones **brings** the equipment to lab. ■ Professor Jones **brought** the scales last week. ■ Professor Jones **has** always **brought** the equipment to lab.
Pronoun case shifts: Pronouns change forms to indicate whether they are subjects or objects in a sentence.	■ **He** came to the demonstration. ■ **We** gave the demonstration to **him.** ■ He is the lab assistant **who** has the problem. ■ She is the assistant to **whom** we deliver our reports.
Sentence fragments: Make sure every sentence has a subject and a verb and is not a dependent clause.	■ *Incorrect:* I am delivering the notes. Which you requested. ■ *Correct:* I am delivering the notes, which you requested. ■ *Incorrect:* The Second Law being the appropriate one here. ■ *Correct:* The Second Law is the appropriate one here.
Sentence-combining errors: Avoid comma splices and fused sentences.	■ *Splice:* The results were poor, however they will improve. ■ *Fused:* The results were poor however they will improve. ■ *Correct:* The results were poor. However, they will improve. ■ *Correct:* The results were poor, but they will improve.

the errors in a typical document. So they significantly improve the chances that you will be able to catch *all* your errors—the goal that independent writers set for themselves. Spell-checking devices have saved many an author from embarrassment. However, because of the same kind of laziness that causes misspelling, too few people take the time to run a spell check. The language edit is the perfect opportunity to devote the time and the care that professionalism requires in this area.

One final problem needs to be considered in your language edit: the need to use inclusive language—language that is not sexist, discriminatory, or offensive to the increasingly diverse social and ethnic groups entering the contem-

GUIDELINES AT A GLANCE 19
Editing sentences

Guideline	Example
Use active voice instead of passive voice whenever possible.	■ *Passive:* The data are now recorded. ■ *Active:* The machine now records the data. ■ *Passive:* The measurements were taken twice. ■ *Active:* We took the measurements twice. ■ *Passive:* The results were attributed to the unusual cold. ■ *Active:* The results deviated because of the unusual cold.
Use action verbs rather than linking verbs as often as possible. Especially avoid overusing the various forms of *to be* (is, are, was, were, have been, etc.).	■ *Weak:* The system is operative in the coldest weather. ■ *Stronger:* The system operates in the coldest weather. ■ *Weak:* The trouble is due to a bad experimental design. ■ *Stronger:* The trouble results from a bad experimental design.
Don't overuse indirect sentence beginnings.	■ *Indirect:* It is well known that mollusks readily absorb the poison. ■ *Direct:* Mollusks readily absorb the poison. ■ *Or:* The literature shows that mollusks readily absorb the poison. ■ *Indirect:* There are 203 new plants in Peoria. ■ *Direct:* Peoria has 203 new plants.
Use parallel structures for items in a series and for coordinate constructions (especially watch sentences with *either . . . or, both . . . and,* and *not only . . . but also*).	■ *Not parallel:* The contract is poorly written, unethical, and breaks the law. ■ *Parallel:* The contract is poorly written, unethical, and illegal. ■ *Not parallel:* The contract is both unethical and it is illegal. ■ *Parallel:* The contract is both unethical and illegal. ■ *Not parallel:* The contract is not only unethical; it also breaks the law. ■ *Parallel:* The contract not only is unethical, but also breaks the law. ■ *Or:* Not only is the contract unethical; it also breaks the law.
To increase readability, convert sentences with parallel items into lists, especially if you plan to discuss the items further.	The contract is: ■ poorly written ■ unethical ■ illegal.

GUIDELINES AT A GLANCE

porary workforce. As we suggested in Chapter 9, inclusive language is now the norm in technical communication because it reflects a new set of attitudes as well as the new diversity in many areas traditionally dominated by white American males. Nondiscriminatory language is required by all government agencies and companies working on government contracts and has become the standard for most professional organizations as well.

So learn to use inclusive language when you write, and during your language edit, remove all accidental instances of noninclusive language. Guidelines at a Glance 22 on page 287 can help.

GUIDELINES AT A GLANCE 20
Checking punctuation

Rule	Example
Use a **comma** after a phrase or clause that introduces an independent clause.*	■ Because of the cold weather, the instruments worked slowly. ■ When we completed the readings, we recorded them in our notebooks.
Use a **comma** before a coordinating conjunction connecting two closely related independent clauses.	■ We measured the outcroppings, and we took further readings on our way back to camp. ■ We took the readings several times, but we could never get them to agree with the ideal measures.
Use a **semicolon** between two closely related independent clauses that are not joined by a coordinating conjunction.	■ We measured the outcroppings early in the day; we took further readings on our way back to camp. ■ We took the readings several times; however, we never got them to agree with the ideal measures.
Use a **comma** between items in a series.	■ We measured the outcroppings, ran the tests, and took samples. ■ The contract is poorly written, illegal, and unethical.
Use a **semicolon** between items in a series *only if* any of the items already contain commas.	■ We accounted for three sets of variables: A, B, and C; Q and R; and G, H, and I. ■ The team divided into three smaller groups: Jones, Smith, and Aziz; Harris and Rampart; and the rest of us.
Use a **comma** to set off nonrestrictive elements in a clause (those which could be handled in a separate sentence without changing the meaning), but do not set off restrictive elements.	■ *Nonrestrictive:* The river, **which we followed for some ten miles,** kept us busy all day. ■ *Restrictive:* The river **that flows between Cairo and Lake Victoria** is called the Nile.
Use a **colon** to introduce a list that follows a sentence.	■ The following measurements were conclusive: . . . etc.
Use an **apostrophe** to indicate possessive nouns. Possessive pronouns usually do not take an apostrophe. Pronouns with apostrophes are usually contractions.	■ *Correct:* researcher's measurements (singular possessive noun), researchers' measurements (plural possessive noun ending in *s*), women's diseases (plural noun that does not end in *s*), today's world (means "the world of today") ■ *Incorrect possessive:* he's, it's (really means "he is" and "it is") ■ *Correct:* his, its (means "of him" and "of it")

*A clause is any group of words containing a subject or a verb. An independent clause can stand alone as a sentence, whereas a dependent clause (indicated by the presence of a subordinate conjunction like *because* or *when* or by a relative pronoun like *who* or *which*) must be connected to an independent clause in a sentence. A phrase is a group of words that act together to modify another part of the sentence.

The edit for mechanical style comes after the language edit in level 2. It involves checking conventions of usage that may vary slightly from field to field and from company to company but usually follow general patterns for such matters as capitalization, hyphenation, use of numbers and symbols, use of

GUIDELINES AT A GLANCE 21
Checking word usage

- *Use technical terms carefully.* Terms such as *quark, central processing unit,* and *isomorphism* have special meanings determined by the usual discourse in specialized fields of study and practice. Make sure you use them in the way your audience uses them. When you use technical terms for audiences of outsiders, define the terms and explain their significance.

- *Avoid using technical terms (jargon) when familiar terms are more appropriate.* Don't say "input" when you mean "advice." Don't say "interface" when you mean "cooperate."

- *Avoid unbroken strings of abstract words.* Like *truth, justice,* and *beauty,* most technical terms are highly abstract. Many other abstractions are merely vague. Keep your writing as concrete and specific as you can. Don't just say that the water was warm; say that it was 42 degrees Celsius. Don't just say that there were five men in the village; say that there were four adult males in warrior attire and one chieftain preparing to conduct the cleansing ritual.

- *Be careful not to confuse words that have similar sounds or meanings.* Most handbooks list such troublesome words as "effect" and "affect," or "continuous" and "continual." Your field of study and professional practice will largely determine which groups of such words will trouble you the most. Keep an eye open for such trouble spots.

- *Don't use more words than you need.* Avoid redundancies such as "yellow in color" or "tentatively seems to suggest." Use simple pronouns to take the place of overblown phrases such as "in terms of" and "with particular reference to."

GUIDELINES AT A GLANCE 22
Using inclusive language

Noninclusive Language	Examples	Edited Examples
Sexist or gender-biased language	<ul><li>Mankind has made great progress.</li><li>Educated men want better jobs these days.</li><li>The company was required to man the post every morning.</li><li>Every student should bring his lab book.</li><li>In addition to reason, human beings use intuition to solve problems, especially women.</li></ul>	<ul><li>Humankind has made great progress.</li><li>Educated people want better jobs these days.</li><li>The company was required to staff the post every morning.</li><li>Every student should bring his or her lab book. *Or:* All students should bring their lab books.</li><li>In addition to reason, people use intuition to solve problems.</li></ul>
Discriminatory or ethnically insensitive comments, cultural over-generalization	<ul><li>Because death is an everyday occurrence, people in Third World countries do not have the same respect for life that we do.</li><li>Latin peoples tend to be enthusiastic and hot-blooded.</li><li>American Indians are closer to nature than most whites.</li><li>Blacks and whites understand each other better in the South.</li></ul>	*Avoid such comments altogether.*

287

GUIDELINES AT A GLANCE 23
Internal citations in APA style

Type of Citation	Example
Indirect reference to source in literature	Climate models suggest that the projected increase of global temperature has already begun (Schneider, 1989).
Indirect reference to source in literature with multiple authors	Climate models suggest that the projected increase of global temperature has already begun (Walker, Hays, & Kasting, 1981).
Indirect reference to multiple sources in literature	Climate models suggest that the projected increase of global temperature has already begun (Walker, Hays, & Kasting, 1981; Schneider, 1989).
Direct reference to source in literature	Schneider (1989) argues that greenhouse warming has already begun.
Direct reference to source in literature with multiple authors	Walker, Hays, and Kasting (1981) discuss feedback mechanisms that may result in long-term stabilization of the global climate.
Direct quotation of source in literature (with page numbers)	Schneider (1989, p. 13) admits that the rate and size of the projected change "is still controversial."
Reference to source without author	The U.S. Office of Technology Assessment (1991) suggests a number of options for dealing with the effects of global climate change.

italics and underlining, spacing, and presentation of data. There is a surprising degree of inconsistency in these matters. To deal with them, you will probably need to cross-check your general English handbook, your professional handbook (such as the CBE or APA), and an up-to-date dictionary (such as those by Random House or American Heritage).

Another convention to check for is your documentation style for references and your bibliography. In technical communication, author–date systems are usually preferred to the footnote or endnote systems that are still used widely in some academic disciplines.

In Guidelines at a Glance 23 and 24, we follow the author–date system of the American Psychological Association (APA) style, which is used in many of the social sciences and in education. It is not far removed from the styles that are used in most scientific and engineering disciplines. Guidelines at a Glance 23 shows you how to cite the source in the body of your document. Guidelines

GUIDELINES AT A GLANCE 24
Listing references in APA style

Type of Reference	Example
Book	Schneider, S. H. (1989). *Global warming: Are we entering the greenhouse century?* San Francisco: Sierra Club Books.
Article in a specialized journal	Walker, J. G. G., Hays, P. B., & Kasting, J. F. (1981). A negative feedback mechanism for the long-term stabilization of earth's surface temperature. *Journal of Geophysical Research 86,* 9776–82.
Chapter or essay in a book of collected papers	Redish, J. C., Battison, R. M., & Gold, E. S. (1986). Making information accessible to readers. In L. Odell & D. Goswami (Eds.), *Writing in nonacademic settings* (pp. 129–53). New York: Guilford.
Article in a newspaper	May, C. D. (1988, August 21). Pollution ills stir support for environmental groups. *New York Times,* p. 20Y.
Article in a magazine	Anderson, W. T. (1990, July/August). Green politics now come in four distinct shades. *The Utne Reader,* 52–53.
Government document	U.S. Office of Technology Assessment. (1991). *Changing by degrees: Steps to reduce greenhouse gases: A summary report.* Washington, DC: U.S. Government Printing Office.
Company publication	Federal Express Corporation. (1989). *Annual Report.* Memphis, TN: Federal Express.
Personal interview notes[4]	Palmer, J. S. (1998). [Transcribed notes from interview with R. Cleminson, Director, Center for Environmental and Energy Education, Memphis State University, Memphis, TN]. Unpublished raw data.

at a Glance 24 shows you how to present the source in the list of references given at the end of the document.

You should check with professors in your major field to see which style most researchers and scholars follow in your area of study. Also remember that in industry the style may change from company to company. As a technical communication student, you should get into the habit of looking up the style requirements and conforming to the demands of each writing situation.

---

4. Personal interviews or telephone conversations that have not been transcribed are considered unrecoverable data and should not be included in the reference list. However, they can be cited in the text as follows: E. Hampton (personal communication, December 2, 1995).

As you begin your edit for language and mechanical style, use the review in Guidelines at a Glance 25 as a general checklist. Figure 10.3 reproduces a page of the final manuscript for *Information in Action,* showing the copy editor's suggestions for language and mechanical style changes. Even seasoned authors are grateful for such expert advice.

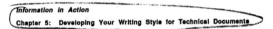

Information in Action
Chapter 5: Developing Your Writing Style for Technical Documents   5-21   279

HC **Use Effective Modifiers to Interpret and Recommend**

BTX Although one of the goals of technical communication is to present information in a fair and balanced way, another is to give other people the advantage of your expertise. So when you are reporting, don't just provide neutral information; tell your users what effect it could have on their actions. Connect facts with previous actions or with future possibilities (or both). Don't just say, "Our profits fell by 5% last year." Say, "Because of mismanagement in our shipping operation, our profits dropped 5% last year." Or "Because our profits fell by 5% last year, we may be forced to lay off a number of workers in the next quarter."

Notice that these are not simple sentences. Modifying clauses such as "because our profits fell by 5% last year" are necessary for effective explanation and interpretation. These dependent clauses, groups of words that, like sentences, have subjects and verbs, but cannot stand on their own, serve as important qualifiers by setting conditions, describing, defining, and explaining. Evaluative adjectives such as "effective" and "correct," and adverbs such as "adequately" and "poorly" can serve a similar interpretive function.

Notice how in the following sentences the adjectives, adverbs, and dependent clauses in italics help the author to explain or interpret facts and events. If these modifiers were removed, how much less would the sentences communicate?

LN   1. The proposed dump site cannot *adequately* meet the *best* standards of waste disposal *because it lacks the recommended clay base.*

2. *Whereas Department A functioned at a profitable margin,* Department B failed to achieve its minimal goals.

3. *If you neglect the calibration in the initial heating,* the process will fail to produce *good* results.

4. The blue crystals, *which are mined in Albania,* are *very difficult* and *expensive* to replace.

5. We must be sure not to alienate the Albanians, *who are crucial to the mission as proposed.*

BTX Two cautions on the use of modifiers are in order. First, make sure that you use the main clause of a sentence, not its dependent clauses, to make the point you want to emphasize. In Sentence 2

FIGURE 10.3

Sample copyedited manuscript page.

GUIDELINES AT A GLANCE 25
Review of language and mechanical style edit

Language Edit

- *Chunking of information:* Does a new section or subsection start every three to five paragraphs?
- *Paragraph length:* Are the paragraphs kept to a 100- to 200-word range in length?
- *Topic sentences:* Does each paragraph start with the topic sentence?
- *Clear logic:* Is the organization of the paragraphs logical?
- *Weaving of paragraphs and sentences:* Do the sentences and paragraphs move from old information to new information?
- *Sentence grammar:* Does each sentence work with Guidelines at a Glance 13 and 14?
- *Sentence punctuation:* Does each sentence follow Guidelines at a Glance 15?
- *Word usage:* Does the document use words as effectively as possible, according to Guidelines at a Glance 16?
- *Spelling:* Is every word spelled correctly?
- *Inclusive language:* Does the document refer to people and ideas in language that is fair and based on real knowledge rather than assumptions?

Mechanical Style Edit

- To check the document's use of *capitalization, spacing, and related mechanics,* refer to a good handbook or dictionary.
- To check the document's use of *references,* refer to Guidelines at a Glance 18 and 19 for APA style, or use the preferred documentation style for your particular area of study.

Level 3: Format and Copy Clarification Edit

Having completed your edits for language and mechanical style, you can now move to the level that deals with format and copy clarification. The format edit aims for consistency, readability, and usability in typography, layout, and graphics—topics we covered in our discussion of document design in Chapter 7. Assuming that you are using a fairly advanced form of word-processing or desktop publishing software, Guidelines at a Glance 26–28 offer help in these matters. Once again, professional style guides (such as the *Chicago Manual* or the APA *Publication Manual*) go into far greater detail, as do most company or agency style manuals.

In the copy clarification edit, the next step, you should amend any copy that remains unclear or illegible for final production in the published document. If you are working in a professional setting, a production copy editor will mark the copy for photocompositors, using codes that correspond to design elements.

GUIDELINES AT A GLANCE 26
Typography

- *Use fonts and styles consistently.* Set up a system and follow it rigidly throughout on matters of type style. For example, use 18-point, bold Helvetica for first order headings, 14-point bold Helvetica for second-order headings, and 12-point Times for main text, with bold for emphasis. Reserve italics for book titles and special terminology. Avoid underlining and fancy styles (such as shadow or outline).
- *Don't overuse typographical variation for emphasis.* For example, don't underline headings that are already bold and enlarged, and don't use italics or bold type too often within the same paragraph.
- *Avoid words in ALL CAPITALS,* even in titles and headings. All caps are harder to read than initial caps. They are leftovers from the days of typewritten documents, when more readable forms of variation were not available. (The same is true of underlining.)

GUIDELINES AT A GLANCE 27
Editing layout

- Double-space all reports, proposals, manuals, and other technical documents unless you are specifically instructed to do otherwise.
- Single-space correspondence (letters and memos; see Chapter 13).
- Leave at least one-inch margins on every side of the page.
- Avoid justification of right margins unless you have a good hyphenation option that allows you to achieve high-quality word spacing within the text. An unjustified right margin (known as "ragged right") is preferable to big gaps of space in the text.
- Place page numbers in the upper right-hand corner unless you are instructed to do otherwise.
- Keep the headers at the top of each page simple but clear [such as "Morgan, Report on Superconducting Super Collider (SSC)"].
- Vary the text with graphics, chunks, and white space. In general, try to increase the amount of white space on the page. Try to have no more than 60 percent of the page taken up with written text.

As an author, you will have a role here, too. You must provide special instructions for the editors or artists who will clarify your work. For example, we often have our students make notes on their copy to indicate that certain photographs or graphics need to be redone before final publication. This way, they can show us that they recognize the need for a professional treatment without having to go to the expense of actually producing it. You should check with your instructor to find out what she or he may require of you at this stage of editing.

GUIDELINES AT A GLANCE 28
Editing graphics

- Place titles and captions above tables.
- Place titles and captions below figures (including graphs, charts, drawings, schematics, maps, renderings, etc.).
- Number tables and figures in separate sequences.
- Provide a list of tables and figures at the beginning of your document.
- Integrate graphics into the text unless you are instructed to do otherwise. That is, keep them as close as possible to the part of the written text to which they refer.
- Place tables flush against the left margin.
- Center figures on the page.
- Be sure to refer to all tables and figures in the text, providing sequence numbers and page numbers for reference.

Level 4: Screening and Integrity Edit

Just as the substantive edit at Level 1 provided one last opportunity for the author and editor to check content, the screening and integrity edits provide final opportunities for an editorial check. The screening edit turns a proofreader's eye on matters of language and format. The integrity edit makes sure all parts of the document relate correctly to all other parts.

Guidelines at a Glance 29 and 30 give the steps for each edit at this level.[5]

GUIDELINES AT A GLANCE 29
Items to check during screening

- All words are spelled correctly.
- Subjects and verbs agree.
- All sentences are complete.
- There are no incomprehensible statements or missing words or graphics.
- Figures that are intended to be camera-ready (not to be redone at production stage) have no handwriting or other flaws on them.
- All parts of graphs and charts are clearly labeled.
- All tables and figures have titles and/or captions and are correctly placed in the text.

---

5. These tables draw heavily on Van Buren and Buehler's treatment of screening and integrity edits. See *The Levels of Edit*, pp. 15–17.

GUIDELINES AT A GLANCE 30
Items to check during an integrity edit

- The wording in the table of contents matches the chapter titles and section headings.
- The wording in the list of tables and figures matches the captions in the text.
- Every table and figure is cited by the correct title and number in the text.
- It is clear where every table and figure begins and ends. All formats for tables and figures are consistent.
- There are no incorrectly numbered sequences (tables, figures, sections, lists, etc.).
- All cross-references (such as "see Section 5") make sense.
- Titles used in headers at tops of pages match titles used in the text.

Level 5: Policy and Coordination Edit

Just as the early levels of edit overlap with revision, this last level overlaps with matters of production, professional practice, management, and ethics, many of which we discussed in Part One.

The policy edit ensures that the document adheres to company policy on official publications. At the Jet Propulsion Laboratory, for example, all reports must have, at the very least, a cover and title page, a table of contents, an abstract, page numbers, figure captions, and table titles. In addition, the document must use at least two levels of headings. All documents must conform to NASA standards and to certain ethical and political standards: Authors must refrain from "derogatory or otherwise inappropriate judgmental comments . . . that would reflect adversely on private companies, government agencies, other investigators, or subdivisions within JPL." In addition, they must not "advertise, endorse, or promote the products or services of a company."[6]

The guidelines that we covered in Part One concerning communication ethics and the politics of information provide a solid grounding for a policy edit. Guidelines at a Glance 31 gives a summary checklist based on that earlier treatment.

The coordination edit is really more of a system of management, a way of planning and directing documents through the final stages of production. At this stage, you can use your management plan (discussed in Chapter 3) as a checklist to see whether you've succeeded in meeting your goals. You should be able to justify any departures from the original plan.

The issue of management includes collaboration with others. In editing, collaboration can really work to your advantage. It is notoriously difficult to

6. Van Buren and Buehler, *The Levels of Edit*, p. 15.

GUIDELINES AT A GLANCE 31

Ethics check

- Have I represented the facts of a case as completely as possible, accounting even for data that may damage the position I take?
- Have I fully acknowledged my sources of information?
- Have I obtained permission to quote or reproduce data if necessary, avoiding violations of confidences or copyrights?
- Have I carefully read and interpreted all pertinent documents before summarizing or using the information they contain?
- Have I competed unfairly by misrepresenting or suppressing information?
- Have I violated the premise of truth in advertising?
- Have I unfairly represented any other person or organization or written anything that could bring harm to others?
- Have I promised more than I can deliver?
- Have I obscured important information that will affect the actions of my readers?
- Have I made clear what consequences readers can expect from the actions I recommend or direct?
- Will anything I have written damage anyone's personal reputation or physical well-being?

be objective about your own work, especially as you near completion. That's why, during the final rush of production, you are likely to need some help. If you are working on a group project, do not leave all the editing to a single person; share the work. If you are working independently, recruit one of your peers or classmates—or better yet, someone in your target audience—to read your paper for errors and other possible offenses.

With the policy and coordination edit, we have come full circle in the process of document production. We're back to the issues that concerned us in Part One. It is useful to remember that document production, like all production, is a cycle. We get to the end only to begin again. Keeping the process going, the wheel of production turning, is one of the key responsibilities of technical communicators.

In each of the following chapters we begin the process again each time we come to a new genre of technical documents. In Chapter 12 on reports, for example, we will discuss preparing an audience-action analysis, writing the core document, preparing graphics, giving an oral briefing, and finally designing, writing, revising, editing, and delivering the full document. You'll need to refer back to these early chapters with each new cycle, but you should also build on what you know, moving closer with each turn of the cycle to creating the document that meets the needs of a specific audience facing a specific task.

EXER CISE
10.1

Part 1. Editing a business letter.   The substantive edit should help to ensure that you have included the appropriate information to enable your reader to make an informed decision or take action. Consider the following letter, then respond to the questions that follow concerning the value of its information to a hiring manager.

December 29, 1994

To Whom It May Concern:

I would highly recommend Ed Hankins for any position commensurate with his excellent technical, sales, and managerial skills. After he successfully covered our Tarrant County sales territory for more than three years, we sent Ed to resurrect the failing Houston branch office. Our investment in Ed was well rewarded; he turned the branch around in less than a year.

Ed's ability to close sales and maintain strong working relationships with clients is impressive. He is a team player: dependable, trustworthy, and effective.

Through his active involvement in professional organizations and his numerous speaking engagements, Ed has represented Morris Industries to hundreds of industry professionals. After working with Ed Hankins for over five years, I am confident that he will be a great asset to any company.

Sincerely,

Robert J. Patterson
Managing Partner
Southwest Division
Morris Industries

If this recommendation were mailed to your personnel office, would you be able to answer the following questions about Ed Hankins and his professional abilities?

- What kind of sales has Ed done? What products or services did he sell?
- How large was the staff he managed, and how extensive was the budget?
- Does Ed have strong communication skills?
- Do you get enough information about Ed's employment history with Morris Industries? If not, what else would be helpful to your hiring decision?
- Do you get too much information about any aspects of Ed's employment history with Morris Industries? If so, what aspects are covered more than necessary?
- Most important, does this letter give you a clear impression of what kind of employee and manager Ed is?
- Would you consider hiring Ed on the basis of this recommendation? Why or why not?

Overall, what kinds of changes (if any) would you make to this letter in a substantive edit? Consider the context of use for this document to decide whether and how it should be changed. If you decide that it should remain unchanged, explain what information it conveys to its target reader and why that is the appropriate information.

Part 2. Editing your own writing.   Using one of the longer assignments you have completed for this class, conduct a substantive edit. Keep the intended reader in mind as you work through your document, checking the information you present for its effectiveness in the context of use.

Developing a Communication Policy.   Review the material on ethics in Chapter 2, and come to class prepared to develop a course policy on what constitutes unacceptable attitudes, subject matter, or style in your technical documents. As a class, try to reach consensus on five to ten policy points. Apply the policy in a Level 5 edit of the documents you write in the class, keeping a list of violations you encounter in each draft for later discussion.

EXERCISE **10.2**

Applying levels of edit to a document fragment.   Sample 10.1 on page 298 is a page from a fictitious technical report. Edit the document several times, going through a new level of edit with each pass. For the policy edit, use either the Communication Policy you developed in Exercise 10.2 or the policy of the Jet Propulsion Lab given in this chapter. If you can't figure out the intended meaning in any passage, write a brief query to the author to try to determine the exact meaning. Keep queries to a minimum. When you are finished, discuss the changes you have made with a small group of your classmates or with the class as a whole.

EXERCISE **10.3**

Recommendations for Further Reading

Barry, John A. *Technobabble.* Cambridge, MA: MIT Press, 1991.

Buehler, Mary Fran. "Defining Terms in Technical Editing: The Levels of Edit as a Model." *Technical Communication* 28.4 (1981): 10–15.

CBE Style Manual Committee. *Scientific Style and Format: The CBE Manual for Authors, Editors, and Publishers.* 6th Edition. Bethesda, MD: Council of Biology Editors, 1994.

The Chicago Manual of Style. 13th Edition. Chicago: The University of Chicago Press, 1982.

Coggin, William, and Lynette Porter. *Editing for the Technical Professions.* New York: Macmillan, 1993.

Introduction: What's missing these Days

In the study of wolves and their habitats, what's missing these days is a objctive study that don't hamper the bias of the author but just gives the facts. In Yellowstone, some people say, there wouldn't be any wolves except that the people have brought them in from Canada. Nobody can tell yet whether the brought-in wolves are mating and reproducing (Jackson and Smith, 1997). The prupose of this report is to establish once and for all that the wolves in the regions surrounding Yellowstone Park are reproducing and so dispel that "significant doubt" mentioned by Jackson and Smith (Jackson and Smith, 1997, page 5).

Problems in the Literature

Jackson and Smith's (1997) study is clearly wrong and based on data that has been figeted. Observations were conducted over the same territory he studied and found that reproduction evidence coefficients were greatly enhanced over the data reported in their work. Their population counts amount to no more than 175 animals in the district of study, however better estimates are that as many as 350 woves roam the prairies and woods of the Yellowstone. How can we explain this discrepancy.

The reason is simple. Jackson and Smith's (1997) study was funded by the Conservation Group who's main purpose was just to put the local ranchers worries to bed about too many wolves in the region. This is no way to do scientific range management. . . .

SAMPLE 10.1   Excerpt from a fictitious technical report in need of editing

Gorrell, Donna. *A Writer's Handbook from A to Z.* Boston: Allyn and Bacon, 1994.

Nelson, Vee. "Sweat the Small Stuff—Editing for Consistency." *Techniques for Technical Communicators.* Carol M. Barnum and Saul Carliner, eds. New York: Macmillan, 1993: 291–304.

Publication Manual of the American Psychological Association. 4th Edition. Washington, DC: American Psychological Association, 1994.

Rook, Fern. "Remembering the Details—Matters of Grammar and Style." *Techniques for Technical Communicators.* Carol M. Barnum and Saul Carliner, eds. New York: Macmillan, 1993: 274–290.

Rubens, Philip. *Science and Technical Writing: A Manual of Style.* New York: Holt, 1992.

Rude, Carolyn. *Technical Editing.* 2nd edition. Boston: Allyn and Bacon, 1998.

Strunk, William, and E. B. White. *The Elements of Style.* 3rd Edition. New York: Macmillan, 1979.

U.S. Government Printing Office Style Manual. Washington, DC: U.S. Government Printing Office, 1984.

Van Buren, Robert, and Mary Fran Buehler. *The Levels of Edit.* 2nd Edition. Pasadena, CA: Jet Propulsion Laboratory, 1980. Available in reprint from Society for Technical Communication.

Williams, Joseph. *Style: Toward Clarity and Grace.* Chicago: The University of Chicago Press, 1990.

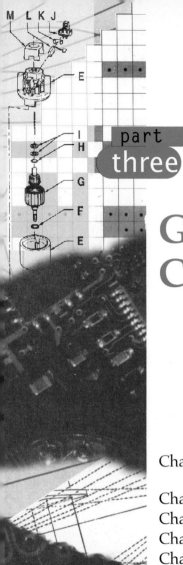

Genres of Technical Communication

▼▼

Our learning suffers from disconnected knowledge—
knowledge disconnected from purposes, models, structure,
or argument.

—*D. N. Perkins,* Knowledge as Design

▼▼▼▼▼▼▼▼▼▼▼▼▼▼▼▼▼▼▼▼▼▼▼▼▼▼▼▼▼▼▼▼▼▼▼▼▼▼

A document is rhetorically effective when its readers are
persuaded that they share a community of logical and affec-
tual understanding with the writer; the document communicates its ability to promote the "community" of the author
and the audience.

—*Scott P. Sanders*

▼▼▼▼▼▼▼▼▼▼▼▼▼▼▼▼▼▼▼▼▼▼▼▼▼▼▼▼▼▼▼▼▼▼▼▼▼▼

Today managers need to know more than what someone is
doing or will be doing. They need to know the value of the
work. They need more than facts; they need assessments,
speculations, and judgments to help guide them in their
decision making.

—*Christine Barabas*

▼▼▼▼▼▼▼▼▼▼▼▼▼▼▼▼▼▼▼▼▼▼▼▼▼▼▼▼▼▼▼▼▼▼▼▼

Look upon the user manual as the document which substitutes for your own physical presence, providing everything
in the way of guidance and assistance that you . . . would
provide . . . if you were there.

—*William Skees*

chapter

11

Correspondence:
Making Connections
in Discourse Communities

CHAPTER OBJECTIVES
▼▼▼▼▼▼▼▼▼▼▼▼▼▼▼▼▼▼▼▼▼▼▼▼▼

After you have worked through this chapter,
you should be able to do the following:

- Recognize the importance (and necessity)
  of creating and maintaining strong
  relations within discourse communities
- Recognize and use the main forms of
  technical correspondence—the letter and
  the memorandum
- Recognize the challenges and opportunities
  created by special forms of electronic
  communication—e-mail, electronic
  conferencing, and hypertext groupware—
  for making connections in discourse
  communities
- Explore the use of electronic
  communication for technical
  correspondence, document review,
  and team authorship

I n this chapter we will face squarely an idea that has been implicit throughout the other chapters: Technical communication is an intensely social affair—a public enterprise, a communal interaction—to which individual authors must make constant adjustments.

This chapter deals with correspondence—forms of written discourse such as letters, memos, and e-mail, used to make connections and form partnerships with other people. Letters and memos offer an excellent context for examining close up the kinds of adjustments authors must make to write effectively for different audiences.

Like the core documents introduced in Chapter 5, letters, memos, and e-mail messages usually appear in a compact form that does not require authors to go through the full process of the CORE method described in detail in Part Two. Indeed in industry, core documents such as position papers usually appear first in memos or letters. Because of this connection between core documents and correspondence, and because correspondence often plays an important role in introducing and shaping longer, more complex documents, we are introducing our section on the genres of technical communication with this chapter on correspondence. Many teachers find it useful to demonstrate audience sensitivity and adaptation by teaching correspondence first. Letters and memos are clearly situated within discourse communities, so working with them helps you understand explicitly the kinds of basic social connections that documents in other genres may seem to take for granted.

We begin by sketching the typical rhetorical situation in which technical correspondence arises. Then we show how authors write letters and memos to build and maintain discourse communities, how other genres are represented in letters and memos, and how special genres, such as job application letters and resumes, arise from special situations. We do not offer an exhaustive treatment of how to write letters and memos in business and industry; rather, we provide brief guidelines for writing the typical kinds of correspondence produced by technical professionals. We conclude with a section on electronic communication, adding guidelines for writing e-mail and participating in various kinds of computer-mediated group interchanges.

The Rhetorical Purpose of Technical Correspondence

Everyone lives and works within many *discourse communities,* groups of people distinguished by how they talk, write, listen, and read. Scholars define discourse communities *regionally* (Southern, Northeastern, Midwestern, and so on), *socially* by social class or ethnic origin, or *occupationally.* Doctors, lawyers, engineers, teachers, auto workers, and beauticians all have their special jargon and style of shoptalk. In actual practice, however, every individual represents a unique blend of the influences from many such communities.

The number of distinct discourse communities—and the complexity of an individual's personal style of discourse—tends to increase in technological cultures because of occupational specialization. If you are an engineer, your discourse differs from that of the other professions. If you are a mechanical engineer, your discourse differs from that of other types of engineers. If you are a mechanical engineer specializing in materials development, your language takes on yet another layer of distinction.

One of the great challenges of technical communication is to keep these distinctions from creating a Babel of misunderstanding. As a technical professional, you will have to cooperate and communicate both within your own discourse community and with people from other communities.

Technical Discourse versus Public Discourse

Fortunately, most of us have experience communicating across community boundaries. Human beings in the culturally diverse United States grow up negotiating differences among people. When you graduate from college, you will not only be an engineer, scientist, or educator. You will also be a member of many other social groups: a family, a social class, a club (or many clubs), a regional or ethnic group, and so on.

Even though these multiple memberships add additional layers of complexity to communication situations in the technological world, they also teach you linguistic flexibility. As you cross from one group to another, you learn to use the language of public interaction—the language of politics, the media, and the arts—all of which seek a general audience. This is also the style of discourse that you learned in your English, speech, and language arts classes. Your job as a technical communicator is to reconcile this public language with the technical language in your chosen field of knowledge. You *represent* that field to the world at large, which means that you must also *interpret* that field to others less familiar with it.

The Purpose of Letters and Memos: Reconciling Tensions between Local and Global Discourse Communities

In the world of business and industry you will need to develop your ability in public discourse to meet the needs of two types of discourse communities that exert simultaneous demands on your loyalty: a local discourse community and a global discourse community.

The *local discourse community* is your place of work—say an electronics plant where you serve as a materials specialist. The field of materials engineering itself is your *global discourse community,* with which you identify and share a special knowledge and language. You can assume that your audience has a good command of the same knowledge and language if you are writing up project results for a specialized journal or giving a presentation at a convention of other materials engineers. But in your local community you must communicate differently. You may be the only materials specialist in your division, so in effect you represent that global community to your fellow workers. You must interpret the results for them in language that they can grasp.

Your experience in college—learning your profession among like-minded students and professors in your field—does not really prepare you for the tensions that develop between local and global discourse communities. Your identification with the global discourse community must be solid enough for you to rank as a specialist in the field. But at work in your local discourse community you may find yourself among people who know nothing about the knowledge and specialized culture that you represent. However, they will depend on you to understand their needs and to supply technical information in a form that they can digest and put to work.

In essence, the global discourse community says, "Learn our language so that you may gain our knowledge." The local community says, "Speak our language so that you may join us in putting your knowledge to use." Letters and memos are among the primary tools that technical communicators use to reconcile these competing demands.

Letters and Memos: The Genres of Correspondence

The word *correspondence* is an interesting one. In technical communication it simply refers to the types of documents that are used to communicate directly to a single individual or to a closely associated group of individuals: letters and memoranda.

But the word implies something deeper. To *correspond* to something means to form a special relation to it, to match up, to create parallels, to cooperate with the other thing almost to the point of sharing an identity with it. A point on a map *corresponds* to an actual location in space. A word *corresponds* to its meaning. In this sense, to *correspond* means to develop a close mental connection. For me to start a correspondence with you—that is, to write you letters or memos on a regular basis—therefore implies that we have begun to strengthen our social connection, to build an identification with one another.

Looked at in this way, letters and memos are the tools of social exchange and community formation. If you neglect them or write them carelessly or thoughtlessly, you may miss an opportunity for an important cooperative venture with another person or organization. Worse, you might insult or anger the other person and engender a grudge that could hurt you later on.

Direct Address: The Personal Quality of Correspondence

If anything distinguishes correspondence from other forms of technical communication, it is the style of *direct address.* Letters and memos began as substitutes for one-on-one conversation. In these forms, *I* (the author) address *you* (the reader) directly, and I assume (or try to cultivate) a mutual interest in the information that I present. Often, I actually know (or want to know) the person to whom I am writing.

Although letters and memos do their work at the personal level, they retain a level of formality often missing in modern conversation. When you write, you must watch your manners closely and avoid becoming what the Europeans call "overly familiar" or presumptuous. The goal is to maintain a respectable distance without seeming unconcerned or cold. Unless you know your correspondents very well, you probably don't want to ask about their health or their kids, even though you address them as "you" and show a clear knowledge of their professional lives. It is important to remember that these "personal" communications play a serious role in the world of business, government, and research. Letters often become official documents that illustrate commitments and serve as informal contracts.

Sample 11.1 illustrates a letter that successfully blends a direct form of address with a tone of cordial formality. It also shows the typical parts of the letter format.

Letters versus Memos

The difference between a letter and a memo has to do with both function and format. The functional difference is a simple one: *Use a letter for someone outside your organization. Use a memo for someone inside it.* We write memos to our department chairman and fellow professors. We write letters to professors at other colleges, to publishers, to scholars in other countries, and so on.

Compare the letter in Sample 11.1 with the memo in Sample 11.2. The major difference is that the letter contains more information about the sender and receiver—a return address and an inside address, for example—which the memo simplifies into TO and FROM lines. If the correspondents are in the same company or organization, they have the same address. Note that even though the

The letterhead contains the return address. Authors who do not have access to letterhead put their return address flush left at the top of the page just above the date. In this style of business letter—called the "block style"—all items are single-spaced and flush left with no indentions.

The inside address follows the date and precedes the greeting. The greeting is followed by a colon (not a comma or semicolon).

The first paragraph establishes the context of the letter by referring to previous correspondence and giving information that sets the stage for the rest of the letter.

Paragraphs are separated by an extra space.

The second and third paragraphs answer specific questions raised by the addressee's previous letter. Notice that enough information appears to remind the addressee of her questions. Correspondents often write many such letters, so they find reminders like this helpful.

The style of the letter is conversational. Notice the use of direct address ("you" and "I") and the representative "we," showing that the author is speaking not just for himself but for the organization as a whole. The tone of the letter is polite and direct. The author tries to be encouraging and helpful, but does not promise what he cannot deliver.

The signature line gives the author's name and title.

Brookline Research, Inc.
54 4th Avenue
Greenville, Mississippi 54750

April 2, 1995

Mary E. Quinlin
Department of Materials Engineering
Tennessee Polytechnical College
Shiloh, Tennessee 38888

Dear Professor Quinlin:

Thank you for your inquiry about our work in polymer engineering. Indeed, we have had good success in our recent applications of polymerized materials in the construction of river levees. The material has proven capable of withstanding strong fluid stresses and has offered a nontoxic and therefore environmentally safe alternative to the older forms of concrete and clay.

We would welcome a visit from you and your graduate students in materials engineering any time this summer as long as we have a couple of weeks' notice. We also require that all visitors sign a nondisclosure agreement, since we have several patents pending on our research in progress.

I'm sorry that we do not have any positions available for interns at the moment. However, opportunities could well arise in the near future, so we welcome the chance to get to know you and your students.

Again, thank you for your inquiry. We hope to see you this summer.

Sincerely,

Robert Jones
Director of Materials Research and
Development

SAMPLE 11.1 Letter

Brookline Research, Inc.
54 4th Avenue
Greenville, Mississippi 54750

MEMORANDUM

June 2, 1995

TO: Lab Supervisors
FROM: R. Jones, Director of Materials R/D
SUBJECT: Visit from Tennessee Polytech Students

Next Monday (June 7), a group of five graduate students and
their professor, Mary Quinlin, will be visiting from the
materials engineering program at Tennessee Polytechnic
College. I will be bringing the group to each of your labs and
turning them over to you for about 45 minutes. I will brief
them on general safety and policy issues and provide them
with lab coats and goggles. Here's what I'd like for you to do:

1. Cover any special safety procedures for your lab.
2. Give a brief overview (no more than 15 minutes) of your
operations and current research focus.
3. Introduce at least one of your best people and give the
students a chance to converse with him or her.
4. Take questions. The students will have signed a
nondisclosure paper, so use your own discretion about how
much to say on any topic.

There's a chance that we may recruit some of these students,
so put on a good show and listen for the best questions.

Thanks for your cooperation.

Authors of memos often use their company letterhead. Some organizations provide an abbreviated letterhead for internal memoranda.

The heading MEMORANDUM appears in all caps, followed by the date. An alternative format adds a line for DATE below the SUBJECT line.

The TO line designates the audience. The FROM line gives the author's name and title. The author initials the memo on this line rather than signing at the bottom. The SUBJECT line gives the topic of the letter as specifically as possible. Follow the rules for good titles in writing subject lines.

The first paragraph briefly gives the background and purpose of the memo.

As in the letter, the memo is single-spaced with an extra space between paragraphs. The paragraphs are in block style with no indentions.

The instructions are broken out in list format to ensure easy reading and to bring attention to these important items.

The style is conversational and perhaps a bit more informal than the letter in Sample 11.1. But it is hardly "familiar." It remains an official communication of a supervisor to a group of employees and, as such, represents the official policies and procedures of the company.

SAMPLE 11.2   Memorandum

writer actually knows the people he addresses in the memo, he maintains a degree of formality in his style. Notice also that memos have no closing signature.
If a memo is drafted or typed by an assistant or colleague (a common practice
in the work world), the author acknowledges authorship by placing his or her
initials after the name in the FROM line.

GUIDELINES AT A GLANCE 32
Writing letters and memos

- *Make connections.* If you're writing a letter, try to use a real name in the greeting. Use constructs such as "To whom it may concern" or "To the Director of Personnel" only if no name is available, and be sure that your choice encompasses both genders. Address the reader directly ("I" to "you"). Respond directly to the needs of the reader and to questions or problems he or she expressed in previous correspondence.

- *Follow the rule of firsts.* First positions—the first paragraph of the letter, the first line of each succeeding paragraph, the subject line of a memo—are positions of emphasis. Place the most important information there.

- *Emphasize actions.* Say clearly what you expect the reader to do with the information you provide.

- *Use a conversational style, but seek an appropriate level of formality.* Personal pronouns, short sentences, questions, and contractions are usually acceptable. However, remember that you represent a corporate position and do not speak only for yourself. Maintain a level of formality to match the seriousness of the information you present. Check carefully for spelling, grammatical, and mechanical errors. They can have a devastating effect on your credibility.

- *Use chunking and listing for emphasis and readability.* As in other forms of technical communication, large blocks of text will discourage the careful attention of busy, browsing readers.

- *Provide guideposts.* Use headings if necessary to guide the reader through a lengthy piece of correspondence. For letters and memos that run more than one page, place a heading flush left on the first line of the second page (and all pages thereafter) that briefly gives the author's name, the reader's name, the date, and the page number, something like this: "Smith to Jones, 10/14/95, page 2."

- *Be courteous.* Treat your correspondent with respect, no matter whether the reader is a superior, a client, a colleague, or an employee. Mutual respect is the key to successful cooperation at all levels.

For a summary of general concerns in writing letters and memos, refer to Guidelines at a Glance 32.

Reporting, Instructing, and Proposing in Technical Correspondence

Now that we've looked at the broad purpose that letters and memos serve in technical communication, we will attempt to classify some of the specific ways they do their job of linking discourse communities. Although the format, tone, and style of letters and memos are different from those of reports, manuals, and proposals, correspondence can fulfill the same functions as these three major genres of technical communication. We will examine each function in turn.

Correspondence That Reports

Technical professionals frequently use letters and memos to provide information that guides decisions and actions. As in more formal technical reports, this kind of correspondence usually addresses an audience of managers or other decision makers and their advisors.

Within your own organization you will write memos to report your own activities and advise your colleagues on courses of action. To people in other organizations—clients and those who inquire about your work—you will write letters that provide information and recommendations suited to their particular needs.

A typical kind of memo produced by technical communicators is the *monthly report* or *progress report*. It provides an account of how well you are progressing toward your goals, what unexpected obstacles you have encountered, and how you have solved problems.

Like any good report, such a memo should not be a simple diary or narrative of every action you've undertaken. Nothing is duller than a day-to-day list of someone else's routine activities—often referred to derisively as a "laundry list." Instead, your memo should be a report of your main accomplishments: what you have done to advance the interests of your organization and a statement of what you will need to accomplish ever bigger and better things. See Sample 11.3.

In addition to informing others of your progress and reminding your supervisor of your value, such reports can be used to compile an annual report for your company or project team. If your reports are well written, clear, and accurate, you can sometimes just transfer whole chunks of text into the annual report.

Monthly reports are similar to activities reports (see Chapter 12). When a correspondent calls on your expertise, however, your memo or letter will be more like a report on options, a feasibility study, or a recommendation report (again, see Chapter 12). Most likely, you can provide a brief *position paper* on the topic to satisfy the request. But some inquiries will demand longer responses. Don't let the popular definition of a memo as a quick note confuse you on this score. In many instances, technical memoranda run for many pages.[1] Of course, you should always try to be as concise as possible and to avoid dumping unselected and unrefined data on your correspondent.

---

1. For a sample of memos of various lengths written for a government bureau, see the famous (or infamous) *Pentagon Papers,* ed. Neil Sheehan and others (New York: Quadrangle, 1971).

MEMORANDUM

August 22, 1997

TO: Robert Jones, Director of Materials R/D
FROM: Jill Stevens, Polymer Lab Director
SUBJECT: Monthly Report, July 1997

During the month of July, the staff of the polymer lab solved the brittleness problem that had slowed our progress in June. We are now back on schedule toward meeting the goals set in Contract 48762 for the levee-reinforcement project. If no unforeseen obstacles arise, we should be able to pass our polymer-reinforced concrete (material #N22876) on to the assembly lab by early September.

Other accomplishments include:

- an enhancement of the smoothness properties of material #P78654, which should greatly aid the firm's medical initiative;

- a definitive solution of the problem with chemical stains on material #P89876, which has slowed the constructive materials project (Contract 27912) for several months now;

- an initial solution of the related stain problem for material #N00709, which is not due to assembly until October. We are about two months ahead of schedule on that project because it is so closely related to several others.

We continue to have trouble with our mixing system, which is taxed all the more by our current work on brittleness. We can probably stay on schedule without any new equipment, but a new centrifuge would speed our work.

In annual terms, we remain over budget by about $600 but have stayed within our monthly allowances for the first time this year. We should be able to make up this deficit in the next two months, since Dr. Singh has taken a leave without pay for two weeks and we did not get approval to replace him.

SAMPLE 11.3   Monthly report in memo format

Correspondence That Instructs

One of the most common kinds of internal correspondence is the *directive* or *instructional memo,* which usually informs employees or colleagues of new policies and changes in operating procedure.

GUIDELINES AT A GLANCE 33

How to write an instructional memo or letter

- Establish the problem (as it affects the reader) in the first paragraph.
- Put point sentences first in every subsequent paragraph.
- Clearly indicate actions to be taken, distinguishing *policies* (which say what the reader should do—and why) from *procedures* (which say how to do it).
- Be especially scrupulous about using the active voice and a direct form of address ("you") with imperatives in procedural instructions ("do this; then do that").
- Divide the prose into manageable chunks, using bulleted and numbered lists, headings, and even diagrams where appropriate.

If you want people to read and follow such instructions, it is important to design a memo that closely resembles a good manual. To accomplish this goal, use Guidelines at a Glance 33.

Sample 11.4 shows what can go wrong if these guidelines are not followed. Actions become lost in thick narrative paragraphs, and readability suffers. Sample 11.5 offers an improved version.

Correspondence That Proposes or Promotes

Authors of interoffice proposals often use a memo format, especially in response to a request to "put things in writing." As in all proposals, your audience has the power to grant funding or permission to carry out a project. To accomplish your purpose, you need to do the following:

- Analyze the problem or need.
- Provide a plan to solve the problem and a rationale for that plan.
- Discuss the management of the project.
- Account for costs and project needs (such as equipment, personnel, and time).
- State the benefits to the organization as a whole.

You can sometimes cut corners when discussing personnel matters, since your immediate audience will probably know the people involved in the project, but don't be too quick to assume such knowledge. Remember that your proposal is a "for the record" document that one audience may use to justify projects to a second audience about whom you know little or nothing. Your manager may use your memo to justify the project to upper management, for example; upper management may show it to investors; and so on. Providing more details than you need is better than providing too few.

MEMORANDUM

January 5, 1993

TO: All Employees
FROM: Jay Addington, Director of Computer Services
SUBJECT: Viruses

About a week ago, a secretary in C Department came to me with a problem. Her computer was slow when she started it in the morning, and she kept getting error messages that she had never seen before. I took the diagnostic program down to her office, and, sure enough, she had contracted the virus XYZ, which in its full form can destroy the data on an entire hard drive. This is the kind of thing all employees must watch for.

If this happens to you, or if you have trouble locating information on your hard drive, call Computer Services and we will install the diagnostic program. You can run it very simply by clicking on the word CLEAN in your utilities menu after it is installed, then selecting "scan" or "clean" from the dialog window. The computer takes it from there. The CLEAN program has been effective in taking care of all the viruses we've encountered.

Of course, you ought not to bring disks to work that you've used on your home computer or any other possibly infected computer unless you scan them for viruses before using any office software and especially before getting on the network. In fact, this is company policy.

Good luck!

SAMPLE 11.4   Poorly designed instructional memo

Sample 11.6 on page 316 is a memo proposal following up on a discussion about expanding a computer writing lab.

Sample 11.7 on page 319 is a student memo proposing a topic for a project. You may use this basic structure when requesting approval from your instructor for a topic or for an action plan (such as a group or project management plan).

Even for proposals that go outside your organization, correspondence can sometimes provide the best medium. Some funding agencies—especially foundations and corporate sources—often prefer letters to formal proposals, at least at the "preproposal" stage. These are essentially concept papers in letter format. Sample 11.8 on page 321 provides an example.

MEMORANDUM

January 5, 1993

TO: All Employees
FROM: Jay Addington, Director of Computer Services
SUBJECT: How to Prevent and Eliminate Computer Viruses

The virus XYZ has infected a number of our computers. In full swing, XYZ can destroy the data on an entire hard drive. Please follow closely the standard policies and procedures for dealing with viruses. Here is a brief outline of practices to follow.

Policy

You should not introduce disks into your office computer if you've used them on computers outside the office unless you check them for viruses first.

What To Watch For

Your computer may already be infected with a virus if you observe the following problems:

- The computer is unusually slow in running its normal programs.
- Unusual or frequent error messages appear on the screen.
- Data on your hard drive seem to have disappeared inexplicably.

If you encounter any of these problems, call Computer Services and have the CLEAN program installed.

Using the CLEAN Program

We recommend that you use the CLEAN program on a regular basis anyway, just to be sure. We can bring it to you and install it in about five minutes.

Once installed, the program is easy to run:

1. Click on the word CLEAN in your utilities menu.
2. Select "scan" from the dialog box.
3. If the dialog box indicates disk errors, select the "clean" option, and the program will remove the virus.
NOTE: You can select "clean" even before you "scan," but the program runs more slowly if you do. We recommend that you "scan" first.
4. Select "quit" and return to your regular operations.

Thank you for your attention to this important matter.

SAMPLE 11.5   Revised instructional memo

Longer memos can use double space or space and a half to improve legibility.

The memo is addressed primarily to the person who has the immediate decision-making power.

The subject line functions like a good title for a technical proposal. It is a bit long, but it has appropriate descriptive power.

The distribution list covers people who have an interest in the project.

The introduction provides the context for the decision, refers to initial agreements, and summarizes the overall proposal.

This section analyzes the problem.

Memorandum

22 November 1991

To: John Dinkel, Associate Provost for Computing

From: Jimmie Killingsworth, Writing Director, Department of English

Subject: **Proposal for Expanding the English Department's Computer Writing Center into a Full Macintosh Classroom with Capabilities for Advanced Teaching and Research**

Distribution: Larry Mitchell, Head, Department of English
                    Anthony Aristar
                    Valerie Balester
                    Paul Taylor

At our recent meeting, Anthony Aristar, Paul Taylor, and I discussed with you the possibility of improving our computing facilities in the writing and linguistics programs. We landed upon a short-term plan to expand our 15-station computer writing center so that it can serve a dual function—as both an open-use writing center, designed to help students from all disciplines produce and improve their written communication, and as a computer classroom that will assist the English department in enhancing research and development in such fields as computer-assisted classroom management, information storage and retrieval, and computational linguistics.

The result of the expansion would be a networked Macintosh classroom with 26 stations. Ten of these would be relatively high-grade machines to allow smaller groups (such as graduate seminars) to pursue more advanced work. One would be a very high-grade machine, to be used by advanced graduates for software development. The Mac classroom, like the writing center itself, would be open for use by classes from all disciplines, though English would have the "inside track" for class scheduling each semester.

Following are some of the details of our plan, which we present for your approval. We hope the project can be funded as soon as possible, though we understand that summertime is probably the earliest the new machines can be delivered.

The Need

With the arrival in the Department of English of two new professors specializing in computer applications to the study of language (Aristar and Taylor), our need for computing facilities has intensified. Our current writing center, with 15 networked Macintosh Classics, served by yet another Classic, is inadequate not only for advanced work (such as computational linguistics, the software for which cannot be supported) but also for current efforts to experiment with synchronous conferencing in networked groups of students. The weak server causes the student's computer-mediated interactions to proceed at a snail's pace and also creates a distorted picture of what is possible in such an environment. The pace frustrates students and teachers, and the distortion makes good research impossible.

Even to use the facility for courses, we must now limit each section to 15 students. In these tough budget times, we cannot afford too many small classes, so we cannot achieve optimum use of the facility.

SAMPLE 11.6   Interoffice proposal in memo format

The Proposal

We propose to expand the center so that a full-sized undergraduate writing class can be accommodated. The required 25 seats will be available if we add 10 new, more powerful computers—Mac IIs instead of Classics. The required power for the network can be achieved with a yet more powerful server. The equipment listed below will give us the environment we need to serve undergraduates effectively and to carry out advanced study in linguistics and writing.

The English department will supply the space and the staff for the expanded center. Dr. Mitchell, the Department Head, has assured me that we will have the support needed for room renovations and staff adjustments.

Hardware Needs

   10   Mac IIsi
          8 MB RAM
          80 MB hard disk
          High-resolution color monitor
          Ethernet card and cabling

    1   Mac IIfx
          12 MB RAM
          300 MB hard disk
          High-resolution color monitor
          Ethernet card and cabling

    1   Mac Quadra 900 (server)
          12 MB RAM
          300 MB hard disk
          Monochrome monitor
          Ethernet cabling (Card comes with machine)

The ten Mac IIsi workstations will provide the computing power necessary to run sophisticated applications in rhetoric, linguistics, and technical writing. In order to run memory-intensive languages like LISP, a minimum of 8 MB of memory is required. MacLisp—the programming environment we will be using in this instance—requires itself a minimum of 5.5 MB of RAM to run adequately. The single Mac IIfx is intended to provide one powerful computer which advanced students can use, and where more demanding programs, such as text-oriented applications, can be developed. Since it will be the only machine available to such students, and language-oriented—and especially phonetic-oriented—software requires large amounts of available storage, it needs considerable disk space. We chose a Mac Quadra to function as our server, for it is approximately twice as fast as a Mac IIfx but is only slightly more expensive. The 900 series is preferable to the 700, for as a server this machine must be adequately expandable, and the 700 series is limited. Such a machine would serve our needs for the foreseeable future.

Some of the computer programs we plan to use require a network; several other applications can run on individual machines, but we will load them on the network server simply because there won't be room on the local hard disks. Since large programs and large databases must be accessed simultaneously by multiple users, simple

This section makes the pitch for the new equipment as the solution to the problem, provides assurances on issues of personnel and facilities, and gives a detailed list of the equipment needed. Costs are not provided because they are unknown. Retail costs are irrelevant because the group that provides equipment in this organization saves money by buying in bulk and at special educational rates. In short, the audience will know more about costs than the authors of this memo proposal.

This paragraph provides a rationale for the choices of equipment requested.

SAMPLE 11.6 continued

PhoneNet cabling will not provide an adequate response time. Even a relatively small program, such as a word processor, requires more than five minutes to load when a dozen people start it at the same time across our current PhoneNet network. Consequently, the workstations will need Ethernet cards and cables for the students to work together during class. For the same reasons, the network server will need a large hard disk (for programs and shared data) and extra RAM (for disk caching to improve response time).

Some additional equipment would benefit the classroom but is not as crucial as that listed above. For example, replacing one or two of the color monitors with dual-page monochrome monitors would greatly help students in technical editing and document design courses. A laser printer would enable students in those same courses to evaluate printed output at the time the document is completed, rather than having to print outside of class and return later with the hard copy; a printer in the classroom would also relieve the strain on the laser printers in Room 133, which sometimes run more than an hour behind. A CD-ROM player would make available to students a great deal of information which it is simply not feasible to keep on a hard drive.

Software Needs

> Macintosh Common LISP
> MacProlog
> Think C

These are basic programming languages and are intended simply to provide the environment in which pedagogical work can proceed. Much language-oriented software is written in these three languages, and it is often available free. Having these languages available therefore saves considerable incremental investment. Further, these languages are the environment in which students can and should be educated if they are to become effective professionals.

Benefits

With the new computers in place, we can continue experiments now under way with classes like English 104 (freshman composition), 203 (introduction to literature), 320 (technical editing), and 341 (advanced composition).

Even more significantly, we can begin to use the facility as a once-a-week lab in courses like 301 and 210, our technical writing classes, for which, as you know, there is more demand than perhaps any other course on campus. We think that our technical writing program will benefit greatly if we have a good computer environment to make clear the obvious connections between technical discourse and computer technology. With more powerful computers, we will be able to do more in the area of document design and production than we are now capable of handling.

In addition, with the new Mac IIs, we can do research and development with smaller, more advanced groups of students. We can develop, for example, graduate seminars in computational linguistics and communication in alternative media (hypertext and other such systems).

In short, we will be making our first big step toward the development of our Center for Innovative Technology in English (CITE), the white paper for which we shared with you in our meeting.

Please let me know if you need further information or help with carrying this project forward. Thank you very much for your cooperation.

The benefits section focuses on the advantages for students, who ought always to be the number one consideration in university decisions.

SAMPLE 11.6 continued

MEMORANDUM

Date: April 7, 1998
To: Dr. Jackie Palmer, Project Supervisor
From: Group B Production Team:
          B. J. Egenes
          E. K. Harrigan
          B. R. Hignight
          A. N. Jones
Subject: Approval for reference manual topic

We would like your approval for our reference manual topic, how to buy a diamond engagement ring. This is a topic that will benefit our audience, the students of the English 210-524 class, since many, if not most of them, may one day face this task.

In this memo, we list problems we may address in the manual and propose an approach for researching our audience. We also briefly discuss tasks the project may require and request specific feedback from you, our project supervisor. We close with a list of sample sources we have found for locating content information for our project.

Manual Topics

Buying an engagement ring is a big event in a couple's life. To buy a quality ring, the couple needs to understand diamonds and how they are priced. We plan to prepare and help class members find the best deals for their money and their needs. We will include the following sections, unless our audience survey finds they are not needed:
- Recognizing diamond quality
- Getting the best deal
- Differentiating between diamonds and imitations
- Judging craftsmanship
- Financing the purchase

Audience Survey

Our audience will be the students of the English 210–524 class. To give the students a useful product, we plan on conducting a survey to discern information they need or want to know. Using the survey results, we will be able to research specific areas that will most benefit our audience. Before we administer the survey, we would like for you to review it for form and content.

Project Requirements

The largest cost of our project will be the time spent on research, both in the library and in conducting interviews. There will also be time spent typing, editing, and illustrating the final project. Our project needs include finding time to set up interviews with jewelers and developing, administering, and analyzing a survey for our audience.

continued

SAMPLE 11.7   Student memorandum proposing a project topic for instructor approval

Manager's Actions

We request that you act as manager for our group. Specific actions that we request include topic approval and suggested revisions for each of the following project components: survey, management plan, audience-action analysis, core document, oral presentation, and final product.

Supporting Research

We include the following APA-style annotated bibliography as evidence of our ability to find sufficient resources to complete this project.

Diamond information center. (1997, December). [Online].
Available: http://www.adiamondisforever.com/.
> This web page analyzes each type of diamond cut based on shape, weight, color, clarity, depth, table, and price.

The diamond source. (1998, March 26). [Online].
Available: http://www.diamondsource.com.
> This web site contains basic information on ordering an engagement ring, such as how much to spend, what kind of diamond to buy, when to buy a diamond, and why select a diamond over other precious stones.

Newman, R. (1995). *The diamond ring buying guide.* Los Angeles, CA: International Jewelry Publications.
> This comprehensive book describes how to determine the value of a ring as well how to recognize rip offs. It also explains the clarity, color, carat weight, and cut of a diamond.

Stone, C. Personal interview with owner of House of Big Rocks. Bryan, TX.
> We plan to interview Mr. Stone to get information from a professional in the field. In this way we can hear first hand what a jeweler looks for in a diamond, what pricing he thinks is reasonable, and what consumer questions he considers typical.

The wedding. (1988, February 14). *New York Times,* pp. 65–72.
> This article, a special section on weddings and the changes society has made to adjust to more modern times, states that one unchanging thing about weddings is the solitaire diamond engagement ring. The article explains the tradition of the solitaire and the increasing trend of women shopping for their own rings.

Thank you for considering our topic for approval. If you have any questions or require further clarification on any of the above issues, please contact B. J. Egenes, our project team leader.

Immediate Action Requested: Topic approval and consent to act as our project team manager.

SAMPLE 11.7 continued

Department of Biology
Texas A&M University
College Station, TX 77843
(409) 845-5555

October 10, 1992

Jane Sims, Research and Development Advisor
Public Relations
Apple Computer Company
Dallas, TX 77777

Dear Dr. Sims:

I enjoyed talking with you at the recent research and development workshop in Dallas. I am writing, as you advised, to present a brief summary of the Texas A&M project to develop software for the scanning of ocean floors using video technology and Macintosh computers. By donating or lending computer hardware, Apple will potentially open up new markets of computer users in oceanographic biology and environmental impact assessment.

The first paragraph establishes the context by reminding the audience of the initial contact with the author.

Current techniques of surveying marine life are expensive, time-consuming, and unreliable. The only method now widely practiced is a time-consuming manual counting process. A team of divers lays a grid over the portion of the sea floor under investigation and then, by hand, takes an inventory of the organisms present within each section of the grid. The process is obviously slow, expensive, occasionally dangerous, and often open to inaccuracies and inconsistencies.

Our system uses a VHS video recorder in conjunction with a Macintosh IIsi microcomputer equipped with a video capture card to record, store, and statistically analyze data. Using this system, scuba divers do not have to spend long hours beneath the surface manually counting and recording specimens. They simply take video images of the sea bottom. Once the images are digitized, the computer saves even more time and energy by automatically analyzing the data. The method is faster, less expensive, more reliable, and less likely to disturb marine organisms and environments.

To refine the technique, we need more powerful hardware. At present, we have exhausted the capability of the Mac IIsi. Thus we are requesting the donation or loan of a Macintosh Quadra 950 microcomputer. With this additional power, we believe we can perfect the method and promote its wide use in the field.

The image-analysis approach is likely to become the method of choice in several areas of scientific and technological investigation. Its promise for environmental impact analysis

SAMPLE 11.8 Preproposal in letter format

continued

is particularly strong. A measure of its promise is the interest that Mobil Oil Company has shown in the project. The company has welcomed us to use its oil platforms as a testing ground and plans to use the information gathered to determine the environmental impact of oil spills.

By investing in the project, Apple can make a major contribution to a methodological breakthrough that not only benefits science and industry, but also opens the way for new applications of high technology in the booming field of applied ecology.

Thank you for the opportunity to explain our project. We would be happy to provide further details or a full proposal. I hope you agree that the project represents a potentially lucrative investment for Apple.

Sincerely yours,

M. K. Bell
Research Assistant

The last paragraph creates the opportunity for further communication.

SAMPLE 11.8 continued

Gaining Entry to a Local Discourse Community: Letters of Application and Resumes

Now we turn to one final function for the letter form: seeking a place in a specialized discourse community—or applying for a job. We will also discuss the specialized form of report known as a resume.

Letters of Application

A *letter of application* is, in many senses, a proposal. It considers the needs of the reader and offers to fulfill those needs; as in the proposal, the author of the document is directly involved in the proposed outcome. The letter of application says, "You appear to have a need, and I have the ability to meet that need. Let me persuade you that I can help." The letter offers selected support for your argument; it is usually accompanied by a *resume* which presents a more complete personal profile and confirms the assertions of the letter.

The key to writing a good application is to think of yourself as the missing element in the organization's context of production. Without overstating your

GUIDELINES AT A GLANCE 34
Procedure for developing application letters

1. Before you begin, review the job announcement thoroughly, and use the reference resources in the library, as well as your personal contacts, to find out as much as you can about the organization to which you are applying.
2. Review your own credentials (perhaps by preparing a draft resume as described later in this section). How do your interests dovetail with those of the company? How can you help to enhance their productivity?
3. Draft the first paragraph, referring to the job announcement or position advertisement and summarizing your credentials in a sentence or two. Find a good way to say briefly how your interests coincide with the company's current needs.
4. Draft a second paragraph that focuses on your accomplishments, selecting those that fit the special requirements of the job. If you have secondary interests that also apply, add another paragraph.
5. Draft the final paragraph, summing up your theme of how you believe you can make a solid contribution to the organization and thanking the reader for his or her attention to your application. At this point, you should also refer directly to the enclosed resume and, if necessary, say when you are available for an interview and how to get in touch with you.

credentials you must convince your prospective employers that, with your abilities, you can fill a gap in the productive process and improve output.

There are many possible strategies for drafting letters of application. You can use and adapt the procedure given in Guidelines at a Glance 34 for many different audiences. Sample 11.9 on pages 324–325 and 11.10 on page 326 follow these guidelines. Sample 11.9 is a letter responding to an announcement requesting instructors and assistants for a summer institute in biotechnology, while Sample 11.10 is a letter applying for a full-time engineering position with a city sanitation and safety department. Let's use Guidelines at a Glance 34 to trace the development of the letter of application for the summer institute (Sample 11.9).

The author is responding to the following position announcement:

Instructors and Assistants Needed
for Summer Teachers Institute in Biotechnology

The Department of Educational Curriculum and Instruction at Grundy Polytechnic University seeks several instructors and assistants to work with elementary school teachers in an intensive four-week summer institute in biotechnology. Institute participants will be first, second, or third-year teachers interested in incorporating biotechnology topics into their teaching programs. Instructors will assist university professors in adapting presentations for the teachers and will help develop classroom activities. Assistants will help prepare for and clean up after laboratory exercises. Minimum educational levels: Instructor Position, Bachelor's Degree; Assistant Position,

April 17, 1998

Myrth Ileana
24 Salem St.
Winston, NC 88888

Dr. Dolly Darling
Director, Summer Biotechnology Institute
Department of Educational Curriculum and Instruction (EDCI)
College of Education
Grundy Polytechnic University
Midwest, Ohio 33333

Dear Dr. Darling:

In response to EDCI's advertisement posted on the departmental bulletin board, I am interested in the instructor position for the July Summer Biotechnology Institute. I feel I would bring to the Institute the following strengths:

- seven years elementary and secondary teaching experience,
- experience as a teacher liaison for a series of cross-disciplinary teacher seminars,
- experience working with teachers in professional development,
- experience in developing curriculum for elementary mathematics and science education teachers and students, and
- a history of working collaboratively across colleges and disciplines.

After receiving my bachelors degree in biology in 1989, I taught elementary science education for three years and secondary science education for four. During part of this time, I worked with Project 30, a national cross-disciplinary effort promoting collaboration between faculty of the Colleges of Arts and Sciences and Education. As teacher liaison for the Project 30 Steering Committee, I helped plan and deliver a series of four after-school seminars on different themes. For each theme, university geology, geography, biology, chemistry, and physics professors demonstrated its importance to their respective disciplines.

From the classroom, I moved to a position with a regional educational laboratory. In this capacity, I conducted research, provided technical assistance, developed materials, and collaboratively designed and delivered state, regional, and national conferences, forums, and workshops to engage a wide range of constituents in discussions related to improving teaching and learning. With a strong institutional focus on assisting at-risk students, we carefully and deliberately integrated multicultural perspectives and approaches into every workshop, forum, or activity. We also developed curricular activities for both students and teachers and offered alternative approaches to assessing student understanding.

I am very interested in working with the Biotechnology Summer Institute this summer. In assisting university professors teaching the courses and in helping teachers find ways to use the information in their classes, I will continue to explore

SAMPLE 11.9 Letter of application for summer instructor position alternatives to traditional educational structures that promote student understanding and support collaborative professional learning communities devoted to quality education for all.

My resume is attached. Please feel free to call or e-mail me to further discuss specific contributions I might make to the Institute. Thank you in advance for considering me for this position.

Sincerely,

Myrth Ileana, B. S., Biology

SAMPLE 11.9 continued

college senior. Preferred fields: Education, Biology, Chemistry, Pharmacy. Position requires 30 hours/week during the Institute, which runs July 6–31. Salary range $500–$1000/month. Send letter of application and resume (postmarked by April 15) to Summer Biotechnology Institute, EDCI, Grundy Polytechnic University, Midwest, Ohio 33333.

Before submitting her letter of application, the author looks up the telephone number of the department, calls to find the name of the person in charge of the Institute, and telephones her to find out exactly how many instructor positions are open and what specialty areas are important for those positions. She also reviews a university catalog found at the library and accesses the university's website for more information about the department.

Next, she updates her resume, a relatively simple task, since she keeps a current copy on her computer, adding publications and experiences as they occur. However, she rearranges sections within her resume, then adds a summary page highlighting skills, experiences, and achievements that the search committee might find of particular interest. This summary page goes at the front of the resume. The completed resume appears later in this chapter (Sample 11.12).

She then plans her letter of application. The intent is to convince the director of the Institute to hire her. However, to accomplish this, she must present information honestly, focusing always on the needs of the Institute. The tone should be respectful and mindful of the fact that the director may have dozens or even hundred of other letters to mull through. It is not enough to simply request that she be hired. Rather, as in all forms of technical communication, she must offer evidence to convince her audience that she is, in fact, the best person for the job.

Evidence supporting letters of application usually takes the form of skills, talents, experiences, or accomplishments, but may also include recommendations from respected people in the field of interest. *Skills* are normally acquired through training (e.g., typing, computer skills, laboratory skills, wood working skills, writing skills), while *talents* are considered inherent. However, there is

2443 West 2nd Street
Tartan, Utah 66822

23 February 1993

Chester Thomas, City Engineer
Department of Sanitation and Safety
P. O. Box 2222
Salt Lake City, Utah 65666

Dear Mr. Thomas:

I am writing to apply for the job of environmental engineer advertised in the January issue of *Urban Engineering*. I have served as Tartan's City Engineer ever since I took my degree in environmental engineering at the University of Utah three years ago. This experience, combined with my specialized study and year's work as a co-op student in your department, gives me the knowledge to make a strong contribution to Salt Lake City's engineering department.

In Tartan, I have been able to apply a number of the techniques I learned while working in Salt Lake City. I have used the microbial leaching techniques for the clean-up of lake margins and perennial wetlands and thus have been able to study on a smaller scale the same kind of work that has occupied the environmental wing of your department for nearly a decade now. As an undergraduate student, I studied with Dr. Richard King, who developed a number of the techniques, and recently I've been able to apply some innovations in my work at Tartan that have produced a number of good results. I would very much like to see if the same methods will work on the larger scale at Salt Lake City.

My work as City Engineer in Tartan has given me valuable insight into the overall operations of an engineering department, and it has allowed me to polish my ability as a report writer and technical advisor to city government. But I would prefer, before going higher in management, to do more work in my field of specialization. Environmental engineering has been my first commitment all along, and I would welcome the opportunity to go deeper into the field.

If you think, as I do, that my credentials and the current needs of your department make a good match, I would enjoy talking with you about the job. My address and phone numbers appear on the enclosed resume, which also contains further details about my education and professional career.

Thank you for your consideration of my application.

Sincerely,

Mary Beth Samsom
City Engineer, Tartan

enc.

SAMPLE 11.10   Letter of application for engineering position

GUIDELINES AT A GLANCE 35

Reviewing the application letter

- Does it project enthusiasm about the job prospect?
- Does it give a brief but full account of the applicant's relevant accomplishments?
- Does it say how the applicant expects to contribute to the productivity of the organization?
- Does it use direct address but maintain an appropriate level of formality?
- Does it observe the conventions of effective style and correct grammar?
- What more would you want to know about the applicant if you were the employer?
- Is anything included that you feel is unnecessary?

often little difference between skills and talents that have been developed or honed to a high level. *Experiences* may be indicated by educational level, courses taken, or jobs held (paying or non-paying). Volunteer service often indicates to a prospective employer not only experience in a given area, but high moral character (a personal commitment to pursue worthwhile unreimbursed activities). *Accomplishments* are formal acknowledgments by others of things you have done (such as publications, awards, high grades, or programs, if you competed for the privilege of participating in them).

The author then drafts her first paragraph, stating the position for which she is applying (since the Institute has more than one opening) and acknowledging that she is responding to a specific advertisement. (Some companies publish several versions of an ad, soliciting people from different backgrounds. It helps the reader to know from what discourse community you originate.) The author then establishes links between skills she possesses and the needs expressed in the ad (or discerned from her research). If she had been able to find out more about the specific job through telephone calls with contacts, this is where she would have put this information to use.

In the ensuing paragraphs, the author describes strengths and accomplishments that she feels make her especially qualified for the job. Notice that her use of technical educational terminology indicates her familiarity with current issues in the education field. She can safely assume that those wanting more detailed information can refer to her attached resume. However, the reverse is not true. She can not assume that the person reading her resume will also have read the cover letter. In many organizations, resumes are detached from all other materials submitted and are reviewed separately.

In her final paragraph, the author states how she can help strengthen the Institute. She provides contact information and closes with a reference to her attached resume.

You can use Guidelines at a Glance 35 to evaluate the effectiveness of letters of application.

GUIDELINES AT A GLANCE 36
Procedure for developing resumes

1. Decide upon the relevant categories of information, such as EDUCATION, WORK EXPERIENCE, AWARDS AND HONORS.
2. Type the categories into a computer file.
3. Begin to fill in short notes about items to include under each category.
4. Check the items for completeness, and add descriptions of positions where necessary.
5. Arrange the items for maximum effectiveness, usually in *reverse chronological order.*
6. Begin to experiment with effective page design, creating a document that allows browsing readers to locate quickly the categories of information that are most important to them.
7. Edit the language items in the resume for consistency and conciseness, watching especially for grammatical parallelism and other good practices of list-making and clearing away any unnecessary words. Check carefully for errors in grammar and style.
8. Polish the design, and produce the document with the best printer you have access to.
9. Show your resume to several friends, or to technical professionals if possible, and follow their suggestions for revisions. You will probably have to revise *several* times to get your resume into final shape.

Resumes

Writing a good resume can be a time-consuming task. Thousands of people make their living doing it for others. The resume is really a cross between a report and an advertisement. Like a report, it records factual data and accomplishments. But it also has to "sell" those achievements by using an accessible, eye-catching format that highlights important information. Though you must give a complete record of your major accomplishments, you must also find ways to distinguish yourself in what is likely to be a large pool of job candidates. In the context of use, resumes tend to run together in the minds of busy readers going over a huge stack of applications. Use the letter of application as a way to open the prospective employer's mind to your distinguishing qualifications; then use the resume to reinforce those distinctions and influence the final decision.

In Guidelines at a Glance 36 you'll find a procedure for developing effective resumes.

Your resume should be as short as possible, preferably a single page if you've had less than ten years of work experience. As you progress in your field, you will probably need to create a one-page *summary resume* to accompany a fuller listing of your accomplishments that runs for several pages.[2]

2. The longer professional profile is sometimes called a *vita,* or *curriculum vitae,* especially in academic circles (as the Latin suggests).

Sample 11.11 gives a completed resume designed to accompany the letter of application for the engineering position shown in Sample 11.10.

This resume takes a minimalist, no-nonsense approach. It presents the author's key qualifications, gives any distinctions she may have earned that make her different from other job candidates (these are emphasized in the accompanying letter), and drastically reduces or omits information about her personal history that is not directly relevant to the job search.

Here are a few points that distinguish this design from others you may have seen or want to create for yourself:

- It omits the "career objective" line that is often included at the top of a resume. Ms. Samson could have written something like "To secure employment as an urban environmental engineer." But this objective is obvious in the application letter and content of the resume.
- It omits a "personal data" section, which usually gives a few details of the person's background, such as marital and health status, hobbies or additional abilities, and even height and weight (if they are flattering). Critics of this approach think that this information is irrelevant and that presenting it borders on the unprofessional. Advocates see it as a way to show that you are a multifaceted individual, not just a work machine.
- It places the experience section before the education section. This is a common practice for people who have worked in at least one job after earning a college degree. If you are just getting out of school and have worked only part-time (or not at all) in your field of choice, you may want to put education first, since your schooling constitutes your leading qualification.
- It divides the page into three vertical columns. The design is joined at the top by a pyramid structure of first one central column (name and title), then a double column with home and work addresses. In the three-column section the headings, dates, and central items such as job titles and degrees get positions of privilege (with plenty of white space) in the first two columns. In the third (right-hand) column the type size is reduced for more detailed descriptions. The prospective employer can get a quick view of the candidate's work and education history by scanning the first two columns without having to dig out dates and positions from descriptive information.
- It uses no fancy fonts, type styles, or paper. Instead, it uses an attractive serif font (Times Roman) that allows the greatest number of words per page while remaining readable and clear.

This is certainly not the only possible design for a resume. The style and format may vary according to your background and field of interest.[3] Sample 11.12 (pages 331–334) is a longer resume designed to accompany the letter of application for biotechnology summer institute discussed earlier and shown in

---

3. For some good variations on resumes for a variety of occasions, see Tom Jackson's *The Perfect Resume* (Garden City, NJ: Doubleday, 1981) and Elizabeth Tebeaux's *Design of Business Communication* (New York: Macmillan, 1990).

Mary Beth Samson
Professional Engineer

2443 West 2nd Street
Tartan, Utah 66822
(101) 678–9012 (home, messages)

City Planning and Engineering Office
City Hall, Room 424
Tartan, UT 66821
(101) 677–8901 (voice) 699–2121 (fax)
e-mail: samson@citywide.gov

Experience

1991-present	City Engineer Tartan, Utah	Responsible for feasibility studies, proposals, and quality control of all projects involving city construction, safety projects, and sanitation. Chief accomplishments include new wetlands protection system, renovation of water treatment plant, and planning study for citywide recycling program.
1989–1991	Research Assistant, Department of Environmental Engineering, University of Utah	Worked with Dr. Richard King on projects for reclaiming wetlands with the use of microbial leaching systems. Co-authored proposals and reports to National Science Foundation. Assisted in the direction of undergraduate design projects.
1985–1990	Assistant City Engineer Salt Lake City, Utah	Worked several semesters in university co-op program. Participated in sanitation planning study, analyzed water samples, and wrote routine reports to City Council.
1982–1985	Various Student Jobs	Worked several jobs in Provo, Utah: part-time server at McDonald's, sales clerk at Dillards Department Store, and clerical assistant at Jones Engineering.

Education

1989–1991	M.S., University of Utah, Environmental Engineering	GPA 3.85. Thesis: "The Efficacy of *E. Vacini* 2010 for Use in Microbial Assisted Reclamation of Wetlands" (directed by Richard King)
1985–1989	B.S., University of Utah, Civil Engineering	Graduated with honors. GPA 3.34. Concentration in Environmental Engineering. Engineering Honors Society, Society of Women in Engineering.

Honors

1991	Society of Women in Engineering Award for Graduate Research
1985–1989	Society of Women in Engineering Fellowship for Undergraduate Education

References   *Available upon request*

SAMPLE 11.11   Resume for engineer (accompanies Sample 11.10)

Sample 11.9. This resume is preceded by a sheet summarizing skills, talents, accomplishments, and experiences that the author feels especially qualify her for this position. It also demonstrates how clip art can be used effectively in designing resumes. Publications in this sample use APA documentation style.

Your school's guidance and placement office may also keep a file of sample resumes that you can check for other possibilities. You can also find good examples—and bad ones—on the World Wide Web. To determine the potential effectiveness of sample letters and resumes, use the evaluation questions in Guidelines at a Glance 37. Whichever format style you select, remember that the most important thing is the content—your accomplishments. Write carefully and select the information with the idea that this document represents *you* as a technical professional.

MYRTH ILEANA, B. S., BIOLOGY

HIGHLIGHTS

245 E. 54th St.
New York, NY 10001   e-mail: whozit@aol.com
(212) 000-0000 (home)

EDUCATION

1986 - 1989   University of Missouri, Kansas City, MO
                    Bachelor of Science, Biology (B. S.), May, 1989

1985 - 1986   Metropolitan Junior College, Kansas City, MO

SUMMARY OF QUALIFICATIONS/EXPERIENCES

Teaching experience

  3 years elementary, grades 5 and 6; 4 years secondary, 7–8 life science

Planning and providing professional development and technical assistance

  2 years (curriculum development, assessment, standards-based mathematics and science education)

Publications

  2 refereed articles; 1 paper presented at a professional conference

Conferences and workshops coordinated and/or facilitated (5)

Honors

  Summer Teachers Honors Workshops: 1989—NASA, Johnson Space Center,
  Houston, TX, Space Science; 1990—Los Alamos National Labs, Los Alamos,
  NM, Chemistry; 1992—University of Colorado, Boulder, Radiation Biology;
  1993—University of New Mexico, Albuquerque, Astronomy

SAMPLE 11.12   Resume for instructor (accompanies Sample 11.9)

MYRTH ILEANA, B. S., BIOLOGY

CURRICULUM VITAE

245 E. 54th St.
New York, NY 10001   e-mail: whozit@aol.com
(212)000–0000 (home)

 PROFESSIONAL POSITIONS

1996 - Present   Eastern Educational Development Laboratory (EDL), NY, NY
                  Senior Training/Technical Assistance Associate

1992 - 1996   Santa Fe Public Schools, 610 Alta Vista, Santa Fe, NM
                  Teacher of Life Science

1989 - 1992   Kansas City, MO Public Schools, Kansas City, MO
                  Teacher, grades 5–6, Southwest Elementary School

 EDUCATION

1993, Summer   University of New Mexico, Albuquerque, NM

1992, Summer   University of Colorado, Boulder, CO

1986 - 1989   University of Missouri, Kansas Cit, MO
                Bachelor of Science, Biology (B. S.), May, 1989

1985 - 1986   Metropolitan Junior College, Kansas City, MO

 OTHER RELEVANT WORK EXPERIENCE

1994 - 1995   Project 30, University of New Mexico, Albuquerque, MO
                Teacher Liaison, College of Education Dean's Office

1991, Summer   Camp Adventure, University of Missouri at Kansas City, MO
                Environmental Education Instructor, Health & Kiniseology

 PUBLICATIONS

Articles

Ileana, M. (1992, December/January). The garbage game. *Science Scope*.

Ileana, M. (1988, December). Three decades of recombinant DNA. *The Science Teacher 52*(9):15–19.

Conference Proceedings

Killingsworth, M. J. and Ileana, M. (1992, February). How to save the Earth: The greening of instrumental discourse. In Oravec, C. L. and Cantrill, J. G. (Ed.), Proceedings of the Conference on the Discourse of Environmental Advocacy. Salt Lake City, UT: University of Utah Humanities Center.

SAMPLE 11.12 continued

MEETINGS COORDINATED AND/OR FACILITATED

Title	Locale	Date	Time	Attendees	Role
Alternative Assessment Follow-up Workshop	OK City, OK	4/21/97	6 hours	24 teacher trainers	Coordinator & co-facilitator
Promoting Instructional Coherence Case Study Design Advisory Committee meeting	Dallas, TX	4/11/97	6 hours	2 each: teachers, principals, district super-intendents	Joint coordinator & facilitator
Professional Development Awards Training-of-Trainers Winter Meeting	Austin, TX	2/2/97– 2/6/97	3 days	28 principals, 32 project directors, & staff	Joint coordinator & facilitator
Alternative Assessment Follow-up Workshop	Albuquerque, NM	11/21/96	6 hours	15 teacher trainers	Coordinator
Promoting Instructional Coherence Regional Conference	Albuquerque, NM	10/3/96– 10/4/96	1½ days	100 regional constituents	Primary coordinator & facilitator

 HONORS

Summer Teachers Honors Workshops (Selected Participant):

    1993 University of New Mexico, Albuquerque, Astronomy

    1992 University of Colorado, Boulder, Radiation Biology

    1990 Los Alamos National Labs, Los Alamos, NM, Chemistry

    1989 NASA, Johnson Space Center, Houston, TX, Space Science (NEWMAST)

State Teaching Certificates:

    Missouri, life; New Mexico, expires 1999

Dean's Honor List:

    University of Missouri, Kansas City, MO, Fall of 1987 and Fall of 1988

SAMPLE 11.12 continued

GUIDELINES AT A GLANCE 37

Reviewing the resume

- Does it give sufficient information about the candidate?
- Is it designed for legibility and effective emphasis of the most important information?
- Do any features of the design—headings, typography, placement of information—detract from the overall appearance or effectiveness of the document?
- Is the document well edited? Could the style be improved?
- What else would you like to see on the resume?
- Should anything be omitted?

 MEMBERSHIPS

Member, Association for the Education of Teachers in Science (AETS), 1996–Present

Member, National Science Education Leadership Association (NSELA), 1996–Present

Member, National Science Teachers Association (NSTA), 1988–Present

 SERVICE ACTIVITIES

Member, Dewey Elementary School Site-Based Management Team, New York, NY, 1996–Present

Co-Leader, Dewey Elementary Brownie Troop 516, New York, NY, 1996–Present

Sponsor, Science and Math Club, Southwest Elementary School., Santa Fe, NM, 1995–1996

Co-moderator of Student Council, Santa Fe Middle School, Santa Fe, NM, 1993–1994

Science Fair Judge:

   New Mexico State Science Fair, Santa Fe, NM, 1994, 1995, and 1996 (1996 Judging Committee Chairperson)

   Santa Fe Public School District, Santa Fe, NM, 1993 and 1994

   Southwest Elementary School Science Fair, Kansas City, MO, 1990

 SKILLS AND HOBBIES

Hiking, camping, swimming

Macintosh Microsoft Word, FileMaker Pro, PageMaker, Adobe Photoshop, ClarisWorks

SAMPLE 11.12 continued

Making Electronic Connections: E-Mail and Beyond

Throughout this book we have seen how electronic communication, particularly the use of the computer, has changed the context of production for technical communicators. Electronic communication gives you a number of options for correspondence. Of course, you can use the telephone—an instrument that, a generation ago, transformed the way people conducted business on a day-to-day basis. But today you can also use a computer network to send an e-mail message to an individual correspondent or post your information on an electronic bulletin board for broader dissemination.

E-Mail: A New Medium for Technical Correspondence

To understand fully the opportunities and constraints that the emerging media have introduced, it's a good idea to think about how electronic communication relates to older modes of communication such as personal conversations, telephone calls, and written correspondence. Consider the relative advantages of each mode.

During a personal conversation you can watch your correspondent's eyes and bodily movements, checking for signs of attention, understanding, and emotional responses. You can use gestures and alter your pitch and tone for emphasis. You can assume a mutual understanding of the background of the conversation and can condense your speech accordingly. If the correspondent doesn't understand, he or she can ask a question. You are right there to answer it.

What do you lose in a telephone call? Instead of having a physical being before you, you have only a voice transmitted electronically, so you miss gestures and body language. Still, you can convey information to your partner quickly in a condensed form because he or she can always stop you to ask a question. Again, you can vary your tone of voice and listen for oral cues from your partner in conversation. Moreover, you can increase your efficiency. You can sit at your desk and have conversations with hundreds of people without having to go out and meet each one.

With a letter, not only do you lose the correspondent's physical presence; you also lose the feedback provided by an immediate voice response. Now you must calculate more closely the effects of your words and information. You must analyze the knowledge that your audience is likely to possess, cover your topic's background extensively (or at least remind the reader of previous contacts with the information), and provide full and clearly stated reasons for all of your interpretations.

In other words, shifting to the written mode requires that you provide more *context* for the information you present. You imagine the kinds of questions your readers might ask and the uses to which they will put the information. Then you write in a way that anticipates their questions. The result is that you take longer to explain any point than you ever would in a conversation.

For example, in a face-to-face conversation with a colleague you might say, "I disagree. We need to keep the funding for labs and travel separate. You remember what happened to X Division, don't you?" If your correspondent still looks doubtful, you might add, "And the situations are closer than you may think!" On the telephone, the conversation would run about the same, except that you would have to interpret a nervous silence on the other end of the line rather than a doubtful expression on the face. But in a letter you would probably have to spell everything out much more fully. Your explanation might go something like this:

> I'll have to disagree with the notion that we can combine research and travel funds, which you suggested in last Monday's meeting. Even though research often depends on travel, the two categories should remain separate because the legislature funds them separately. To mix the two types of money would border too closely on misappropriation of funds. You may remember the charges against X Division, which arose under similar circumstances. They, too, wanted to mix funds from legislatively mandated categories (in their case, clerical salaries and equipment funds).

If letters require so much more time and energy, why do we still write them? Because written text continues to offer a number of advantages that are missing in telephone calls:

- *A letter allows you to state your full position on a topic without interruption.* This process can help you clarify your thoughts. It also offers your correspondent a chance to study your position carefully and then frame a response.
- *A letter creates a written record,* a "paper trail" to which you can refer later and that increases the legal power of the exchange. In some circles—in the international diplomatic corps, for example, or in organizations concerned with trademarks and patents—letters may even have the force of a contract. You can tape your conversations, of course, in an effort to achieve the same effects, but tape recording involves a number of legal and ethical uncertainties and usually must be accompanied by some written record (signed permissions or transcripts) before it has the same effect as written correspondence. A letter signifies a seriousness of purpose that is missing even in "on the record" conversations. A letter "puts it in writing."

Electronic mail, or *e-mail*—the transmission of messages between correspondents on a computer network—offers most of the advantages of the written medium while recovering some of the advantages of the phone call and the face-to-face conversation. It provides a written record and allows for reflective reading and writing. It also permits a speed of transmission that puts regular mail to shame—even expensive overnight mail.

Whether you are transmitting internal "memos" within your company using a local area network or sending "letters" halfway around the world via networks such as the Internet,[4] you can transmit in a matter of seconds detailed messages or even whole documents uploaded onto the network. Your correspondent can download and print the document. Then, almost instantaneously, your reader can send you a response, keeping the printed document or the e-mail file for a record of the transaction. This form of communication virtually amounts to a *written conversation.*

Our hypothetical conversation with a colleague might begin pretty much as it did in a memo, perhaps with a bit more informality, since most people do not see e-mail as having the formal power that a memo on official letterhead has:

> Your comment last Monday about combining research and travel funds worries me. Research and travel are related, true enough, but the two categories should remain separate because the legislature funds them separately. Could mixing the categories be interpreted as misappropriation of funds? Remember the charges against X Division, which arose under similar circumstances? They, too, wanted to mix funds from legislatively mandated categories (in their case, clerical salaries and equipment funds).

Your colleague might then respond:

> I understand your worries about mixing categories, but I don't think we have to be concerned about the plan I'm proposing. Where X Div. went wrong was messing with salaries. That category is sacrosanct to the state auditors. But there are precedents for mixing research and travel. Should I fax you the notes I have from the last audit?

You could come back, including a little of the context because you know that while he can read the response almost instantaneously, he may not have the opportunity to read and respond right away:

> Yes, I'd like to have a look at those notes on the last audit. They could be useful in the future. But what you've told me about the mixing of categories satisfies me that you're doing the right thing now.

Given the advantages of e-mail—the way it offers the advantages of writing letters and telephoning while minimizing the disadvantages of both—you may wonder why anyone would use the other media at all any more. But e-mail still has some problems, such as the following:

- *Access is not universal.* Your correspondent may not be connected to a network compatible with yours. You will feel seriously frustrated if your screen

---

4. On the variety of networks and bulletin boards and their respective histories, functions, clienteles, and advantages and disadvantages, see Howard Rheingold, *The Virtual Community: Homesteading on the Electronic Frontier* (Reading, MA: Addison-Wesley 1993).

displays "message returned" or "message undeliverable" after you've spent a long time carefully crafting a message, which you now must download and reformat to print as a letter.

- *Your correspondents may have divergent attitudes and expectations about e-mail.* Because it is a new medium, users are still a bit unsure about how much they can depend upon their correspondents to read e-mail messages and give them the same consideration they give to written correspondence. After all, conventions for writing and reading letters have evolved over several centuries, whereas e-mail has come into wide use only in the last decade or two. Some people treat e-mail like a friendly, informal conversation; others think of it as being more like a letter.

- *Some users clog transmission lines by using e-mail for publishing, creating a genre now widely known as "junk e-mail."* A message sent to more than a few correspondents may read like a form letter. Most readers expect e-mail, like letters, to be crafted for their particular needs. They expect e-mail to be addressed personally to them, not distributed by means of a long general mail list on which their address appears, perhaps without their knowledge or permission.[5]

Table 11.1 summarizes some of the advantages and disadvantages of e-mail compared to the other media for correspondence.

Some of the problems of e-mail could be prevented if authors treated the medium as a serious means of correspondence worthy of all the care and attention they give their regular letters. With this goal in mind, you may find it helpful to think of e-mail as an evolving process with a still fluid, but increasingly clear, set of conventions, which are summarized in Guidelines at a Glance 38.

Exploring and Supporting Electronic Discourse Communities: Mail Lists, Newsgroups, and Discussion Groups

Once you have an e-mail account and an address, you can do more than communicate with one person at a time. You can also set up or join mail lists and participate in other forms of group communication.

Mail lists, or bulletin boards, are groups of users who agree to share their e-mail addresses with one another. Once you have subscribed to a mail list—which you can usually do just by sending mail to the home address of the list—you receive every piece of mail posted to the list, and you can post to everyone else. The lists can include members of a department or a company, or they can provide a "meeting place" for a larger group from anywhere in the world. Mail lists can support either local or global discourse communities.

---

5. The "virtual journal," an innovative attempt to use e-mail networks as an avenue for serious publication, represents a possible exception to this critique. In this instance, subscribers are actually solicited and not merely placed on a mail list without their permission.

TABLE 11.1

Advantages and disadvantages of media for technical correspondence

Medium	Advantages	Disadvantages
face-to-face conversation	Partners can use body language, gestures, and variations in vocal delivery; feedback is immediate; they can condense speech and background information.	It leaves no written record, may be treated less seriously than written correspondence, and is subject to inaccuracies. It is also obviously limited by considerations of space and time.
telephone conversation	The advantages are similar to those of face-to-face conversation except for absence of body language and gestures; but telephoning is even faster and easier to enact on many occasions and across great distances.	It leaves no written record, is subject to inaccuracies, and may be taken even less seriously than face-to-face conversations; it is also subject to malfunctions of technology.
written letters or memos	Written correspondence offers an opportunity for extended reasoning on a position, allows for reflective writing and reading, and leaves a written record of the transaction.	These forms lack the personal warmth and informality of conversations; they require more time and effort for composition and transmission; feedback is slow; and they are subject to malfunctions of technology (word processors, mail delivery, etc.).
e-mail	E-mail allows for full presentation of positions and reflective reading and writing; it provides a written record (or computer file) of the transaction; it permits fast feedback and response.	It can be difficult to edit, difficult to transmit to certain correspondents, and subject to malfunctions of technology. Also, there's some uncertainty about how correspondents will respond; conventions of composition and response are still evolving.

Newsgroups are simply mail lists that provide information on a particular topic on a regular basis, like a newspaper or magazine. Discussion groups are similar, but they allow more interaction among subscribers.

Almost all academic disciplines and subdisciplines, as well as many hobbies and political groups, have their own newsgroups and discussion groups. Exploring these electronic or "virtual" communities can be an excellent way for a novice member of the group—say a sophomore in mechanical engineering, a junior chemistry major, or a graduate student interested in biodiversity—to get a sense of the prevailing discourse in his or her field of interest. You can participate as a "lurker," someone who simply reads the postings, or as an occasional poster (who, for example, asks questions of the other participants), or, depending on your ability, as a full participant. Advanced participants who set the agenda for discussion are sometimes called "catalysts"; others who monitor or filter messages to be sure of their relevance are called "moderators." However, the nature of these roles is evolving as the medium matures.

GUIDELINES AT A GLANCE 38
Using e-mail for technical correspondence

1. Begin your e-mail correspondence with a fairly formal, detailed delivery—as you would for a letter or a position paper. Your first transmission should supply context for your reader and present a coherent set of points for discussion.

2. The correspondence may then evolve into something more like a conversation, thanks to the rapid transmission of responses and feedback. Your correspondent may ask questions or provide a quick counterpoint, so something like a dialogue begins. Or your correspondent may reflect for a while before replying, and then send a longer, more carefully considered response, which demands the same from you. The interchange will then become more like an exchange of letters. The beauty of e-mail correspondences is that they provide the flexibility to support both kinds of communication.

3. E-mail may also evolve from a correspondence into a conference. When you share e-mail from one correspondent with other correspondents, be sure to secure permission from the original correspondent and to provide the necessary context for "over-the-shoulder" readers. As in any good technical communication, give shape and meaning to your transmissions, interpreting and reducing the data to clear action-oriented information. Don't just dump files over the network, hoping that someone will get something out of your mass of data. Develop the message so that it accommodates the needs of the people who are operating in the context of use.

Whether you are sending e-mail to an individual correspondent or participating in some form of electronic group communication, you should always practice good technical communication ethics. Guidelines at a Glance 39 offers some special considerations for ethics and etiquette on an electronic network.

Electronic Conferencing

Electronic conferencing, like a telephone conference call, can involve more than two participants in an exchange of information and ideas. But like e-mail, it occurs on an electronic network.

Conferences may be asynchronous or synchronous. *Asynchronous conferences* occur regularly on mail lists and bulletin boards. A writer posts a message; then, at different times, other users respond to the message or respond to the responses of others. Supported by e-mail networks or some other kind of groupware, the "conversation" can extend over a few days or even weeks and months.

Synchronous conferences, by contrast, occupy users for a single "conversation," during which time all the users are "tuned in" at once. A variety of software exists to support synchronous conferencing, each program offering different features. But almost every program has a few similar features to sup-

GUIDELINES AT A GLANCE 39
Ethics and etiquette for an electronic network*

- Observe all considerations of copyright and intellectual property that you would if you were dealing with print literature. Acknowledge your sources of information, and do not borrow without permission.

- Do not take advantage of your anonymity. On the electronic networks in recent years, there have been some unfortunate instances of verbal abuse, racism, threatening suggestions, and disrespect of various kinds. Avoid such behavior scrupulously, and do not tolerate it in others.

- Avoid "flaming"—rude or angry responses that, while stopping short of verbal abuse, may create ill will among users. Because you can respond quickly, you may be tempted to use e-mail to vent your frustrations or other emotions. Think about the effect of what you write before you post it. It is neither good ethics nor good rhetoric to alienate your readers with careless displays of emotion.

- Do not add people to a regular mail list without their permission. Include only interested parties. Everyone else will think you are imposing on them with junk mail.

- Don't post advertisements. If you do, users will not be sympathetic to your products, since it's a point of general agreement that the Internet is reserved for non-commercial uses. You may, of course, discuss your work and offer suggestions to interested parties about how to use or obtain your products.

- Check the rules of any user group before you post any messages. There may be restrictions on who can post and what kinds of information are allowed. Even after you know the rules, you ought to get used to the particular style and rhetorical conventions of the group before you begin to post.

- Most groups have lists of *frequently asked questions (FAQs)*, which you ought to obtain before you waste everyone else's time by repeating the same old queries.

- Don't include a copy of the original message when you write a response. This practice wastes the correspondent's time in scrolling or rereading. Instead, add a line summarizing the original message.

- Don't send messages in ALL CAPS. As you know from Chapter 6, they are very hard to read. In many online discourse commmunities the use of all caps is considered the same as shouting.

*Many of these suggestions come from Levine and Baroudi, *The Internet for Dummies,* and Rheingold, *The Virtual Community.*

port users who are linked by a local area network. Writers usually compose messages on a section of their computer screens that only they can see. Once the writers transmit them, the messages appear on a portion of the screen that allows general viewing. Everyone reads one another's messages and writes responses, creating a text file that roughly resembles the dialogue of a play.

Figure 11.1 shows a computer screen that would appear to users in a fairly typical synchronous conferencing program: Interchange® by Daedalus, a program developed for classroom use in college writing classes.

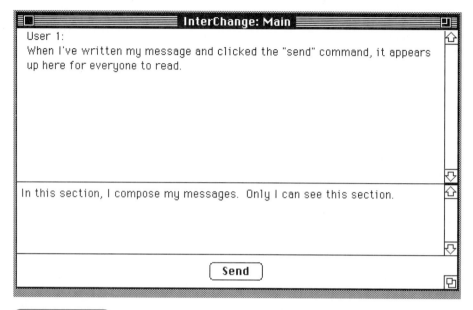

FIGURE 11.1

The main screen in a program for electronic conferencing
(Interchange® by Daedalus).

You can probably see why a group of communicators would choose this
manner of electronic conferencing over a simple discussion around the conference table:

- Users can carefully compose their contributions and present their positions
  more thoroughly.
- They can think about and visually compare the positions of others before responding.
- They can hold a conference without leaving their offices, or they can gather
  around a conference table equipped with networked computers and add a literate dimension to the discussion.
- They create a written record that they can review later and even incorporate
  into reports.

Sample 11.13 shows the kind of record an electronic conference produces.
Following the convention of online "chat rooms," these users chose to employ
code names to maintain anonymity.

The interchange is not quite as smooth as a conversation, as you can see. It
takes a while to compose and transmit your message, and during that time the
discussion continues without you. People get used to these slight dislocations

Earth Mother:
How are your papers coming along? My presentation is on Wednesday. Any ideas for graphics I could use?

Godzilla:
I have no numbers, etc. that I can graph. What do I do??? (My topic is self-esteem in school children.) My presentation is on Friday, AHHHHHHHHHHHHHH!

Earth Mother:
Godzilla, you could show an outline highlighting main points. I am going to use one large graphic to explain my topic.

Godzilla:
O. K. Outline for one, four more to go. We've got to do 5 you know. Maybe I could find a cartoon showing a bunch of depressed school kids and put it overhead.

Cliff Sides:
What are your sources of information? They may have some figures or surveys that could be graphed.

Godzilla:
I don't have any good graphing info yet, I need to go to the library this weekend and find something.

Earth Mother:
Would it be wise for me to show the economic, political, and environmental advantages of biomass as opposed to oil in one chart or 3 charts?

Godzilla:
Maybe you could make three little comparison graphs/charts and put all of them on one overhead.

Earth Mother:
Good idea!

SAMPLE 11.13   Output from part of a small-group synchronous conference on Interchange®

fairly quickly, though, and they are offset by the advantages of having conversations that produce written responses and records.

One further advantage of computer conferencing is that it tends to involve a greater number of participants. Members of the group who might be cautious or shy about presenting their ideas in open discussion may be more willing to put ideas forward if they have time to compose them carefully and then present them electronically rather than in their own voices.

GUIDELINES AT A GLANCE 40
Participating productively in synchronous conferences

- Work on your typing speed and accuracy. The faster you can type, the faster you can get your ideas out. But transmission of too many errors will probably lead to embarrassment if not disgrace (even though most participants are fairly forgiving in these sessions).

- Keep your messages as brief and crisp as possible, and transmit them frequently. Break the issues and topics into parts, dealing with one part in each message. If you send too large or too complex a mass of information over the line at once, you run the risk of being ignored by readers who are eager to be typing themselves. More-over, if you tie up other users by making them read long messages, you will reduce their opportunities to give you advice and other responses. Simply put: Practice *chunking* (see Chapter 7).

- Don't monopolize the reading space by sending too many messages. Use some form of the polite conversational technique of *turn-taking.* If you seem to be trans-mitting more than anyone else, back off and take some time to read what others are writing. The idea here is to benefit from the ideas of others, not simply to dispense your own words of wisdom to the world.

- In general, apply the skills of good conversation, letting others "speak," "listening" carefully, and responding fairly and completely.

Moreover, several conversations can develop within a general discussion. Discussion groups splinter off and then merge back into the general session as they wish. Sometimes this generates an extended discourse on a special issue that involves only two or three of the participants. They can confer freely without worrying about interrupting the main flow of discussion.[6]

As with e-mail, the conventions and etiquette for holding electronic conferences are still evolving. At present, the level of formality varies widely, depending upon the context. And, as with any medium of communication, some users are more adept at presenting their ideas than others. It's usually advantageous to include as many participants as possible in the discussion. Guidelines at a Glance 40 covers some ways to become a productive participant in synchronous conferences.

Hypertext for Collaborative Writing

E-mail and electronic conferencing have created exciting new possibilities for technical correspondence and interpersonal communication. But they have barely scratched the surface of innovation in electronic communication. Some newer programs not only allow free-form communication over computer net-

---

6. We are indebted to our friend Paul Taylor, the author of Interchange®, for many of these observations about electronic conferencing. Professor Taylor has studied many synchronous conference sessions and is in the process of publishing his research.

works but also actually provide "intelligent" guidance for users. By creating arrays and forms, they help conferees to define their positions in clear and logical ways to compare them with competing positions.[7]

Other programs involve productive innovations in *computer-supported cooperative work* (CSCW). They actually bring writers together to become more effective coauthors.

The basic medium for most such innovations is *hypertext,* a form of computer-based writing that builds a weblike text by linking nodes of information in a database. It is the basic form of the World Wide Web. Authors of hypertext create both nodes and links, adding a new dimension to the idea of chunking and relating information. They replace the single-hierarchy or "linear" approach of most paper documents with a more open organization. Then their readers can browse in many directions and even manipulate the hypertext to create new links and add new nodes of information.

On the screen, the reader views windows (or "cards" or "texts") that represent the nodes ("items" or "objects") in the database. The links that the author has created appear as special parts of the text in the window. They are identified by icons or by typographical variation; they may be words that are boldfaced or underlined, for example.

Figure 11.2 shows a structural diagram of a hypothetical hypertext. The screen presents windows A and B, which correspond to nodes A and B in the hypertext database. The bolded lowercase letters in the window (such as **b** and **g**) correspond to links the author has created. The user reading node A can explore linked information by pointing to (or clicking on) any of the lowercase bold letters. This action—pointing to the bold **b,** for example—activates a new window (in this case, window B).[8] (The links in the database diagram are not in normal alphabetical order—an illustration of how usual hierarchical orders tend to disintegrate in hypertext.)

By selecting a particular option, then, the reader opens a field of text to read or to interact with. Within that text may be further options. So the user can keep exploring a particular aspect of a subject in greater and greater depth.

Hypertext products hold great promise for cooperative work among authors. A few of the more obvious applications include the following:

- Researchers can build coauthored databases and explore together relationships among various bits of information by creating alternative links and nodes.
- As they begin to compose documents, they can use the hypertextual writing environment like a complex storyboard, keeping the main "board" relatively

7. See, for example, Jeff Conklin and Michael L. Begeman, "gIBIS: A Hypertext Tool for Exploratory Policy Discussion," *ACM Transactions on Office Information Systems* 6.4 (1988); Catherine C. Marshall et al., "Aquanet: A Hypertext Tool to Hold Your Knowledge in Place," *Hypertext '91 Proceedings,* 261–275.

8. This explanation is based on that of Jeff Conklin, "Hypertext: An Introduction and Survey," *Computer-Supported Cooperative Work: A Book of Readings,* ed. Irene Greif (San Mateo, CA: Morgan Kaufmann, 1988), 423–475.

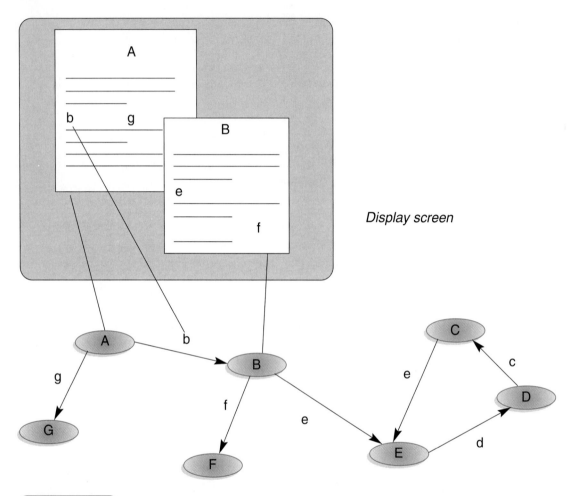

FIGURE 11.2

The structure of a hypertextual database.

Source: Jeff Conklin, "Hypertext: An Introduction and Survey," *Computer-Supported Cooperative Work: A Book of Readings*, ed. Irene Greif (San Mateo, CA: Morgan Kaufmann, 1988) 425.

simple, but using links to allow coauthors to read related documents and sources.

- They can draft and redraft the document online before producing a more traditional paper document.

Technical communicators have only just begun to exploit the new technology of hypertext, so we are just now beginning to understand its peculiar strengths—and weaknesses. One problem, for example, is that many hypertext

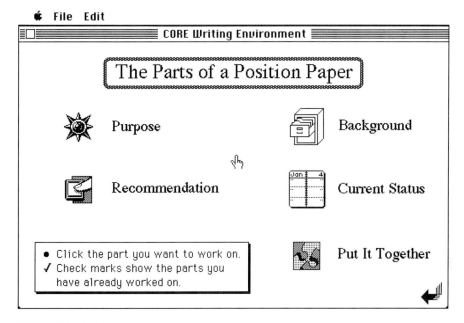

FIGURE 11.3

Screen from CORE Writer.

authoring programs allow you to create only online documents, so you can't easily transfer the collaborative work in your database over to your final paper document.

Another problem is the tendency of users to get lost in the hypertext environment. Authors of hypertexts must be very careful to provide sufficient "maps" for their users, guides that help users know where they've been and where they are going. The increasingly popular addition of "artificially intelligent" elements—devices that ask the user guiding questions about the specific information needed or entered and that help with the storage and organization of new subject matter—will allow users to track their own work and gain new information about how they reason through a problem.[9]

For a concrete example of how a writer can use this tool, consider Figure 11.3, a screen from a hypertext that introduces users to the CORE method of technical writing. The name of the application is "CORE Writer."

---

9. See Patricia A. Carlson, "Artificial Neural Networks as Cognitive Tools for Professional Writing," *Sigdoc '90* (ACM), 95–110.

This section of the hypertext contains instructions for writing a position paper. The user can choose to work on any part of the paper in any order.

By placing the little pointing finger icon in the middle of the screen on the purpose icon, for example, readers can get guidance on writing a purpose statement—a set of questions that will help them to formulate the statement. As Figure 11.4 shows, a place to actually do the writing will appear in the activated window.

If the user is uncertain about exactly what a purpose statement is, help is available. Clicking on the eye icon will open a window that presents a definition and example of a purpose statement, as shown in Figure 11.5.

Now the user can return to the second screen and type in the statement and then return to the first screen to work on the other components of the position paper. Once all the parts are composed, the user can select "Put It Together," and the program will assemble the parts in proper order. At that point, the user can download the position paper to a word processor, revise it if it seems too formulaic or mechanical, and print it out for class.

In addition to the advantages for the user, hypertext offers some advantages for producers, especially collaborative writers. Since the basic unit of design is the *screenful* rather than the text section (as in reports and proposals) or

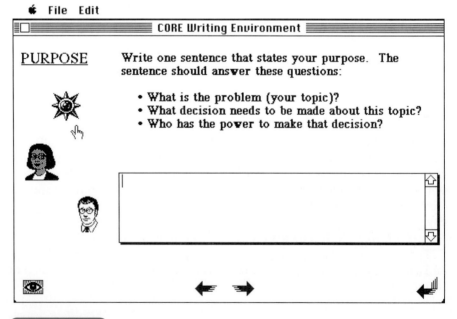

FIGURE 11.4

Second screen from CORE Writer.

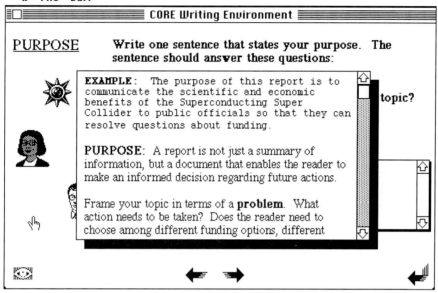

FIGURE 11.5

Third screen from CORE Writer.

the page (as in manuals), the best way to organize the information is by dividing it into *modules,* or blocks of text like the one in Figure 11.6. Because of their visual orientation and size, screen modules are well suited to storyboarding and

As your interest in your task intensifies, you tend to lean forward in your seat, moving your face closer and closer to the computer screen. The resulting posture places great strain on your lower back and neck. So, despite the natural inclination to do so, you should resist the temptation to lean in like this. Design your work station and develop work habits that allow you to keep your back straight, with ears aligned fairly closely with your lower back.

FIGURE 11.6

A text module.

other techniques for group authoring. They also reinforce other trends in technical document design, especially chunking.[10]

As we move to our discussions of the main genres of longer documents, remember that any of these may be produced not only as written texts but as hypertexts. It is becoming increasingly common to post reports, scientific papers, and technical manuals on the World Wide Web. Faced with an assignment to create a web project or to transform a print text to a web text, remember that the shift in medium requires you to rethink your organization and overall design very seriously.

In many ways, the web offers the ultimate form of technical correspondence. It collapses the conceptual distance between correspondence and other genres, as well as local and global discourse communities. Now a local community can post a report on the web and within days receive feedback from numerous, if sometimes random, sources from around the world.

EXERCISE
11.1

Using Guidelines at a Glance 32, draft an appropriate letter or memo in response to one of the following cases. Discuss your draft with a group of your fellow students, then revise it according to the feedback you get. Submit your final draft to your instructor for evaluation.

1. As president of your college's philanthropic organization, you must write to the Dean of Students to get her approval for your annual fundraiser. This fundraiser, a haunted house, will replace last year's "Game Day for Kids." You're hoping that the new idea will reap higher profits and reduce the potential for injuries and other problems, but getting permission from the dean is the first hurdle. In addition to approval for the fundraiser itself, you are requesting the use of a campus building for the event, which will run for four days. The dean's opinion of your organization is relatively positive, but she has expressed concern about the injuries at last year's Game Day.

2. A recent snowstorm caught employees of your company off guard because they didn't know where to turn for information about office closings. Your job description as vice president of human resources puts the responsibility for this policy on your shoulders. In a memo addressed to all levels of the company's employees, you must apologize for the inconvenience caused by the previous lack of a standard procedure and inform them of the new policy that the board of directors recently approved. The company president will make the decision about office closings and inform a specific local radio station and television station by 6:00 a.m. on the workday in question. Weather-

---

10. For more on modular design and related issues, see the excellent article by Stephen Bernhardt, "The Shape of Text to Come: The Texture of Print on Screens," *College Composition and Communication* 44 (1993): 151–175.

related closings will fall into one of three categories: 1) complete shutdown for the workday, 2) opening delayed until 10:00 a.m., 3) opening delayed until 1:00 p.m. Closings will not be announced for more than one day at a time.

3. For the past six months, you have been part of a team of chemical engineering consultants working at a job site 200 miles from your home office. The project has been proceeding very well so far, but the client company's policies necessitate the naming of an official project leader. While you have been acting in that capacity for the duration of the project, your home company prefers to keep project teams on a more equal footing, in hopes that eliminating unnecessary hierarchies will improve cooperation. Because of the client company's repeated requests for an official project leader, however, you are afraid that the entire project may be in jeopardy with the client company. The client company is frustrated by the lack of a designated leader because the final responsibility for the project does not rest with any one person. The client sees this as irresponsible and uncooperative and is ready to pull the plug on the project. Your colleagues have unanimously agreed that you should officially be named their leader. In a letter to the president of your company you must explain your reasons for requesting that he disregard company policy, and you must nominate yourself for the position, requesting that the president personally contact the client company with the new hierarchy.

4. The soft drink distributing company with which your restaurant has been doing business for the past six years has recently mishandled several of your orders, and, to make matters worse, they have overbilled you twice for regular shipments. Luckily, your accountant was able to catch these mistakes and correct them, but you are considering the possibility of giving your business to a different distributor. Your relationship with the local distribution manager has always been good, but she was recently promoted to regional manager. The local manager who replaced her has not handled your concerns well, even sending subordinates to deal with you. You feel that you are an important customer, putting in regular orders and always paying on time, but the new local distribution manager is giving you the run-around. You expect to be treated like the valuable customer that you are, and you are concerned about these recent problems. You decide to write to the regional manager, since you believe that she should know what has happened and because you know that she would handle the situation effectively. However, one of your employees has just informed you that the new local manager is her protégé, and she chose him specifically as her replacement.

5. On a cross-country flight, you are seated next to your state representative. You strike up a conversation about a specific piece of legislation, and the two of you agree wholeheartedly. By the end of the flight, you have developed a strong rapport, and he tells you that he might have an internship available for you in the state capital. Elated, you thank him and tell him that you will write to his personal secretary first thing in the morning to get more information, and he promises to recommend you highly. Once he has left, though, you remember that you still have two months before you graduate, and you

cannot take the job right away. You write him a letter, telling him that you have not yet graduated but emphasizing your interest in the position. You hope that he will let you begin the application process immediately so that you can secure the position as soon as possible, even though you cannot start for two months.

6. Your company's annual performance awards dinner will be held in a few weeks. The awards committee has chosen winners in all six categories. You must write a memo to the supervisors of each of the winners in which you invite them and their guests to the black-tie dinner and inform them of their role in the evening's events. Each of the supervisors will be expected to give a short speech in praise of his or her winning employee. Since some of last year's speeches went on for close to twenty minutes, the awards committee is hoping that you will urge the supervisors to keep these remarks brief. Keep in mind that two of the supervisors to whom you'll be writing were among those who gave long speeches last year.

EXER**CISE**
11.2

Evaluating and revising letters that report, propose, or instruct.   Samples 11.14, 11.15, and 11.16 are examples of correspondence that attempt to perform one of the major functions of technical communication. None of these samples succeeds very well. In your discussion group, or on your own, make a list of strengths and weaknesses in one of the letters, using Guidelines at a Glance 35 and the ones included in the discussion of memo reports, instructions, and proposals. Then try revising the sample. Compare your revision with those of others in your group or class, or submit the revised letter to your instructor for evaluation.

EXER**CISE**
11.3

Part 1. Evaluating letters of application and resumes.   Consider the letters of application and resumes in Samples 11.17, 11.18, and 11.19 (beginning on page 358). In your discussion group, use Guidelines at a Glance 37 to evaluate each document. Make a list of each document's strengths and weaknesses. Compare your list with other groups' lists.

Part 2. Writing letters of application and resumes.   Draft a letter of application for a job in your field advertised in the local paper or in one of your professional publications. Include a well-designed, one-page resume. In a group of fellow students, discuss your drafts. Decide which of the letters and which resume is best suited for the job the author seeks. In the whole class, present your top choices, pointing out their merits over those of the other samples in your group. Drawing upon this discussion, revise your application, and submit it to your instructor for evaluation.

PEORIA POINT
Apartments

Dear Residents: June 12, 1995

COMMUNITY POLICIES

OFFICE HOURS:
Monday thru Friday   8:00 a.m. to 5:00 p.m.
Seasonal Leasing Hours   9:00 a.m. to 6:00 p.m.
Weekend hours will be posted
Office hours are subject to change

LAUNDRY ROOM
Open every day 8:00 a.m. to 10:00 p.m.
Please report any equipment problems to the office. Please
do not use laundry trash containers for your household
trash. We appreciate your help in keeping your laundry
facilities clean.

REGARDING YOUR SAFETY
The property owner and the owner's representative do not
provide, guarantee or warrant security. Owner and owner's
representative do not represent that the dwelling or
apartment complex is safe from criminal activities, by other
residents or third parties. The existence of nighttime walk-
thru or drive-thru services or other systems are not a
guarantee of your personal safety or security and they are
not a guarantee against criminal activity. Clever criminals
can defeat almost any kind of crime deterrents. Walk-thru
services or drive-thru services and owner's representatives
cannot physically be every place at every moment of the day
or night. Owner and owner's representatives assume no duties
of security. Owner reserves the right to cancel personnel
listed above at any time. Any security-related personnel
must <u>not</u> be relied upon by residents to work all the time.
Criminals can circumvent almost any system designed to deter
crime. Remember to please call the police <u>first</u> if trouble
occurs or if potential crime is suspected. Please read and
follow all recommendations in the "<u>PPA</u> Guidelines for
Residents Security" which has been furnished to you.

SWIMMING POOL RULES
1. WARNING NO LIFEGUARD WILL BE ON DUTY. CHILDREN SHOULD
    NOT USE POOL WITHOUT ADULT SUPERVISION. DIAL 911 FOR
    EMS OR POLICE EMERGENCY.

How might this letter be revised?

SAMPLE 11.14   Reporting letter in need of revision

continued

continue
revisions

2. Persons using pool facilities do so at their own risk. Owners assume no responsibility for accident or injury.

3. Residents and guests should be especially careful in supervising and watching their young children at the pool.

4. Residents and guests shall be responsible at all times to make sure that small children do not get out of the apartments unnoticed and that they do not wander into the pool area alone. Remember to use night latches, deadbolts, and window latches when small children are inside.

5. The gate into the pool may not be propped open or otherwise rendered inoperable, even temporarily.

6. No children under the age of 13 will be allowed in the pool at any time, unless accompanied and supervised by a parent or guardian or a person over the age of 18 years who has been given written authority by the parent or guardian to supervise the child and who has assumed responsibility for such supervision.

7. The pool may be used only between 8:00 a.m. and 11:00 p.m.

8. The pool may be used only by residents and their guests. No more than 2 guests of an apartment unit may use the pool at any one time without owner's express approval.

9. No person who is ill may use the pool.

10. No food may be served or eaten in the pool areas at any time without owner's express approval. No glass containers are allowed in the pool area.

11. Any person who is, in the sole judgment of owner's representative, under the influence of alcoholic beverages may be excluded from the pool.

12. No running, horseplay, fighting, dangerous conduct or noise which is disturbing to the other residents is allowed in the pool area. No diving in the shallow part of the pool is permitted.

13. Radios, record players, or other instruments may be used only at the pavilion area. They are not allowed in the pool area.

14. No toys, inner tubes, or any other objects will be allowed in the pool at any time.

15. Residents and guests will place their own towels over pool furniture when using suntan oil or other lotions.

16. Owner is not responsible for articles which are lost, damaged or stolen.

17. Those using the pool shall dry themselves off before leaving the pool area.

SAMPLE 11.14 continued

18.   Safety equipment is not to be used except in case of emergency.

19.   Resident shall be responsible for payment of clean-up expenses, repair costs, and damages caused by resident and resident's guests.

20.   Residents should feel free to ask others to cease any violation of these rules. Residents are requested to immediately notify owner or owner's representative of violation of these rules by others.

21.   Parents, guardians or custodians of children are totally responsible for compliance with these rules. These rules apply to residents, occupants, guests and their children.

22.   VIOLATION OF THESE RULES WILL ENTITLE OWNER TO TERMINATE RESIDENT'S RIGHT OF OCCUPANCY.

DISTURBING THE PEACE

Residents of our communities are wonderful people, but occasionally a few must be reminded that loud noise and loud music will not be tolerated in the community! Call the answering service to report any disturbances after regular office hours.

continue
revisions

BEING NEIGHBORLY

Our community is a shared living experience. So, when it comes to your neighbors, please respect their parking spaces, noise levels and yard areas. Let's all be good friends and follow the Golden Rule.

If you have any question or suggestions for our management team, please don't hesitate to call us at the on-site management office.

Sincerely,

Peoria Point Management

SAMPLE 11.14 continued

How might this letter be revised?

Memorandum

To: All Staff on the Platinum Project
From: Bill Smith, Project Manager
Subject: New Policy on Paper Work
Date: April 24, 1994

I had occasion recently to talk with upper management about the biggest problem we've got. You know what it is: The paperwork. The people over in Group B keep whining about the monthly report forms in particular and want them simplified. Well, these things have worked ok for 10 years, but since you're so bent on changing, management has made us a new policy. This ought to simplify your work load.

What you do is this: Just write up a simple memo after each job. Make it no more than a page long, and include the following information: personnel involved, purpose of work, statement of work, project outcome, amount budgeted, amount actually spent. When you list the personnel, just give everybody's last name. Don't give more than a sentence for the statement of purpose. Your statement of work ought to be a list of activities completed—in list form (1, 2, 3, etc.) with a sentence or two explaining how you did each thing. At the end of the memo, justify any funding that went over the amount budgeted. That ought to do it.

Then, at the end of the month, just staple all these memos together and that will be your monthly report. Don't submit each one separately.

I hope you all appreciate this new policy. I know I do.

SAMPLE 11.15   Instructional memo in need of revision

EXERCISE 11.4

Part 1. Setting up and using an e-mail account.   If you have not already done so, go to your college's computing center and set up an e-mail account. Log on, and begin a correspondence with one or more of the other members of your class. Make notes about any difficulties you had setting up your account and transmitting messages. Also record your observations about the experience. Discuss your notes in class. Talk about ways to make your work on e-mail more efficient and successful.

Part 2. Role-playing an e-mail interchange.   Using your e-mail accounts, conduct an e-mail interchange with one or two other members of your class. As your point of departure, use one of the problems in Exercise 11.1. Take note of how your letters and memos differ from your e-mail transmissions. In your e-mail group, decide who will play the role of original correspondent and who

How might this letter be revised?

The Profit Company—Your Service Giant
44 Tenth Street
Dead End, Texas 79787

April 16, 1994

Tom Winston
% The Davis Company
200 Cattle Street
Ft. Worth, TX 77777

Dear Tom,

The meeting was very informative, relaxed, enjoyable, and always good to see the group.

The lunch was delicious and "Thanks."

I have always gained much information from our meetings and I have been thinking of some way you can better the meetings.

We smaller store don't gain the knowledge the larger stores do. They could be more helpful to us in which line we should check out if shown in Dallas Market.

I think we should be more helpful to new store managers and buyers that hasn't had the opportunities of buying as some of us. I'm speaking of myself. The first meeting or market I went to after the one you sponsored was very difficult for me—because I was not used to buying off-price merchandise and I didn't know where to find it—even though I had gone to market for many years and no one offered to help me.

The last market I took Bill Jones, Lori Small and Susan Joseph with me and they did appreciate this so much.

Sincerely yours,

Josey Smith

SAMPLE 11.16   Letter that proposes (in need of revision)

will be the respondent(s). Agree on a time for the role-playing, then transmit the messages as quickly as you can. Take note of how much time the interchange takes. Discuss your experience in class.

How might this letter be revised?

Marc W. Hammack
999 W. Villa Maria, #1107
Bryan, TX 77803

October 6, 1994

Mr. Richard Sampson
Manager, Human Resources
Ingersoll-Rand Company
P.O. Box 462288
2100 N. First Street
Garland, TX 75046-2288

Dear Mr. Sampson:

I am writing to express my interest in Ingersoll-Rand's Engineering Management Training Program. The information presented to me from our visits both this summer and at Texas A&M's engineering career fair have strengthened by interest in this position.

As you know, I am currently a Mechanical Engineering Major at Texas A&M University, and I expect to receive my Bachelor of Science Degree in December of 1994. I feel that my academic standing as well as both my work experience with Ingersoll-Rand this summer and with Keystone Valve in the past makes me a suitable candidate for a position in this program.

As requested, a copy of my resume is enclosed for your review.

Thank you for your time and consideration. I look forward to hearing from you in the future.

Sincerely,

Marc Hammack

enclosure

SAMPLE 11.17   Letter of application to be evaluated

Marc W. Hammack

UNIVERSITY ADDRESS:
999 W. Villa Maria, #1107
Bryan, TX 77803
(409) 821-6281

PERMANENT ADDRESS:
2826 Oak Street
West, TX 77439
(317) 913-9626

How might this resume be revised?

EDUCATION:

1990–present   Texas A&M University, College Station, Texas
Degree: Bachelor of Science
Major: Mechanical Engineering
Date of Graduation: December 1994
GPR: 3.1, Passed Fundamentals of Engineering
Examination, April 1994

EXPERIENCE:

Summer 1994   Ingersoll-Rand Co.: Rotary Drill Division, Garland, Texas
Development Engineering Department:
Conducted experiments for the redesign of the blast-hole
drill product line's break-out wrench. Responsibilities
included test organization, materials purchase, apparatus
design, experimentation, data collection, and analysis and
documentation of results and conclusions.

Product Support Department:
Observed and performed quality checks on blast-hole
drills. Responsible for the research and development
of the blast-hole drill product line's extended warranty
policy. Edited hydraulics schematics for training
manuals. Acquired product specific knowledge along
with extensive experience with the manufacturing
computer system.

Summer 1992   Keystone Valve: Division U.S.A., Houston, Texas
Research and Development Department:
Assisted in the process of redesigning the high
performance butterfly valve product line. Duties
included benchmark testing of competitors' valves,
extensive design sketches and analysis, prototype
materials purchase, supervision of prototype
assembly, and prototype testing. Developed
interpersonal and group skills as well as fluid
mechanics and material selection techniques in the
design process.

continued

SAMPLE 11.18   Resume (keyed to Sample 11.17)
to be evaluated

continue
revisions

Summer 1991	<u>Keystone Valve</u>: Division U.S.A., Houston, Texas Engineering Department: Input data using various engineering software packages in the process of converting engineering files to a new computer system. Acquired the ability to quickly learn and use a variety of computer software.
Summer 1990	<u>Weisser Engineering</u>, Houston, Texas Worked in a group as an entry level surveyor. Acquired interpersonal skills as well as a basic understanding of surveying techniques.

ACTIVITIES AND HONORS: Member of American Society
of Mechanical Engineers
Named to National Dean's List
Received Barnes and Noble scholarship
Participated in College Station Softball League and
Intramural sports
Placed in various volleyball tournaments
Most Valuable Player and All-District Baseball at
Katy High School

REFERENCES: Available upon request.

SAMPLE 11.18 continued

EXERCISE 11.5

To explore the possibilities of joining or establishing electronically supported discourse communities, do one of the following, then discuss the experience in class:

1. Set up a mail list of everyone in your class and transmit a copy of your latest short paper. Ask everyone for a brief critique of your work, then compile and print out the responses.

2. Join an active newsgroup or discussion group in your major or on a subject in which you are interested. Print out recent samples of the postings, and bring them to class for discussion.

3119 Mason Road   Katy Texas 77449   (713) 579-2079

Computer Hardware and Software Contracting

Personal Information:

I have lived in the Houston Metro area for over 10 years, working in the Computer Industry. I graduated from Eastern Illinois University in 1982. I am currently married, and have two children. My interests involve the use of computers in education, astronomy, and electronic bulletin board (BBS) systems. My active hobbies include Tae Kwon Do, Chess, writing science fiction, and extremely amateur Astronomy.

Educational Background:

I received my Diploma when I graduated from High School in 1975 at Lisle Community High School, in Lisle, Illinois.

I attended College of DuPage in Glen Ellyn, Illinois and graduated in 1977 with a Degree (A.A.) in Computer Science.

In 1982, I received my Degree (B.S.) in Computer Management upon graduation from Eastern Illinois University in Charleston, Illinois.

Employment History:

Contract Software Development

From August 1992 to Present, I have been President of PC Discount Masters. I have been primarily involved in Hardware Sales and Upgrades, and Software Development on a Contractual basis. The software component of my projects have included implementing PICK-based systems on UNIX platforms (programming in C and PICK BASIC), Windows Multimedia (programming in CA-Realizer with Accusoft and Autodesk extensions), and UNIX (a graphics-based BBS system used primarily for auctions). The hardware component has involved upgrading clients' machines with high-performance '486 systems, most of which I built myself.

Senior Systems Analyst

From October 1990 to August 1992, I worked under the IBM AIX operating on an IBM RS/6000 7013, programming software both in the C, and the VMARK UniVerse languages. Systems included Real-Time Truck Dispatching, Warehouse inventory stocking and purchase control, as well as Payroll, Account Receivable, and Accounts Payable.

Senior Programmer Analyst

From August 1989 to October 1990, I developed VIDEOTEX applications for the U.S. Videotel network, including on-line financial systems, entertainment, as well as educational services.

How might this resume be revised?

SAMPLE 11.19   Resize to be evaluated

continued

continue
revisions

While attending college, I worked my way through school, as a short-order cook, steel-foundary bull operator, and computer operator and programmer, along with several odd jobs to pay for tuition.

Software Development Manager

From 1987 to 1989, I was responsible for the development, maintenance, and customer support of Hair Care Salon industry management software. Modules included Point of Sale (POS), Payroll, and Multi-Site Inventory Control.

Computer Operating Systems and Languages Summary

Unix/XENIX/AIX/HP-UX   Windows/MSDOS   3-COM/Novell

C, Realizer/BASIC, PICK/BASIC, Pascal, RPG, Cobol, Fortran

Intel 80×86, Zilog Z80, Intel 8080, and 6502/65816 Assembly

SAMPLE 11.19 continued

Recommendations for Further Reading

Branscomb, H. Eric. *Casting Your Net: A Student's Guide to Research on the Internet.* Boston: Allyn and Bacon, 1998.

Clark, Carol Lea. *Working the Web: A Student's Guide.* Orlando, FL: Harcourt, 1997.

Greenly, Robert. "How to Write a Resume." *Technical Communication* 40, 1 (1993): 42–48.

Jackson, Tom. *The Perfect Resume.* Garden City, NJ: Doubleday, 1981.

Levine, John R., and Carol Baroudi. *The Internet for Dummies.* San Mateo, CA: IDG Books, 1993.

Matalene, Carolyn B., ed. *Worlds of Writing: Teaching and Learning in Discourse Communities at Work.* New York: Random House, 1989: 203–221.

Rheingold, Howard. *The Virtual Community: Homesteading on the Electronic Frontier.* Reading, MA: Addison-Wesley, 1993.

Additional Reading for Advanced Research

Bolter, Jay David. *Writing Space: The Computer, Hypertext, and the History of Writing.* Hillsdale, NJ: Lawrence Erlbaum, 1991.

Easterbrook, Steve, ed. *Computer-Supported Cooperative Work: Cooperation or Conflict?* London: Springer-Verlag, 1993.

Greif, Irene, ed. *Computer-Supported Cooperative Work: A Book of Readings.* San Mateo, CA: Morgan Kaufmann, 1988.

Killingsworth, M. Jimmie, and Michael K. Gilbertson. *Signs, Genres, and Communities in Technical Communication.* Amityville, NY: Baywood, 1992.

Killingsworth, M. Jimmie, "Discourse Communities: Local and Global." *Rhetoric Review* 11 (1992): 110–122.

Zuboff, Shoshana. *In the Age of the Smart Machine: The Future of Work and Power.* New York: Basic Books, 1987

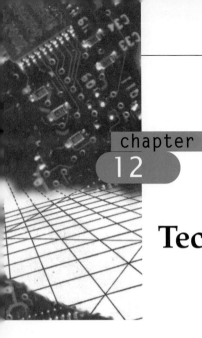

chapter

12

Technical Reports

CHAPTER OUTLINE
▼▼▼▼▼▼▼▼▼▼▼▼▼▼▼▼▼▼▼▼▼▼▼▼▼▼▼▼▼

Step One: Doing an Audience-Action
Analysis for Your Position Paper
and Technical Report
Step Two: Writing the Technical
Position Paper
Step Three: Developing Graphics for the
Technical Report
Step Four: Planning and Presenting an
Oral Briefing
Step Five: Developing the Technical Report

CHAPTER OBJECTIVES
▼▼▼▼▼▼▼▼▼▼▼▼▼▼▼▼▼▼▼▼▼▼▼▼▼▼▼▼▼

After you have worked through this chapter,
you should be able to do the following:

- Perform an audience-action analysis
  toward a short position paper and a full
  technical report
- Write a position paper on a technical topic
  and participate in peer reviews of position
  papers
- Plan and design graphics for your technical
  report
- Plan and present an oral briefing based on
  the graphics and position paper
- Plan, produce, and review the full technical
  report

A s you saw in Part One, one of the great problems in technical communication—perhaps *the* great problem—involves reconciling the needs of information producers with those of information users. How can we find out what our readers need so that we can craft an information product to satisfy those needs?

The CORE method is one response to this question. In Parts One and Two, we discussed each step in the process at a general level. In Chapters 12–15 we will take a closer look at how to apply this method to the major genres of technical communication: the report, the proposal, the manual, and their variations. You will learn to write on topics of your choice in each of these genres and to review and evaluate each type of document. We begin in this chapter with the technical report.

To apply the CORE method to report writing, take the following steps (see also Figure 12.1):

1. *Do an audience-action analysis and write a management plan.* For most technical reports, you will provide researched information and recommendations for an audience of decision makers who use the report to direct future actions.

2. *Write the core document: a position paper of no more than two pages.* You should share the position paper with trial readers to get a sense of the kind of information your audience will need to make their decision. The more readers, the better. The closer the readers are to your actual audience, the better.

3. *Develop graphics to support the main points of your position paper.* Try to cover as much of your information as you can in graphic form, using graphics to display and interpret information and to provide impact and vividness to your arguments.

4. *Give a short oral report, or* briefing, *to the trial or actual audience.* This provides a second occasion for engaging the audience, this time integrating the graphics with the arguments of the position paper.

5. *Develop the full written report.* Here you amplify the core document and original graphics, using the feedback you've gathered to develop an integrated document that provides the reader with everything needed for an informed decision.

This chapter guides you through each step, beginning with the audience-action analysis and the position paper. As you will see, the position paper is essentially a short version of the report that grows out of your understanding of your task as a writer. If you discover that your understanding of the communication situation is flawed, then you can rewrite with little trouble, having invested only enough time and energy to write a short draft.

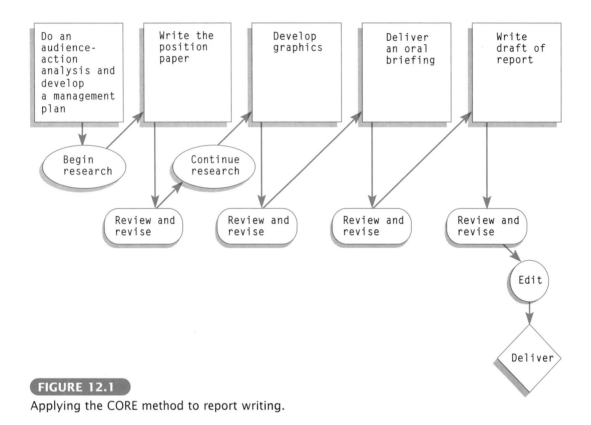

Applying the CORE method to report writing.

Step One: Doing an Audience-Action Analysis for Your Position Paper and Technical Report

As the name suggests, the position paper requires that you develop a *position*, or a place from which an action can proceed. In technical writing, this position takes the form of a claim or recommendation based on technical information. In essence, technical reports are tools that allow users to relate information to operations and interests in the context of use. In fact, we can view every technical report as a document that answers the same big question: *Who is supposed to do what with the information I provide?*

Technical reports are not merely "informative." Indeed, contrary to what many rhetoric books teach, there is no absolute division between "informative" (or "objective") writing and "persuasive" (or "subjective") writing. Technical reports must inform *and* persuade. They must use the best information available to convince people to take rational courses of action.

On the basis of these premises, we can define a technical report as follows: It is a presentation that provides *information* directing a specific *audience* toward a *decision* about future *action* within a specific *context of use.* As part of your audience-action analysis (see Part One) you should be able to write a set of planning notes that identifies each of these components and outlines how they will work together, first in your position paper and later in your full-length report. These notes will keep you on track as you draft both documents.

The *information* that you need for a technical report comes from past experience. It may be your own experience (experiments or observations), or it may be experience encapsulated in an information base (technical literature, a company's files, etc.). Most likely, you will use a little of both. Like a news reporter, you will focus on information as *a difference that makes a difference*[1]—news that affects how we understand the world and how we act. But as a technical reporter, you will direct your news toward very particular groups of people who must make specific decisions and take specific actions.

These *audience* groups may be your peers, clients, or superiors. Professional engineers, government workers, scientists, and professional researchers in industry rarely write reports for other fully informed technical experts. That luxury—the freedom to write for a group of appreciative peers—is reserved mainly for academic researchers. Instead, technical specialists must communicate their findings either to busy nonspecialists who have the power or the money to put the information to work or to other technical specialists who need to be brought up to date on new information so that they can advise decision makers. Typical audiences include:

- managers at various levels within companies and organizations
- government officials (ranging from city council members to executives and legislators in the highest echelons of the federal government)
- technical advisors who lack the special expertise or knowledge that the author of the report possesses

These groups have in common the power to implement actions that will solve technical problems. The technical report opens access for them to new information, shows them how this information relates to a particular decision they need to make, and recommends the best course to follow.

In your report, the *decision* that the audience needs to make should appear in the form of a *recommendation.* This is an argument, based on past experience, for an action in the future. It should sum up your position and tell where things stand now, how they got there, and where they should go.

The *context* of use includes the typical actions, goals, people, places, and objects that affect the decision in question. It includes personal, social, and political concerns as well as natural and technical factors. See Figure 12.2.

---

1. This is the famous definition of information given by anthropologist Gregory Bateson in his book *Steps to an Ecology of Mind* (New York: Ballantine, 1972).

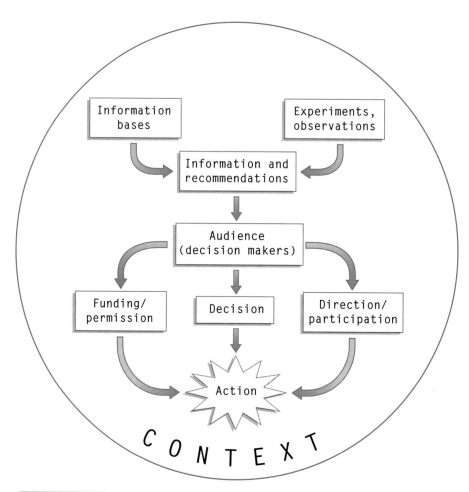

FIGURE 12.2

The report provides *information* directing a specific *audience* toward a *decision* about future *action* within a specific *context of use.*

The *action* that you recommend in your technical report may or may not involve you directly, but it will certainly involve your audience. They will either make the action possible (through funding or permission), direct the action (as managers), or participate in the action (as technicians). Think of the action as the *objective* or ultimate outcome of the report. Ideally, the audience will permit or perform the recommended actions after reading the report.

Different Types of Reports

The technical report genre encompasses many different subgenres—progress reports, data reports, activities reports, trip reports, feasibility studies, recommendation reports, empirical studies, and so on—with slight variations in purpose, content, format, and audience. Progress reports, for example, inform management about what is happening on a technical project, how close the work is to schedule, and any problems that have arisen. Feasibility studies give decision makers a technical analysis of the costs and benefits of several options for action and usually recommend the best alternative for the situation under investigation.

Today, these subgenres are becoming ever more complex and various, so instead of trying to elaborate on each one, this chapter groups all of the various subgenres into two general types of reports: reports on *projects* and reports on *options.* Chapter 13 deals with a related genre—the experimental report—and its close relative, the scientific paper.

Managers, government officials, and other nonspecialists are the typical audience for both types of technical reports. These decision makers often lack the expertise of technical authors, but they have the power to act on information. Unlike a scientific paper addressed to technical peers, the purpose of which is to advance the research program of a scientific discipline, a technical report tries to solve more immediate problems of human existence. For this reason, technical reports are considered less "objective" and more "persuasive" than scientific reports. To be an effective tool in decision making, a technical report must make a clear argument for certain policy options. People in power may choose to accept or reject the viewpoint, but if the argument is based on reliable information, it is not really any more "subjective" than a good judicial decision.

As Table 12.1 shows, the two types of reports described in this chapter include many subgenres. But this flexible classification also leaves room for variations that have not yet been classified in practice.

TABLE 12.1

A simple classification of technical reports

Reports on Projects	*Reports on Options*
activities reports	recommendation reports
progress reports	feasibility studies
trip reports	impact reports
data reports	cost-benefit analyses

Most reports you will write as a technical professional in industry and government fit comfortably into one of these two categories, which are defined by their dominant purpose:

- Reports on projects tell about the significance of something you have done.
- Reports on options present surveys, analyses, or evaluations of alternatives for action, usually with a recommendation for the best option.

The two categories are not mutually exclusive but allow for different emphases. Reports on projects may well consider options for further action in the introduction or conclusion and will likely offer recommendations. Reports on options may present the author's experiences as part of the information. The difference is mainly one of focus, which leads to a difference in organization.

Both reports have, of course, a beginning, a middle, and an end. The beginning (introduction) brings the audience up to date by giving the background and outlining the issues covered. The middle (technical section, or discussion) reports on recent changes and new information that has come to light and brings the old way of doing things into question. The end (conclusion and recommendations) points the way toward future action. Both kinds of reports move from past to present to future in their overall flow. But in the middle section there is an important difference: The report on projects focuses on a single project activity; the report on options focuses on an issue or problem along with a number of related activities.

Using the general guidelines presented in this chapter, you should keep in mind that every reporting situation is unique. You should be ready on every occasion for communication to do an audience-action analysis, looking closely at the particular requirements of your audience and rhetorical situation. Always analyze the relative needs of the contexts of production and use, keeping in mind the big question: Who is supposed to do what with the information I provide? If the requirements of the project do not fit neatly into any report category you've learned, you may have to invent a new form. The two categories given here are merely intended to get you started thinking about typical situations.

Step Two: Writing the Technical Position Paper

Because it is brief, the position paper cannot present a totally convincing case. But you can cover the major parts of your argument, providing the audience with a scaled-down version that is strong enough to convince your readers that the position is worth arguing in a longer report.

To get a sense of the big picture you are trying to show your reader, you can begin with a basic outline. This working outline can also be useful in guiding your research on the topic (see Chapter 4). As you gather information, you can refine and even change your position.

Organizing Reports on Projects

In a report on projects the author may be writing either about his or her own project or activities (as in an engineer's trip reports and monthly reports on activities) or about the actions of others (as a project manager, lobbyist, or company spokesperson). In either case you can use the same basic outline to get started:

1. Introduction
    1.1. Statement of Purpose and Position
    1.2. Background (the Past)
    1.3. Brief Overview of the Present and Future Leading to Section 2
2. Discussion of Recent Findings and Current Activities
3. Conclusion
    3.1. Significance of Findings and Activities within Context of Use
    3.2. Recommendations for Further Action

One student in one of our technical writing classes, Michele Morgan, selected as her topic a scientific project in her hometown of Waxahachie, Texas: the controversial Superconducting Super Collider (SSC). The SSC was a multi-billion-dollar, federally funded effort to discover subatomic particles that exist in theory but have not yet been located experimentally. The project was funded by Congress, the site was selected, and the construction of the facilities was begun. Then, in 1992, during tough economic times, the House of Representatives voted to cancel the project. Its funding was saved only by a relatively close vote in the Senate. The funding was canceled again in 1993, then won back in another close Senate vote, and then lost yet again in 1994.

Michele embarked on her project in the middle of this dispute. After reading the available nontechnical literature on the SSC and interviewing scientists, government officials, and interested citizens, she decided to frame her paper as a technical report to congressional staff members.

The purpose of the report was to give members of Congress and their advisory staffs the kind of information they would need to make a good decision about funding the SSC. Michele decided to recommend that the project be continued with full funding because, aside from its contribution to scientific knowledge, its potential technological and economic benefits justified its costs. Obviously, such a report was needed in this case, but a convincing one never appeared. We can only wonder whether better technical communication could have saved the SSC.

For the rest of this chapter we will trace Michele's steps as she applies the CORE method to write her report. If you have followed the research guidelines in Chapter 4, you should be able to follow Michele's work step by step, thinking through your own strategies for composing your report and actually writing as you go.

We'll start with the working outline she developed (Sample 12.1).

While your working outline provides a general structure, remember that the focus of a technical report should not be solely on the information you have collected in your research, but on what your audience is expected to *do* with the information. As such, it should be both informative and persuasive. This means that each section of your position paper has a specific purpose:

- The introduction should present the position and give the background necessary for the reader to understand the implications of the project.
- The discussion should *make an argument* for the position, giving enough details about the project and its intended outcomes to support the position.
- The conclusion should clinch the argument by showing how the project relates to the overall values and activities of the constituencies represented by the audience of decision makers.

Of course, not all projects will have such global implications, but even simple progress reports need to show how the activities reported relate to a broader context. A manager at a major medical products firm—who was himself trained as an engineer—once complained to me that the monthly activities reports of his engineers were no good. He said that they simply rambled on about what they were doing. What he wanted was not an "activities report" but rather an "accomplishment report" or a "significance report." He wanted to know how the reported activities furthered the goals of the engineer and the company. Sample 12.2 illustrates the difference between the two kinds of reports by giving paragraphs from each.

Because Michele's paper on the SSC argues that the project has a place and a purpose in a larger context, it too is an "accomplishment report." Although she provides decision makers with enough technical details to make the project real in their minds, she focuses their attention on clearly argued positions that they can use to defend the project.

Her full report appears in Sample 12.17 at the end of this chapter.

Organizing Reports on Options

Like the project report, the report on options is a problem-solving document for decision makers. But because it examines several possible ways of approaching

Title: Why We Need the Superconducting Super Collider (SSC)

1. Introduction
   1.1. Purpose—To give background and benefits of the
        Superconducting Super Collider (SSC).
        Position—The potential benefits of the SSC are great
        enough to justify full funding
   1.2. What is the SSC?
        • The SSC will be the world's preeminent facility for
          high energy physics research.
        • It is located in Waxahachie, Texas.
        • Physical facilities consist of an underground
          tunnel 54 miles in circumference with supporting
          buildings.
        • Superconducting magnets focus and guide two beams
          of protons in opposite directions around the
          tunnel. Collisions between the two proton beams
          will create new subatomic particles.
        • Expected results include a new understanding of the
          fundamental nature of matter and energy.

2. Discussion
   2.1. Current Status
        • Target dates
        • Money spent to date
        • Land acquired to date
        • People employed
        • Degree of progress attained
   2.2. The Benefits
        • Technological
          —Medical: proton beam surgery, magnetic resonance
            imaging (MRI)
          —Transportation: prototype of train with
            superconducting magnets
          —Communications: superconductivity, cryogenics,
            electronics, and computing (high-speed counting
            circuits)
        • Economic
          —Local
          —Global
            >Enhances U.S. position in world marketplace
            >Prevents "brain drain"

3. Conclusion: Basic physics research pays for itself in
   growth of new industries and diversification of old ones.
   Keep the SSC alive.

SAMPLE 12.1 Outline for a project report

Excerpt from an activities report in the diary style:

On November 15, 1994, we visited the Cedar Rapids plant for the start-up of the new pulp processors. Harris, Watson, and Jones accompanied me. The trip went well. Much as we did in our trip to Iowa City earlier in the month, we found that the new design initiated in our latest product model (A48a) is working out great. The first day's operation recorded none of the inefficiencies recorded in the early operation of the model A47.

Excerpt from an activities report emphasizing significant accomplishments:

We confirmed the greater efficiency of Model A48a of the pulp processor in two plant visits this month. After two weeks of operation, the Iowa City processor had an improved efficiency of 22% over Model A47. On the first day of operation, the Cedar Rapids processor had similar results, improving upon the average first-day performance of A47 by 23.5%. The time invested in the new design back in January and February has now paid off.

SAMPLE 12.2   Paragraphs from two activities reports showing the differences between the "diary-style" focus on activities and a focus on accomplishments and significance. Managers tend to prefer the second alternative.

the problem, this kind of report has a more obvious problem-solution focus. As such, it is something like a report and something like a proposal.

Here's a simple generic outline for a report on options:

1. Introduction
    1.1. Statement of problem [Optional: Statement of position]
    1.2. Analysis of the problem
    1.3. Why old ways of solving the problem no longer work
    1.4. Overview of the options for solving the problem
2. Discussion
    2.1. Option 1
    2.2. Option 2
    2.3. Option 3, etc.
3. Conclusion
    3.1. Review of advantages of different options
    3.2. Recommendation of best option (or combination of options)

Again, this outline follows the basic form of the position paper but adds a more detailed discussion section in the middle. Sample 12.3 is an outline for a report on options by Patrick Zuzek, a student in mechanical engineering.

Title: Alternative Fuels for the Future: Research and Development in the
Petroleum Industry

1. Introduction

    1.1. Purpose—To discuss recent research and development in alternative fuels
         and to suggest policy options for the petroleum industry

    1.2. The Problem

        • The limitations of oil as a sustainable fuel resource

        • Political issues
          —environmentalism and public relations
          —dependence on foreign oil

        • Position: The industry needs to diversify; certain alternative fuels offer
          good options.

2. Discussion

    2.1. Existing fuels with special promise for industrial development

        • Propane

        • Alcohol

        • Natural gas

    2.2. Alternatives to be treated as stop-gap measures

        • Electric

        • Solar

    2.3. Hydrogen—The Fuel of the Future

3. Conclusion

    3.1. Industry can benefit directly from pursuing the development of propane,
         alcohol, and natural gas alternative fuel options and can benefit indirectly
         (public relations) by promoting electrical and solar options for
         transportation fuels.

    3.2. As these fuels are used alongside petroleum products, industry should
         invest in major research and development for hydrogen as an alternative
         fuel.

    3.3. Legislative lobbying should cease trying to block the development of
         alternative fuels and turn toward a coherent policy for developing
         alternative fuels.

SAMPLE 12.3   Outline for a student report on options (feasibility study)

Notice that in the outline for his discussion section, instead of merely listing the options and dealing with them one at a time, Patrick divides the material into categories related to his argument: "existing fuels with special promise," "stop-gap measures," and "the fuel of the future." This hierarchical arrangement clearly sets up the argument in the conclusion and shows the audience how the options fit within the context of use.

Sample 12.4 is the outline for a somewhat different report on options, a report on solid waste management by Chris Powers, a student in environmental engineering. Instead of covering a number of options in a fairly even amount of detail the way Patrick did, Chris focuses on a single preferred option: the recycling of biowaste.

While Patrick's work (Sample 12.3) is an outline for what is usually called a *feasibility study,* Chris's outline (Sample 12.4) represents a *recommendation report.* It verges on becoming a proposal but stops short of giving a detailed plan for proceeding with the project. Recommendation reports usually come earlier in the project cycle than proposals. If the idea wins support, management will

---

Title: Biowaste Recycling: An Economic and Environmentally Sound Solution to Municipal Disposal Problems

1. Introduction

    1.1. Purpose—To inform the mayor and city officials of San Diego about biowaste recycling as a way of preventing a waste disposal crisis

    1.2. The Problem
   - Ocean dumping—no longer a solution
   - Landfilling—a "time bomb"

    1.3. The Solution: Through the collection and treatment of biowaste, San Diego could decrease disposal costs and profit from the sale of the treated waste.

2. Discussion

    2.1. Biowaste recycling has worked for other cities

    2.2. Public health is a concern

    2.3. Private industry can profit

3. Conclusion

    3.1. Our options

    3.2. Implementation

SAMPLE 12.4   Outline for a student report on options (recommendation report)

appoint a project team to write a detailed proposal. (To get a better sense of the difference between reports on options and proposals, you can compare Chris's report to the proposals in Chapter 14.)

Sample 12.18 at the end of this chapter gives Chris's full report.

Applying Guidelines at a Glance 7 in Drafting the Position Paper

With a basic outline in hand and your research well under way, you are ready to write your position paper. The following steps from Guidelines at a Glance 7 (see page 126) provide a highly organized structure for the position paper that requires a well-defined audience and purpose. As we walk through the steps, we will illustrate each one, using Michele Morgan's paper on the SSC project.

I. Paragraph 1 (Introduction and Background):
  - *Write a sentence that states your purpose.* Follow this model fairly closely:

    The purpose of this report is to provide information about [a problem— your topic] to [an audience of decision makers with the power to make a direct impact upon the problem] so that [a reasonable action can be taken to solve the problem].

    Here's what Michele Morgan wrote for her position paper on the Super-conducting Super Collider:

    The purpose of this report is to communicate the scientific and economic benefits of the Superconducting Super Collider (SSC) to public officials so that they can resolve questions about funding.

  - *Add a sentence stating your position or recommendation on the issue.* Michele wrote the following:

    The potential benefits are great enough to justify full funding and support for the project.

  - *Explain key terms and orient the audience to the project or problem you are addressing.* Michele wrote the following:

    The Superconducting Super Collider will be the world's preeminent facility for high-energy physics research. Located in Waxahachie, Texas, the SSC is a racetrack-shaped, underground tunnel about 54 miles in circumference. Within the tunnel, superconducting magnets will focus and guide two beams of protons in opposite directions around the tunnel. Collisions between the two proton beams will bring to light subatomic particles that will add to our understanding of the fundamental nature of matter and energy.

GUIDELINES AT A GLANCE 41

Questions for reviewing position papers

- Who is the audience? Who will do what with this information?
- Does the audience have the power to contribute to a solution?
- What is the position of the author? What decision is the author asking the audience to make? What actions should result from the audience's reading of the report?
- What additional information does the audience need to understand the background of the problem and the strength of the author's position?
- What additional information does the audience need to make an informed decision?
- Does the author do a good job of convincing you? What are possible objections to the author's position?

II. Paragraph 2 (Discussion):
- *Bring the audience up to date by giving the current status of the project.* Michele wrote,

> Construction of the massive Super Collider is well under way. Land acquirement is complete, main campus buildings are nearing completion, tunnel construction is progressing, and thousands of people are employed by the SSC and its laboratories. The Super Collider currently has an $8.5 billion price tag. Although several billion dollars have been spent on the project to date, today's difficult economic conditions make the SSC a target for spending cuts by some public officials.

III. Paragraph 3 (Conclusion):
- *Explain your position briefly, with special attention to the benefits of the decision you recommend.* Michele wrote this:

> The Superconducting Super Collider is a project that's crucial to the future of the United States as a leader in science. The technology developed by the SSC will lead to numerous spinoff industries while keeping the greatest scientific minds from around the world working in the United States. The future health, industrial, and economic benefits the Super Collider makes possible will far outweigh the burden of the project's initial cost.

Sample 12.5 gives the completed draft of Michele's position paper. Notice how each section develops a sentence or two that corresponds nicely to the basic outline she wrote (Sample 12.1). Use the questions in Guidelines at a Glance 41 to evaluate Michele's paper.

A Position Paper
on the Superconducting Super Collider (SSC)

Michele Morgan, Public Relations

The purpose of this report is to communicate the scientific and economic benefits of the Superconducting Super Collider (SSC) to the public officials who must decide the fate of the SSC. The potential benefits are great enough to justify full funding and support for the project. Denial of funding at this point would cause the loss of thousands of jobs, several billion dollars, and countless scientific breakthroughs.

The Superconducting Super Collider will be the world's preeminent facility for high-energy physics research. Located in Waxahachie, Texas, the Super Collider is a racetrack-shaped, underground tunnel about 54 miles in circumference. Within the tunnel, superconducting magnets will focus and guide two beams of protons in opposite directions around the tunnel. Collisions between the two proton beams will bring forth new subatomic particles, the study of which will greatly increase our understanding of the fundamental nature of matter and energy.

Construction of the massive Super Collider is well under way. Land acquirement is complete, main campus buildings are nearing completion, tunnel construction is progressing, and thousands of people are employed by the SSC and its laboratories. The SSC currently has an $8.5 billion price tag. Although several billion dollars have been spent on the project to date, today's difficult economic conditions make the SSC a target for spending cuts by some public officials.

The Superconducting Super Collider is a project that is crucial to the future of the United States as a leader in science. The technology developed by the SSC will lead to numerous spinoff industries while keeping the greatest scientific minds from around the world working in the United States. The future technological and economic benefits the SSC makes possible will far outweigh the burden of the project's initial cost.

Side annotations:

The title clearly gives the topic of the paper in a straightforward way.

The first sentence states the purpose. The second sentence states the author's position on the issue. The third sentence gives a brief rationale for the position. It says why the author is taking this particular stand.

The second paragraph gives one type of background—a description of the project and its scientific purposes.

The third paragraph gives another kind of background. It says where the project stands at the moment.

The final paragraph briefly develops the position. It says why the decision for the SSC should be positive.

SAMPLE 12.5 Draft of a position paper and core document for a technical report

Step Three: Developing Graphics for the Technical Report

The third step in the CORE method of producing reports is to plan and design graphics to support the argument you have begun to develop in your position paper. As we saw in Chapter 6, the typical audience for a technical report, like most audiences, will respond positively to graphically rich, mixed-media documents written in the straightforward conversational style. As you expand your core document, you should try, whenever possible, to *show* the reader information in graphic form and *talk* directly to the reader through your writing. As you work on your graphics, you will be working simultaneously on Step Four of the CORE method. You can use the graphics you produce in combination with your position paper—and very little else—as the basis for a short oral report or "briefing."

To develop graphics for a technical report, concentrate on the points in the following list and Guidelines at a Glance 42. (These points are covered at greater length in Chapter 6.)

- *Use graphics and hybrid displays to preview and review your main points.* For oral reports, outlines of the topics and subtopics and lists of keywords in order of discussion are especially useful. For the written report, you can convert the topic outline to a table of contents, and you can retain many of the lists to forecast points at the beginning of each section. Flowcharts and tables can also be useful as forecasters, and along with line graphs and other high-density data displays, they are perhaps the most powerful summarizing devices available to you.
- *Use graphics to orient the audience in space and time, showing readers where and when actions take place.* For both written and oral reports, the best orientational graphics include maps, flowcharts, and diagrams.
- *Use high-density graphics to interpret findings and support claims.* You can use complex tables, line graphs, and other data displays to condense and classify huge blocks of past experience and observation. In written reports especially, these graphics are an excellent means for supporting arguments. In oral reports, however, take care not to overwhelm the reader with too many complex graphics.
- *Use low-density graphics for emphasis and impact.* Strong contrasts come through especially well in pie charts, bar graphs, and short tables. These are excellent for oral presentation because your audience can comprehend them at a glance.
- *Use drawings and photographs for vividness and impact.* To create a strong visual image of your subject—an especially useful technique when you are discussing products and places—use line drawings, sketches, and photographs.

To preview and review your main points, use outlines, lists, tables, and flowcharts.

For a discussion of flowcharts and other forecasting and summary devices, see Chapter 6, pages 140–142.

For detailed summaries, use line graphs and other high-density graphics.

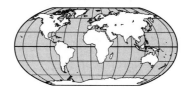

For a discussion of line graphs and other high-density graphics, see Chapter 6, pages 149–156.

To orient the audience in space and time, use maps, flowcharts, and diagrams.

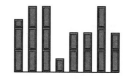

For a discussion of maps and other orientational graphics, see Chapter 6, pages 142–147.

To interpret findings and support claims, use tables, line graphs, and other data displays.

1993	1994	1995	Total

For a discussion of tables and graphical interpretation, see Chapter 6, pages 149–156.

For emphasis, impact, and contrast, use pie charts, bar graphs, and short tables.

For a discussion of bar charts and other low-density graphics, see Chapter 6, pages 156–163.

For vivid images, use line drawings, sketches, and photographs.

For a discussion of line drawings and other depictions of images, see Chapter 6, pages 163–171.

Step Four: Planning and Presenting an Oral Briefing

The fourth step in the CORE method of producing reports is to plan and present a short oral version of your technical report, integrating the material developed in your position paper and your graphics plan. As a way to envision what you will do in your briefing, begin by planning a set of transparencies.

Why We Need the Superconducting Super Collider (SSC)

Purpose - To Give Background and Benefits of the Superconducting Super Collider (SSC).

Position - The potential benefits of the SSC are great enough to justify *full* funding.

1. What is the SSC?

2. How Far Along Is the Project?

3. What Are the Benefits?

    •   Technological

    •   Economic

SAMPLE 12.6   Transparency 1 for a briefing on the Superconducting Super Collider. Michele gives a general outline of the presentation, forecasting the structure of the whole report and stating clearly her purpose and position. For the lettering, she uses a combination of 24- and 18-point Helvetica type, with an occasional use of italics for emphasis.

Creating a Set of Transparencies

To get sense of the task before you, consider Samples 12.6–12.11, the draft transparencies that Michele Morgan developed for her report on the Superconducting Super Collider (the position paper in Sample 12.5). The captions of the samples suggest the author's design strategies.

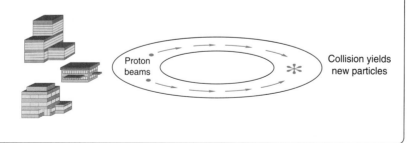

What is the SSC?

- The SSC will be the world's preeminent facility for high-energy physics research.

- Located in Waxahachie, Texas, the underground tunnel consists of 54 miles in circumference with supporting buildings.

- Superconducting magnets focus and guide two beams of protons in opposite directions around the tunnel. Collisions between the two proton beams will create new sub-atomic particles.

- Expected results include a new understanding of the fundamental nature of matter and energy.

Proton beams

Collision yields new particles

SAMPLE 12.7   Tranparency 2 for a briefing on the Superconducting Super Collider. Michele provides a list of key points for the section along with graphical support. The simple diagram aids in her explanation of the project's design. Michele mainly uses easy-to-read, low-density graphics.

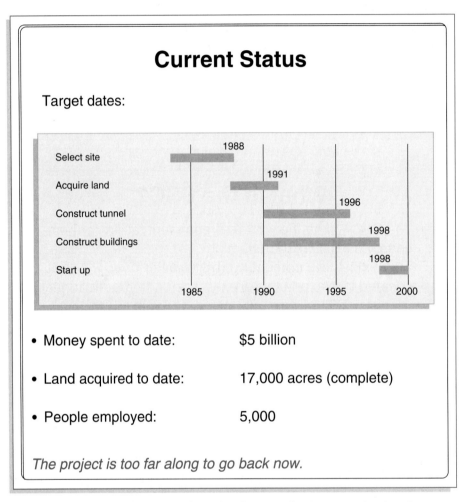

SAMPLE 12.8   Tranparency 3 for a briefing on the Superconducting Super Collider. Following the pattern set in Transparency 2, Michele lists the key points for the next section. She emphasizes her main point—the progress of the project toward completion—with a low-density graphic, this time a Gantt chart.

Michele's transparencies may seem very sketchy, but she was strictly limited to a twelve-minute presentation, so she had to cover a great deal of background information in a hurry. Overall, she followed a good rule of thumb, *allowing a minimum of two minutes for each* transparency, even simple ones like these.

One other rule of thumb that she followed was this: *If the audience were to see nothing but the transparencies, the main points should still come across.* Even a sleepy or uncommitted audience will look at slides or transparencies magnified by a bright light in a dark room.

The Benefits

- **Technological**

  —Medical: proton beam surgery, magnetic resonance imaging (MRI)

  —Transportation: train with superconducting magnets

  —Communications: superconductivity, cryogenics, electronics, and computing (high-speed counting circuits)

- **Economic**

  —Local

  —Global
  >Enhances U.S. position in world marketplace
  >Prevents "brain drain"

  Basic Physics research pays for itself in growth of new industries and diversification of old ones. Keep the SSC alive.

SAMPLE 12.9   Transparency 4 for a briefing on the Superconducting Super Collider. Again, Michele gives a list of points for the section, this time giving more detail for explanations of the benefits—probably the points on which the decision will be made.

Delivering the Oral Report

In preparing to give the briefing, review the general guidelines for oral presentations in Chapter 8. The substance of your report will be an elaboration of the points on your transparencies, which you deliver without further notes. This

• *Proton Beam Surgery*

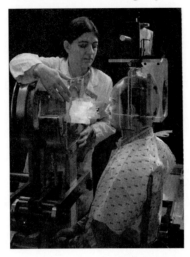

• *Train with Superconducting Magnets*

SAMPLE 12.10 Tranparency 5 for a briefing on the Superconducting Super Collider. Michele selected photographs to add concreteness, vividness, and impact in support of the points made in Tranparency 4. Because of their realism, photographs increase the credibility of claims about future benefits. She found the photographs in brochures provided by the public relations arm of the project.

Source: Photos courtesy of: Lawrence Berkeley Laboratory, University of California (left); Japan Railways Group (right).

requires a strong command of your subject matter and considerable practice. Remember that even if it were conventional to read your report or deliver a written text from memory (and it is not), you would have barely enough time in a ten- to fifteen-minute briefing to read the text of your position paper and present your graphics. So if you have mastered the material in your core document and graphics well enough to explain the main points on the transparencies, you should be in good shape.

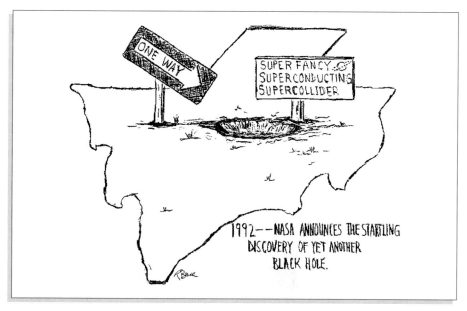

SAMPLE 12.11   Tranparency 6 for a briefing on the Superconducting Super Collider. Michele found this cartoon in a newspaper report and saved it for an optional graphic to be used, if she had time, to end on a light note and create a disarming rhetoric.

The last step is to rehearse and prepare mentally for the presentation. Take note of a few special considerations for the oral presentation of technical reports:

- *Make your purpose and position clear from the start.*
- *Emphasize the information the audience needs to make a clear decision on the issue.* Don't get bogged down in too much detail. But don't err on the side of unsupported generalizations either. Steer a sane middle course between these two extremes.
- *Have additional supporting information at your fingertips* (or better yet, in your head) so that you can cover questions about details with confidence.
- *Avoid a monotonous delivery,* but don't appear overly excited either. Vary your pitch and volume to emphasize key points that lead to responsible decisions. Speak slowly and deliberately to show conviction and seriousness.
- Confidence is the key. Project a tone of calm and knowledgeable assurance. *Rehearsal helps!*
- *Stay within your allotted time at all costs.* If listeners want to hear more from you, they will ask questions or request a more complete report in writing or at a later date.

Guidelines as a Glance 43 summarizes our recommendations regarding graphics and oral presentations.

GUIDELINES AT A GLANCE 43

Checklist for report graphics and oral briefings

1. Make sure your graphics do the following:
   - help your audience sort and structure the information you present
   - orient your audience in space and time
   - interpret information to aid in decision making
   - emphasize the most important points to prevent information overload
2. Make sure the transparencies you develop have the following characteristics:
   - are clear, easy to read, and not overcrowded
   - use primarily low-density graphics to make major points
   - provide effective forecasts and summaries
   - give enough information that if the audience members were not able to hear your oral presentation, they would still get the gist of your report
3. In rehearsing for your presentation, do the following:
   - review your position paper and research notes until you have a strong command of your subject matter
   - practice looking forward, making eye contact with the audience, and maintaining a fluent delivery and an expressive variation of pitch and tone for emphasis
   - time your report carefully, staying within your assigned limits

Step Five: Developing the Technical Report

Together with the outline that you have developed, your position paper and presentation graphics will constitute the basis for your final report. But you will need to write additional material and choose a format for delivering it. Although project reports and reports on options have distinct purposes and organizations, they occupy common ground when it comes to the rhetorical, stylistic, and ethical considerations of final drafting and framing the report. So the techniques discussed below apply equally well to both types of reports. In this stage of your work, you will use most of the critical and productive skills that you developed in Parts One and Two: information development, research, graphic artistry, style, document design, revision, and editing.

Drafting the Report

To draft the report, you should have first of all completed your research. As you work, however, you may find gaps in your argument that will require you to return to your sources of information. Never try to gloss over such gaps. Get the information you need to present a solid argument. Always leave plenty of time for drafting so that you can fill gaps and still have time left for revision and editing.

In writing your draft, you'll need to keep in mind three particular requirements: proper amplification or development of key points, complete and effective documentation, and the ethics of good reporting.

Amplifying Your Work: Expanding the Text and Creating Points of Emphasis   Drafting means, first of all, amplifying the text that you produced in the first four steps of the CORE method. *Amplification* is a technical term in rhetorical studies. In its simplest sense, it means to make a text or speech longer. Amplification involves knowing which information to cover in what amount of detail. It has to do with *elaboration.*

In the world of audio technology, to amplify means to make *louder.* So amplification also relates to the amount of stress you give to certain points through the use of repetition, placement, style, and graphical display. It has to do with *emphasis.*

In writing their reports, the student authors of the samples we have seen amplified the material they used in writing their position papers, graphics, and oral reports. Their original audience-action analyses—their understanding of the particular purpose, audience, and content of their reports—guided their choices about what to expand and emphasize. Michele Morgan chose to emphasize the economic advantages of the scientific project she discussed because the decision she hoped to influence related to funding, costs, and benefits rather than to advancing a more purely scientific agenda. Patrick Zuzek chose to survey a broad field of options because he wanted to lead his audience away from a narrow vision of the petroleum industry. He wanted his readers to open their minds to new alternatives and new approaches to their work. Chris Powers also wanted to introduce his audience to a new way of doing things. But he had a specific method in mind for solving what seemed to be a deadlocked problem.

In writing your own papers, your understanding of the audience, objectives, and contexts in which you are working will similarly guide your choices about what information to amplify and emphasize. As you draft and revise your report, keep in mind the following questions:

- How much information does my audience need to make a good decision?
- Have I given an adequate account of the projects or options I have discussed? From what I have said, can the audience visualize the actions involved and the likely outcomes?

Documentation: Bringing the Reader into the Network of Knowledge   One of the most important tasks in writing the full report is to document your sources carefully. The aim of good documentation and references is to allow readers the opportunity to retrace the paths of information by which you arrived at your position. With good documentation you bring the reader into the network of knowledge about your topic. Give good citations for printed sources and a clear indication of where you got unpublished information (such as dates and places of personal interviews or phone interviews).

Chris Powers's report (Sample 12.18 at the end of this chapter) uses the common author–date system of documentation, crediting his sources when he refers directly to information they provide and listing references at the end of the paper.

GUIDELINES AT A GLANCE 44

Questions to guide ethical reviews of technical reports

- Can you as an author guarantee the accuracy and sufficiency of the research that you cite? Your claims to technical knowledge and expertise—the service you are providing—require this guarantee for the report to be considered ethical. The guarantee does not mean that you must dump all the data into the final report, swamping the reader with repetitive or irrelevant facts. It does mean that you should be able to produce complete data on request.

- Have you developed (analyzed) the information in a way that considers all the available data? Your claim to objectivity—your responsibility to be fair—requires that you not omit troublesome facts or those that don't fit neatly into your chosen categories. Scientist and critic Evelyn Fox Keller defines *objectivity* as "the pursuit of a maximally authentic, and hence maximally reliable, understanding of the world around oneself."[*]

- Have you clarified the reasons why you recommend one position over another? Be sure you do not have any "hidden agendas" in your report. Avoid using rationales that the audience is likely to accept while hiding rationales that may be objectionable.

---

[*]Evelyn Fox Keller, *Reflections on Gender and Society* (New Haven: Yale University Press, 1985), 119.

A full treatment of documentation appears in Chapter 10 on editing. You should review that information before you draft the final version of your report.

Ethical Issues: Meeting Your Responsibilities to the Reader   Providing clear documentation is only one of the many responsibilities you have to your reader. As you amplify the information for your report, consider the questions in Guidelines at a Glance 44, which focus on your duties and responsibilities to your reader. You can use these questions to guide an ethical review of your first draft of the full report.

Sometimes, these ethical issues are not clear-cut. In the report on the Superconducting Super Collider, for example, Michele argues that the SSC should be kept open because of its important contributions to scientific knowledge, to the local economy, and to technological progress in general. While few readers would dispute her claims that the project can advance specialized scientific knowledge, a tough critic might question whether it will have an overall positive impact on the local economy and provide important technology for the future. Moreover, many have grown weary of NASA's claims about the "spin-off" technologies that have come from the space program. Some skeptics view these claims merely as government-science propaganda.[2]

---

2. John Lyne, a rhetorical critic from the University of Iowa, tells a joke that reveals the cynicism that some people feel about these claims: "If the space program gave us Tang," the joke goes, "imagine what kind of fruit drinks we'll get from the SDI project!"

So the report may be slightly unethical in its tendency to overstate the technological advantages while downplaying more basic scientific gains, which actually represent the SSC's greatest potential contribution but may not win as much sympathy on Capitol Hill. In a sense, Michele is selling the project to her audience.

The problem is as old as Plato (who took a hard look at the relationship between ethics and rhetoric in his dialogue *Gorgias*), but it cannot be solved theoretically. We must simply look at each case as it arises and make a judgment on how it fits into a larger context of action. We would consider the SSC report ethical because it is likely to be considered among other reports that will challenge its view of the SSC's contribution. The report does clearly state that the main contribution of the SSC will be in the arena of basic scientific knowledge, even though it goes on to develop a rationale that is more acceptable to an audience of government decision makers.

Reports on projects that recommend a single course of action are more liable to be ethically murky than are reports on options, which typically cover all the possible alternatives for solving a problem. Yet even in reports on options the author must eventually take a clear position, give a recommendation, or make a judgment of some kind. And every decision must be considered potentially unethical and reviewed carefully.

Framing the Report

After the main part of your report is drafted, another consideration arises. How will you frame your report? The answer will depend on the rhetorical situation in which you work. Depending on the circumstances, you can package your report as a memo or as a formal report.

Framing a Memo Report   Memoranda (or memos) are discussed more fully in Chapter 4 on correspondence. Here it will suffice to say that memo reports are internal documents, used by communicators delivering information to decision makers within their own companies, organizations, or agencies.

Memo reports differ from formal reports in appearance and style. They have a different covering format and a slightly different tone:

- The covering format is simply the To/From/Subject/Date form typical of communications in office environments.
- The tone may be slightly more personal than that of a formal report, since memos are always internal documents addressed to an audience within your own organization. You probably know the reader personally, but you also know that the reader may share your report with others whose needs and agendas you may not know. So you cannot lapse into the kind of informality characteristic of some interoffice communications.

Sample 12.12 is a memo report on options, written in response to an inquiry on the topic of global climate change. (See Case 4.2 in Chapter 4 for a scenario

Note the typical format for a memo with to/from/date lines. See Chapter 11 for more about memo formats. The subject line contains the equivalent of a good informative title.

The first paragraph sets the context, using direct address—*I* to *you*.

By paragraph 2, the report proceeds as any report on options would.

As in any report, the section is the basic unit of design (See Chapter 7 on document design.) The headings are partly informative, like headlines in a newspaper, and partly structural (see "Conclusions" and "Recommendations" later in the report). The organization is hierarchical. Each paragraph begins with a point sentence.

<u>MEMORANDUM</u>

April 1, 1992

TO: Regina Ergill, Chairperson,
        Sansville Electric Cooperative

FROM: Jim Keller, Public Information Specialist

<u>SUBJECT: What Should We Do about Global Warming?</u>

I am writing to respond to your memo of March 15 asking for information about the concept of global warming. I agree that it is an important topic, which many of our members may take a strong interest in. I have reviewed the technical and policy literature and have consulted with our staff scientists and engineers. I have also contacted Professor Harry Weatherly, the foremost meteorologist at Downtown State.

The available literature and expert assessments suggest that, while global climate change is a significant topic for current scientific research, fully worthy of continued investigation, not enough is yet known about the phenomenon to warrant major changes in our energy policies. An article in a 1989 issue of *Science* magazine suggests that continued efforts at energy conservation represent the best possible course. Indeed, our co-op may be rather far ahead of other utilities in doing our part to prevent drastic changes in the global climate.

The sections below report on my findings and recommend a stepwise policy based on the best available literature.

<u>The Problem: Science and Policy</u>

Since the summer of 1988, the warmest on record, the public has become alarmed over reports that modern industrial society may have significantly altered the climate of the Earth. The burning of carbon-based fuels over more than two centuries of industrialization and, more recently, the introduction of large quantities of complex carbon gases have resulted in the largest concentrations of atmospheric carbon in the history of the Earth. Carbon gases trap the sun's heat in the atmosphere, creating the "greenhouse effect." Under normal conditions, this heating of the Earth is moderated by the capacity of the seas and other "carbon sinks" to reabsorb the gases. However, the increase of

SAMPLE 12.12   Memo report on options: A response to an inquiry

carbon gases currently in the air correlates with an increase in global temperatures of 1 to 2 degrees Celsius. This fact leads some researchers to argue that industrial society has created an imbalance in nature, favoring the greenhouse effect and an overall trend toward global warming. The ultimate effects could be disastrous.

The problem for ordinary citizens and policy makers divides into two parts:

1. We must sift through the evidence and arguments of the scientific experts to decide whether the greenhouse theory has merit.

2. On the basis of our decision, we must further decide whether and how to adjust our personal and public actions to allow for this new scientific information.

Recent findings suggest that, though we cannot yet be sure about how much human beings have altered the global climate, we do know beyond a reasonable doubt that the Earth has warmed during this century. Further, if policy changes are delayed too long, we may have missed our chance to remedy the situation by curbing actions that have led to greenhouse warming. Thus, we should implement policies that would decrease global warming according to current theories and the best available evidence.

The Greenhouse Hypothesis

The hypothesis of the greenhouse effect has been around for several decades now, long enough to have received full treatment in recent textbooks. It has become a well-established theory in atmospheric science and is widely used, for example, to explain why the surface temperature of Venus is so much higher than that of Mars. Venus has an atmosphere rich in carbon dioxide and other gases. Like the glass panes of a greenhouse, these gases trap the sun's radiant infrared energy so that more heat remains inside than is allowed to escape. The atmosphere of Mars, by contrast, is poor in carbon dioxide and thus releases more of the sun's heat back up into space. While the Martian surface is frigid, the atmosphere of Venus is a "runaway greenhouse," far too hot to support life (Schneider, "Greenhouse Effect," 771).

The controversy arises when scientists apply the theory to human influences upon the Earth. As early as 1827, the

The author interprets the problem by dividing the question into 2 main issues. Each may affect the recommendation and the ultimate action differently. Note the frequent use of lists as an element of document design.

SAMPLE 12.12 continued

French scientist and mathematician Jean-Baptiste Fourier suggested that human activity might alter climate. In 1957, Revelle and Seuss coined what has become a very popular phrase among scientists, "large-scale geophysical experiment," to describe the tendency of human-influenced processes to result in an increase of carbon dioxide in the air (Ramanathan, 293-94).

Normally, a balance is maintained between the introduction of carbon dioxide into the air and the incorporation of carbon dioxide back into the global ecosystem. In the oceans, carbon dioxide is dissolved; in green plants, it is used in photosynthesis, which releases oxygen. But processes like electric power generation, industrial combustion, and the burning of fuel for transportation, cultivation, and the heating of homes and factories for an exponentially expanding population have overloaded the atmosphere, releasing more carbon dioxide than the natural processes of conversion can accommodate (ReVelle and ReVelle, 504).

In a now famous study, C. D. Keeling showed graphically how the increase in global temperature over the last few decades matched very closely the increase in atmospheric concentrations of carbon dioxide, as measured at the Mauna Loa observatory in Hawaii (ReVelle and ReVelle, 505; Keeling).

Future Scenarios

According to the hypothesis, then, the increased level of carbon dioxide in the air produces the greenhouse effect. Eventually the temperature may stabilize, but at some higher level.

The grimmest forecasts suggest that atmospheric alterations could result in a net gain in global warmth large enough to cause substantial melting of the Earth's polar caps and a consequent rise in sea level. Rising sea levels would make important population centers such as New Orleans and the Netherlands uninhabitable. The problems of food supply in the face of overpopulation and famine, which social scientists already rate as a crisis, would grow worse. In areas with the most fertile soils, such as the American Midwest, "the world's breadbasket," decreased rainfall and increased heat could lead to desertification, while temperate climate patterns could prevail in areas with soils less inviting for large-scale agricultural development (New Mexico, for example). There could also be negative contributions from related problems, such as acid rain—also caused by emissions

SAMPLE 12.12 continued

of waste gases from industrial and technological processes
(Pain, 38–40; Lemonick, 36–39).

Faced with these possibilities, a number of scientists have
become concerned about how soil and water fit for human and
animal use can be sustained. Specialists in atmospheric
science, like Stephen Schneider, James Hansen, and George
Woodwell claim that there is a broad consensus among scientists that the Earth is warming up (Schneider, <u>Global
Warming,</u> 23).

But few will claim consensus about the causes, the exact
effects, the rate, or the future of the warming. When Hansen
testified before a Senate hearing in 1988 that he was 99%
sure that the hot summer was caused by the greenhouse
effect, even scientists who basically agreed with him, like
Schneider, criticized the statement (Schneider, "Greenhouse
Effect," 779). Schneider himself has been criticized for
going public with information before a solid consensus has
been reached in the scientific community (Schneider, <u>Global
Warming,</u> 195–96).

Most scientists tend to be cautious in their conclusions and
are generally reluctant to relate their findings to policy
options. A typical comment comes from the University of
Chicago atmospheric specialist, V. Ramanathan: "Observed
records do reveal a warming of the order of about 0.5 K,
but . . . theoretical understanding of the climate system
is by no means complete" (298).

<u>Related Concerns</u>

In the 1980s, three factors have increased concern about the
possibility of human-induced climate changes and have led
more scientists and other observers to take the greenhouse
hypothesis seriously:

1. Mounting evidence shows that chlorofluorocarbons (CFCs) and
   other complex organic gases released during recent industrial processing intensify the effect of carbon dioxide and
   other "greenhouse gases" (Schneider, <u>Global Warming,</u> 21).

2. Evidence proves that these complex gases contribute to
   the destruction of the ozone layer in the stratosphere,
   one of the important chemical buffers that moderates the
   intensity of the sun's rays in the lower atmosphere. In
   1985, a previously unrecorded "ozone hole" was discovered
   over Antarctica (Kerr, 785–86).

SAMPLE 12.12 continued

The conclusions answer the main questions raised by the data. All the conclusions should have some influence over the actions recommended in the next section.

3. In recent years, a pattern of galloping desertification has developed in many areas of the world, which, when coupled with the destruction of rain forests for cattle production and other forms of development, clearly weakens the ability of the Earth's green systems to assimilate carbon gases (Gore, 115-25).

All of these developments increase the danger of "feedback loops" that would cause global warming to increase, not linearly, but exponentially (Gore, 50-55). This danger leads advocates of global warming prevention to urge action now.

Conclusions

1. A trend in global warming over the last century clearly correlates with an increasing introduction of carbon gases into the atmosphere.

2. The causes, effects, and extent of human-induced global climate change remain unclear.

3. Because of increases in complex carbon gases, ozone thinning, and the destruction of carbon sinks, the potential danger of global warming increases to the point that to wait is to risk irreparable damage.

4. Since the risk involved outweighs the alleged lack of closure on the scientific question, some change in current policy is justified.

The recommendations advise directly on decisions about what actions to take. They answer the question raised in the subject line, "What should we do about global warming?"

Recommendations

Observers on the political scene differ widely on how the public should respond. In their books, scientist Schneider and Vice President Al Gore both suggest the need for international legislation comparable to the Montreal Ozone accord, but conservatives argue that the scientific understanding of global warming and its effects does not yet warrant such expensive actions that would regulate industry on a large scale.

Schneider mentions an alternative plan that is likely to be more widely acceptable because it is more moderate. This "tie-in strategy" says quite simply that "society should pursue those actions that provide widely agreed societal benefits even if the predicted change does not materialize" (Schneider, "Greenhouse Effect," 779).

SAMPLE 12.12 continued

Along these lines, obvious directions for policy changes at all levels--from international to personal action--include the following:

1. a strong effort at <u>energy conservation,</u> which would cut down on carbon emissions, and

2. a widespread effort at <u>forest regeneration,</u> which would restore effective carbon sinks.

Despite enthusiastic Department of Energy recommendations, tree-planting alone could not stop global warming (Marshall). The addition of a solid conservation program for fossil fuels would not help much more, according to the worst-case scenarios.

Nevertheless, much would be gained by efforts to restore and protect forests and to conserve energy supplies, even if current theories on global warming turn out to be over-stated. If the theories prove to be on the mark, much stronger policies will be necessary, but the tie-in strategy will have already turned action in the right direction.

For our purposes, then, the literature points to two possibilities for policy adjustment:

1. Maintain and consider extending current initiatives in energy conservation.

2. Consider a program for cultivation and replanting of local forests.

References

Gore, Al. <u>Earth in the Balance: Ecology and the Human Spirit.</u> New York: Houghton Mifflin, 1992.

Keeling, C. D. "Characterization of Information Requirements for Studies of CO_2 Effects." U.S. Department of Energy, Carbon Dioxide Research Division, December 1985.

Kerr, Richard A. "Ozone Hole Bodes Ill for the Globe." <u>Science</u> 241 (1988): 785-86.

Lemonick, Michael D. "Feeling the Heat." <u>Time</u> 2 Jan. 1989: 36-41.

The list of references is a bit formal for a memo report, but is often required. The list uses MLA documentation style.

SAMPLE 12.12 continued

Marshall, Eliot. "EPA's Plan for Cooling the Global Greenhouse." <u>Science</u> 243 (1989): 1544-45.

Pain, Stephanie. "No Escape from the Global Greenhouse." <u>New Scientist</u> 12 Nov. 1988: 38-43.

Ramanathan, V. "The Greenhouse Theory of Climate Change: A Test by an Inadvertent Global Experiment." <u>Science</u> 240 (1988): 293-99.

ReVelle, Penelope, and Charles ReVelle. <u>The Environment: Issues and Choices for Society.</u> 3rd ed. Boston: Jones, 1988.

Schneider, Stephen H. <u>Global Warming: Are We Entering the Greenhouse Century?</u> San Francisco: Sierra Club, 1989.

------. "The Greenhouse Effect: Science and Policy." <u>Science</u> 243 (1989): 771-81.

SAMPLE 12.12 continued

that may have led to such a report.) The author is a technical expert, and the audience is a decision maker. The subject matter is information that may affect an organization's policy—the way people will act. Following the basic outline for reports on options, the author provides both information and recommendations on directions for future policy.

The only things that make this report different from a formal technical report are the framing devices: the format and the more personal introduction. The author addresses the audience directly as "you" and refers to a specific situation that gave rise to the report.

Framing a Formal Report   Formal reports require a bit more framing than memo reports. Depending on the special requirements of your audience, you may need to add any number of additional elements. The full report may include any or all of the parts listed in Table 12.2.

The *letter of transmittal* is a cover letter that accompanies the report, explaining briefly its context and purpose. If it goes into more detail, it may serve as an executive summary, the part of your report that the busiest decision makers will read. Sample 12.13 shows a letter of transmittal for Chris Powers's options report on biowaste recycling. For a complete treatment of the form and content of letters, see Chapter 11.

TABLE 12.2

The parts of a full formal report. The elements in bold type are essential. The degree of formality will determine whether the parts marked "optional" are required.

- A memo or letter of transmittal (optional)
- **A title page**
- A table of contents (optional)
- **An abstract**
- An executive summary (optional)
- Lists of tables and figures (optional)
- **The report itself**—based on outlines for activities or options reports
- A glossary (optional)
- An index (optional)
- An appendix (or several appendices) (optional)

A *title page,* such as the one accompanying the full reports at the end of this chapter, simply lists the content, audience, author, and date of the report. The company or agency for which you write may, of course, require other information, such as project numbers and other forms of identification.

A *table of contents* is useful for longer reports. The table of contents usually lists at least the first two levels of headings and may go as far as the first three levels (such as chapter headings, section headings, and subsection headings). Short reports (under ten pages) should not require a table of contents. Sample 12.14 is the table of contents from a professional report on options.

The *abstract* of a report (sometimes called a summary or synopsis) gives the fullest representation of the content as possible within a very short space and ranges from 100 to 300 words. It is a summary that gives a balanced view of the report. A good rule of thumb for writing the abstract is to write a sentence for every major section in the report, making sure that you clearly state the following:

- the problem addressed in the report
- the report's purpose
- the position supported by the report

Both of the sample reports at the end of this chapter (Samples 12.17 and 12.18) provide abstracts short enough to be included on the title pages of the reports.

An *executive summary* differs from an abstract in its emphasis on recommendations and conclusions. It is directed primarily not to technical staff but to *executives,* who have a special interest in costs and benefits. Position papers, such as those discussed earlier in this chapter, can often serve as executive summaries, as long as you return to them after writing the full report so that you

Office of the City Engineer
42 County Drive
San Diego, CA 77777

October 20, 1992

Honorable John R. Public, Mayor
City Hall
San Diego, CA 77777

The letter is single-spaced in a block format. At the top of the page, it includes (in
order) a return address (without the name
of the sender, which is reserved for the signature line), the date, and the inside
address. All elements of the letter are flush
against the left margin.

The formal greet-
ing is followed by
a colon (:).

Dear Mr. Mayor:

The first para-
graph establishes
the context.

Enclosed is the report you requested from our office on June 16. You asked that we
consider options for dealing with the city's solid waste problems.

The second
paragraph sum-
marizes the key
finding and rec-
ommendations.

Our study concludes that we should follow New York and Los Angeles in a program
to recycle biowaste. We think that this course of action will not only prevent an
environmental crisis but also turn a profit for the city in the long run.

Thank you for your attention to this crucial topic. Please call on us if we can assist
you further.

Sincerely yours,

The signature
line includes the
title of the
sender. The
abbreviation
"enc." indicates
that something
(the report) is
enclosed.

Chris Powers
Assistant City Engineer

enc.

SAMPLE 12.13   Letter of transmittal for a report on options

can revise them to show any changes in information or interpretation that may
have come about during the writing process. Executive summaries should always be relatively short and crisply written.

An *appendix* (or several appendices) may appear at the end of a long report.
An appendix contains information that may be useful for technical reviewers
but that will not be of much interest to most readers. Appendices may include
additional data to support the report's positions and evidence such as documents and letters that do not fit effectively into the body of the report.

Don't underestimate the importance of these framing devices—especially
letters of transmittal, titles, abstracts, and summaries. These may be the only
parts of your report that many people read, and they will certainly be the first

GUIDELINES AT A GLANCE 45
Review checklist for a full draft of a technical report

- Does the report provide enough information to allow the audience to make a good decision? Is there too much information? What should be added or omitted?
- Does the report provide an adequate account of the projects or options discussed? Can you visualize the actions involved and the likely outcomes?
- Are the graphics sufficient to the task? Should any graphics be added? Which graphics could be improved?
- How could the writing be improved?
- How could the document design be improved?
- Does the report need further editing? What type of editing?
- Is there adequate documentation? Could the reader find the documents referred to and trace the sources of information, reconstructing the network of knowledge?
- Are there potential problems with the ethics of the report? (See Guidelines at a Glance 44.) How could these be solved?

thing that all readers see. You want to get off on the right foot, and these devices are the keys to doing so. In addition, be sure that your report is well designed and edited according to the principles discussed in Chapters 7 and 10. Guidelines at a Glance 45 provides a checklist for reviewing your final report.

Contents

SAMPLE 12.14   Table of contents from a professional report on options
Source: Congress of the United States, Office of Technology Assessment, *Changing by Degrees: Steps to Reducing Greenhouse Gases,* Summary Report, 1991.

EXERCISE
12.1

Doing an Audience-Action Analysis.   This exercise will guide you through the process of writing a paper on a topic of your choice. To get started, you need to select a topic and then do an audience-action analysis that defines the components of the rhetorical situation in which you must write.

A. Select your topic.   As you try to think of a topic, consider which special knowledge that you are learning in one of your major courses might be useful to special groups of decision makers. You will probably need to do some initial research before you finally settle on a topic, so you may want to begin with a couple of alternatives in mind.

Here are some sample topics that you can use or adapt for a report:

- Methods for ensuring a plentiful and safe water supply for your region of the country
- Traffic control models—how they could be applied to improve traffic in your community
- Alternatives in public transportation—other systems that might help to balance costs and benefits in your city
- Nuclear power generation—the promise of the latest improvements in design and operation
- Electric power generation—options for public utilities
- Obesity as a concern among office workers—how a corporate fitness program can help
- Disposal of hazardous wastes—dealing with the problem on a local or national level
- Adult computer literacy—community-based resources and programs
- Year-round schooling—costs and benefits
- DNA-profiling of accused criminals—dependability, costs, and advantages over other means of identification
- Comparative advantages of brick versus wood home construction in your region
- Recreational facilities in local municipal parks—adequacy and safety in meeting current demands
- Motorcycles or "donorcycles"—safety design issues
- Mosquito control—disease vectors versus environmental concerns
- Nutrition for homeless children—degree to which shelters are meeting their needs
- Over-the-counter pain relievers—safety and effectiveness comparisons

B. Do your analysis.   As you select your topic and begin to gather information about it, ask yourself the following questions, refining your answers as you work:

- *Who exactly is the audience?* (The city council or school board? The Federal Drug Administration? The local power utility? Upper management in your company?)

- *What action outcome do I envision?* (The city will provide funds to redesign and upgrade local parks? The FDA will require new labels for pain relievers? The power company will commission a study on the feasibility of nuclear power?)
- *What recommendations should I make?* (Construction of wood houses to reduce indoor exposure to radon gas? City regulation of nutritional content of meals served in homeless shelters? The use of one brand of aspirin over another for specific types of pain relief?)
- *What kind of information will my audience need to be able to reach a good decision? What sources should I use?* (Data on the number and type of injuries in local parks? Testimonies of citizens and experts? Explanations of alternatives and model practices? Comparisons of costs and benefits among several options?)

Discuss your short answers to these questions in a small group in your class. Then evaluate one another's responses. Which of the projects seems to be the best conceived? What does it have that the others lack? When you have finished your discussion, ask the author of the best-conceived project to describe it to the whole class. Then compare the merits of the best project from each group.

Part 1. Reviewing Outlines for Technical Reports.   In your discussion groups, select one of the outlines in Samples 12.3 and 12.4 on pages 375 and 376. Discuss the strengths and weaknesses of the structures as planned. How would you structure it differently? Also discuss with the members of your group how your topic fits into one of the two general types of reports. Seek advice on the best way to organize your report.

Part 2. Writing Your Outline.   Using the models in this chapter and the advice you've received in group discussion, draft an outline for a full report on your project. Bring the outline to class, and discuss it in your group. Using the feedback you get, revise it and submit it to your instructor for approval.

EXERCISE **12.2**

Part 1. Evaluating Position Papers.

Evaluation A.   Michele's paper (Sample 12.5, on page 379) is nicely written, but as a technical paper it leaves some things to be desired. In a small group of your fellow students, discuss its strengths and weaknesses. Using Guidelines at a Glance 45 on page 401, try to decide what the paper needs to make it a first-rate technical paper. If this were your paper, would you add any of these features now, at the position paper stage, or would you wait until you develop the full report? Remember that the paper should be short, no more than about two or three double-spaced pages.

EXERCISE **12.3**

Evaluation B.   Sample 12.15 shows Patrick Zuzek's position paper for a project with an engineering slant. Like Michele's paper, this one gives a recommendation on the future direction of research and development. But it uses a slightly different structure and conveys a different tone. Discuss the paper in your group, trying to pinpoint the differences in structure and rhetoric—and in persuasiveness. In what ways is Patrick's paper more effective than Michele's, and in what ways is it less effective? What does Patrick's paper need to be completely successful? Again, begin with the questions in Guidelines at a Glance 45.

Position Paper on Alternative Fuels

Patrick Zuzek

The purpose of this report is to give members of the Petroleum Industry Council information about alternative fuels so that they can develop or influence a fuel policy that will not only sustain current levels of performance, safety, and comfort in cars but also promote economic and environmental interests. In the past, diesel, and to a much greater extent gasoline, have been the exclusive fuels for vehicles. Both of these fuels are derived from petroleum. In recent years, the use of these fuels has been questioned. Environmentalists have pushed continually for reduced emissions and higher mileage since the sixties. These measures reduce air pollution, smog, acid rain, and (theoretically) the greenhouse effect. Environmentalists would also like to end the damage caused by drilling for oil. Energy independence is a national economic interest. At present, America and most other countries are captive to Middle Eastern politics and OPEC's oil price setting. The oil embargo and price fluctuations during the Gulf War provide vivid evidence of this dependence.

Oil and auto industries have an interest in alternative fuels. Since experts now firmly believe that the world oil supply will last for only thirty to eighty more years, industry growth depends upon diversification. Legislation already exists for further reduction of emissions and for mandatory variable fuel and zero emissions (electric or hydrogen) vehicles. Concerned industries should try to point legislation towards a viable solution rather than try to stop or delay it because an alternative will be decided, and without industry input it could be the wrong decision.

Three existing alternative fuels have special promise for development by the petroleum industry: propane, alcohol, and natural gas. These fuels would be used alongside reformulated

SAMPLE 12.15   Position paper for a project with an engineering slant gasoline in the short term to meet existing EPA legislation. The technology is already available, and the conversion is relatively simple and low cost. These fuels are also avail-able nationally to some degree. During this time, new tech-nologies such as solar and electric energy would be developed and used for limited and specialized functions, such as urban commuting, where zero exhaust emissions could also be used to advantage.

However, the real fuel of the future is hydrogen. Although it could eventually replace gasoline, hydrogen still needs extensive research. Hydrogen would be favorable because it is readily available, it's a good source of energy, and it produces very few emissions.

Development of hydrogen fuel while using other fuels in the meantime would benefit everyone. Environmental interests would be pleased with the reduction of oil drilling as well as the annual reduction of emissions until emissions finally become negligible with hydrogen. The auto industry would be pleased by the advantage of more development time as well as reduced short-term costs. The oil and other energy industries would be able to slowly shift production toward alternative fuels, thus providing some stability.

New fuel laws should not be seen as the enemy of industry. The development of responsible legislation could provide a better environment for everyone. It could also provide economic strength and stability to America's economy and two of its most important industries, energy and automobiles.

SAMPLE 12.15 continued

Interventions in Early Childhood
Kate Hendrix and Patricia J. Molloy
Education Development Center, Inc.

[The purpose of this background paper is to provide public health officials with the information they need to educate their constituencies on how to reduce violent behavior in minority youth by applying interventions in early childhood.] The prevention of violence among minority youth has been approached from many different perspectives, but intervention efforts that begin in adolescence may be too late in the process of child development to effect comprehensive, long-term changes in aggressive behavior. Early childhood, then, represents an optimal time for violence prevention intervention.

continued

SAMPLE 12.16 Professional position paper
Source: *Public Health Reports* 106 (May–June 1991): 275–276.

Aggression is learned from a very early age; and as children move through the normal course of development, family and peer relationships, living environment, and media violence all affect the development of aggressive behavior. For minority children, however, an increased likelihood of low socioeconomic status may increase exposure to violence and present greater obstacles to the development of social competence than those confronting children from higher socioeconomic backgrounds. Interventions designed to reduce the development of aggression in early childhood fall into two broad categories: educational-behavioral and therapeutic.

Although not abundant, educational interventions for new parents do exist and are often incorporated into prenatal and well-child health care. These programs focus primarily on increasing care giving and discipline skills, anger management, and social support, and they employ parent aides-mentors and home visitation. Relatively few effective educational intervention strategies targeting very young children exist, but many programs designed for early elementary school children are adaptable to culture-specific day or preschool environments, or both. These interventions commonly use strategies designed to teach or enhance interpersonal problem-solving skills, self-esteem, anger management, communication, conflict-resolution skills, and empathy.

For parents and children already displaying violent or aggressive behaviors, therapeutic interventions are also available, but often less readily so for low-income families. Aside from individual or family counseling, only a few therapeutic alternatives are available for parents. However, some residential and therapeutic day care programs involve both parents and children. In addition, foster care, grief counseling, and conduct-disorder therapy are available for children.

Aggressive behavior is produced by a constellation of factors, and effective intervention programs must be based on a multidimensional, culture-specific approach. Additionally, interventions in early childhood must often include two target populations—both parents and children—because of the enormous familial influence during early childhood. However, early childhood intervention strategies are often hampered by the problem of access. Many families are isolated from social services and quality health care, and many children under age 5 are not enrolled in center-based day care or preschool. Reduced access means that numerous young families, especially families of low socioeconomic status, find that center- or school-based interventions are not readily available. Community-based interventions, then, may offer the most effective means for reaching inner-city families with very young children who are at high risk for violence.

There is clearly a need for further research and evaluation, especially in the development of reliable measures of behavioral outcomes. In addition, a number of critical issues, including cultural norms, socioeconomic status, media violence, children's resiliency, and social policy, must be considered in designing effective interventions. Until racial disparities are removed from society, minority children will continue to be at higher risk than their nonminority counterparts for exposure to violence and possible subsequent aggressive behavior. Interventions in early childhood that encourage the development of social competence may be one of the most effective strategies to prevent violence among minority youth.

SAMPLE 12.16 continued

Evaluation C.   Sample 12.16, beginning on page 405, is a summary of a background paper delivered at a professional conference for public health officials. In its published form, it closely resembles the kind of unpublished position papers that technical professionals regularly produce. We've added an introductory sentence [in brackets], but otherwise the original paper employs the structure we've recommended—from statement of purpose and position to background and on to a brief elaboration of the position.

In your group, discuss what makes this professional paper different from the student examples. Do not assume that, just because it's a published paper, it is perfect. Note ways that the style could be improved to reach a wider audience. Review the guidelines in Chapter 9 if you need to, and note places where jargon could be eliminated and the sentences made clearer. What about the use of passive voice? Is wordiness a problem? How would you revise the paper if it were yours or a coauthor's?

Part 2. Writing Assignment.   After you have completed the evaluative part of this exercise, you should be able to write a position paper yourself. Use the following steps:

1. Using the notes you wrote for your audience-action analysis in Exercise 12.1, draft a position paper on a technical topic. As you write, keep foremost in your mind the question, "Who will do what with this information?" The information that you provide may not solve the whole problem (what to do about global climate change caused by increased carbon emissions, for example), but it may contribute to the overall solution (by cutting down on carbon emissions) or solve a particular local version of the problem (relieving air pollution in a small region).
2. Bring your draft to class and share it with a small group of classmates. After everyone in the group has read all of the drafts, spend some time discussing each one, using the same evaluative techniques that you developed in Part 1 of this exercise.
3. After the evaluation session, revise your paper, drawing on the insights you gathered from the peer review.
4. Submit the revised paper to your instructor for additional comments.
5. On the basis of these comments, prepare a final draft of your paper.

Part 1.   In your discussion group, return to the sample position papers on alternative fuels (Sample 12.15, pages 404–405) and interventions in early childhood (Sample 12.16, pages 405–406), which you evaluated in Exercise 12.3. Make a list of the graphics you would need in a full report based on these position papers. Which would be most appropriate for the oral briefing and which should be saved for the full written report? Compare your list with those of other groups.

EXERCISE
12.4

Part 2.   Now consider your own position paper (developed in the second part of Exercise 12.3). Make a list of graphics you will use first in your oral report and then in your full written report. Share the list with your discussion group and get further ideas. Complete the list, and submit it to your instructor for further advice.

EXERCISE **12.5**

Part 1. Evaluation.   In your discussion group, go over the transparencies in Samples 12.6–12.11. What are their strengths and weaknesses? If you were doing the briefing, what would you do differently? Make a list of improvements, and discuss them with the class as a whole.

Part 2. Preparing and delivering an oral briefing.   Use the following steps:

1. On the basis of the position paper and graphics plan you developed in Exercises 12.2 and 12.3, develop a plan for a short oral presentation. Draft a set of transparencies for the presentation.
2. Show your tentative transparencies to your discussion group. For every group member's transparencies, ask the following review questions:
   - Who is the audience?
   - How is the intended audience likely to respond to each transparency?
   - Do the transparencies clearly convey the author's intended position?
   - Would the audience be able to give a short account of the topic just from reading the transparencies?
3. After the discussion, revise your transparencies and rehearse a presentation based on the transparencies and position paper until you are ready to present an oral report to the class or to a small group from the class.
4. Take turns delivering your briefings. In an informal group discussion, critique each presenter on the basis of Guidelines at a Glance 12 in Chapter 8 and the special considerations discussed above. You may also refer to the checklist in Guidelines at a Glance 42.

EXERCISE **12.6**

Part 1. Reviewing Technical Reports.   In your discussion groups, select one of the sample reports given at the end of this chapter on pages 410–425 (Sample 12.17 or 12.18). After everyone has read the report, discuss it in detail, making a list of features you think could be improved. Use the checklist in Guidelines at a Glance 44 to review the document generally. Use the questions in Guidelines at a Glance 43 to perform an ethical review. Do all of the parts contribute to an effective whole? If not, what does the report need to become a better decision-making tool? Once your list is complete, share it with other groups in a general discussion.

Part 2. Writing Your Report.   Using the information you've developed in research, the notes you've made in your audience-action analysis, your graphics plan as well as the graphics you used in your oral briefing, and a working outline based on the models for project reports or options reports, you can now write your full report. Use the following steps:

1. Draft a formal technical report based on the audience-action analysis, the position paper, the graphics, the oral report, and the outline you prepared in the exercises for this chapter.
2. Using Guidelines at a Glance 44, review your draft with a small group of classmates.
3. After the discussion, revise your paper and submit it to your instructor for an evaluation. Be sure to prepare a title page and an abstract. If you are required to use additional elements or if you think you need more framing, also prepare a letter of transmittal, a table of contents, a list of tables, a list of figures, a glossary, an index, and/or appropriate appendices.

Recommendations for Further Reading

Barabas, Christine. *Technical Writing in a Corporate Culture: A Study of the Nature of Information.* Norwood, NJ: Ablex, 1990.

D'Arcy, Jan. *Technically Speaking: Proven Ways to Make Your Next Presentation a Success.* New York: American Management Association, 1992.

Killingsworth, M. Jimmie, and Michael Gilbertson. "How Can Text and Graphics Be Integrated Effectively?" *Solving Problems in Technical Writing.* Lynn Beene and Peter White, eds. New York: Oxford University Press, 1988: 130–149.

White, Peter. "How Can Technical Writers Give Effective Oral Presentations?" *Solving Problems in Technical Writing.* Lynn Beene and Peter White, eds. New York: Oxford University Press, 1988: 191–204

Documents for Review

The title gives as much information about the content of the report as possible.

In technical writing classes, the title page is a useful device for indicating the intended audience and the perspective of the author.

It is best to keep the text on the title page as short and simple as possible.

<u>The Superconducting Super Collider:</u>
<u>An Investment in America's Future</u>

A Report Submitted to the Research Staff of
Congressman Ben Smith

by Michele Morgan
Waxahachie Research

October 20, 1992

SAMPLE 12.17   Student report on activities (project report)

The Superconducting Super Collider:
An Investment in America's Future

Abstract

The Superconducting Super Collider is a project in high-energy physics centered in a Waxahachie, Texas facility currently under construction. Recently, Congress considered cancelling the project. This report shows why funding for the project should continue. Progress toward completion and start-up is so far along that cancellation would cost thousands of jobs and billions of dollars. Scientific knowledge as well as economic and technological growth would suffer. Benefits relating to applications of superconductivity in electronics, communications, transportation, and medicine would be lost. The SSC creates jobs as it fortifies America's position as a leader in the development of high technology. It is a good investment in the future of the country.

The abstract, like the title, covers as much of the report as it can. Each section of the report is expressed with a sentence or two in the abstract. The abstract is thus a miniature version of the report. The information it gives stays at a fairly high level of generality, though, avoiding details such as statistics.

SAMPLE 12.17 continued

The title is repeated on first page of the report proper.

The unheaded introduction states the problem and the position and forecasts the content and purpose of the report.

The internal documentation uses the author–date system (APA style).

Informative headings reveal the content and structure of each section.

The text refers directly to all figures, which are incorporated into the report as close to the reference as possible.

Documentation placed at the end of the paragraph usually indicates that much of the information given in the paragraph was taken from this source.

Notice that topic sentences appear at the head of each paragraph. If the reader reads nothing but these, the main points would still come across.

<u>The Superconducting Super Collider:</u>
<u>An Investment in America's Future</u>

   Americans have led the world in basic scientific research for most of this century. Much of this success has been possible because this country has strongly supported basic research and construction of scientific instruments that were, in their time, as challenging as the Superconducting Super Collider (Watkins, 1992). However, tough economic times threaten such projects and cause taxpayers and their representatives to wonder if the investment is worth the return.

   The purpose of this report is to communicate the scientific and economic benefits of the Superconducting Super Collider (SSC) to the public officials who must decide the fate of the Super Collider project. The Super Collider must receive full funding and support from government officials. Denial of funding at this point in the project would cause the loss of thousands of jobs, several billion dollars, and countless scientific and technological breakthroughs.

<u>What is the Superconducting Super Collider?</u>

   As the largest and mostly advanced particle accelerator ever built, the Superconducting Super Collider will be the world's preeminent facility for high-energy physics research. The Superconducting Super Collider will be located in Waxahachie, Texas, in a racetrack-shaped, underground tunnel about fifty-four miles in circumference and twelve feet in cross-section diameter (See Figure 1). Approximately ten thousand superconducting magnets will focus and guide two beams of protons in opposite directions around the tunnel. The beams will be accelerated to the speed of light and made to collide head-on at an energy of forty trillion electron volts. Researchers expect the collisions to create new subatomic particles that will be detected and analyzed, adding to the understanding of the fundamental nature of matter and energy (Wylie, n. d.).

   The Super Collider will be the largest scientific instrument ever built (Anonymous, 1990). Super Collider detectors will have to cope with about one hundred million collisions every second. More than seventeen thousand bunches of protons and quarks will race around each ring about three thousand times a second. Opposing proton bunches will cross through each other sixty million times per second at each interaction point. Every time two bunches cross, one or two collisions occur; however, probably only ten to one hundred of the collisions per second will be of interest.

SAMPLE 12.17 continued

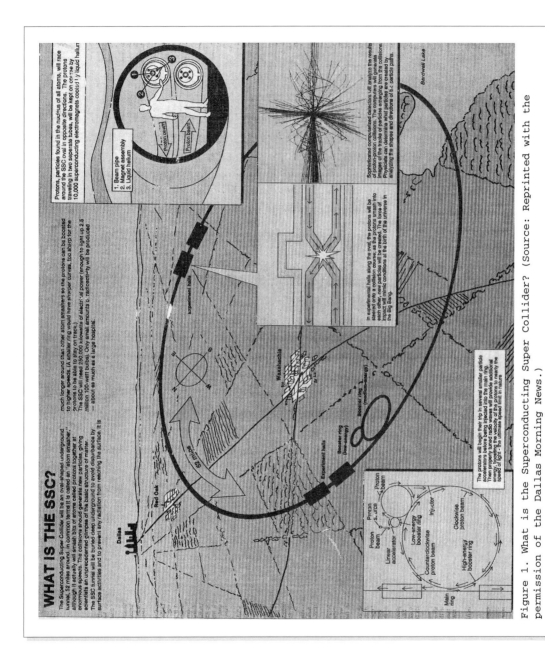

Figure 1. What is the Superconducting Super Collider? (Source: Reprinted with the permission of the Dallas Morning News.)

SAMPLE 12.17 continued

The detection system is equipped with electronics fast and accurate enough to read the data from the various detector elements, select the interesting events, and record the data before the next two bunches cross only sixty nanoseconds later. The Super Collider will produce more particles, with higher energy, than any existing accelerator (Wylie, n. d.).

The goal of high-energy physics research is to understand the basic nature of matter--the fundamental particles and forces that constitute atoms. With the help of particle accelerators and detectors, high-energy physicists have made dramatic advances toward this goal while providing important economic benefits for the nations that have underwritten the projects. Particle accelerators have proven valuable for many purposes unrelated to high-energy physics; accelerators are often built and used for other applications. The scientists and technical experts at the Super Collider will develop innovative solutions to difficult scientific problems while broadening the applications of existing knowledge (U.S.D.O.E., n.d.).

The conclusion of the section summarizes the main point and uses the occasion to reiterate the now-strengthened position.

The Superconducting Super Collider is a project which is crucial to the future of the United States as a leader in science. The technology developed by the SSC will lead to numerous spin-off industries while keeping the greatest scientific minds from around the world working in the United States. The future technological and economic benefits the Super Collider makes possible will far outweigh the burden of the project's initial cost.

Much of the work of this section is done by the graphic. The only details given are the ones crucial to the decision.

Current Status of the Project

Construction of the massive Super Collider is well under way (Figure 2). At this point, land acquirement is complete,

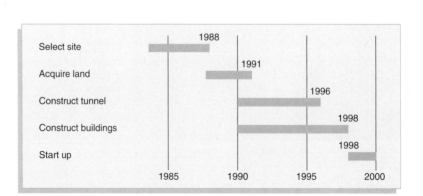

Figure 2. Progress toward completion of the SSC.

SAMPLE 12.17 continued

main campus buildings are nearing completion, tunnel construction is progressing, and thousands of people are employed by the Superconducting Super Collider and its laboratories. The cost of the Super Collider so far has been $8.2 billion.

Economic Benefits of the SSC

Although several billion dollars have been spent on the project to date, today's difficult economic conditions make the SSC a target for spending cuts by some public officials (Jordan, 1992). Still, the Superconducting Super Collider is not a revenue drainer, it is a revenue enhancer. More than 6,000 SSC-related jobs have been created since 1987, and in Fiscal Year 1993 this number will increase to more than 8,000 jobs.

The Super Collider is creating technologies and industries that will expand our economic base and provide thousands of jobs well into the 21st century. These are white- and blue-collar jobs for high-tech workers that will help to offset cutbacks in the defense industry (SSC Central Design Group, 1992). As a writer in the Wall Street Journal puts it, "The Superconducting Super Collider is the single national initiative most likely to spur economic activity, invention, and long-term prosperity. It is pump-priming appropriate to this technological age" (Silber 1992, p. 12).

The Super Collider will be the equivalent of a major "public works project." Construction of the Super Collider will require massive amounts of building materials and over $1 billion in heavy construction. The ongoing building of the physical structure will provide a much needed "jump start" to the United States economy. More than a thousand industries, millions of hours of labor and hundreds of thousands of workers will benefit from involvement in the SSC (SSC Central Design Group, 1992).

Research and development is the backbone of the United States manufacturing success. Innovation through continued scientific research will help to keep the cost of producing durable goods low. Killing the Super Collider would save only a fraction of one percent of the president's budget, but its costs in terms of lost scientific achievement and future technological advances would be staggering (Watkins, 1992).

Technological Gains

The technological and industrial revolution that will follow from inventing and making the SSC's scientific components will eclipse the economic impact of the Super Collider. As the world's leading scientists and engineers work on the challenges of the Super Collider, they will inevitably invent new materials and technologies. The SSC will advance a number

The author prefers to paraphrase throughout the report but occasionally uses a direct quotation to add authority (e.g. the *Wall Street Journal* on economic issues) and to preserve memorable language (e.g. "pump-priming").

The documentation gives a page number for direct quotations only.

The author quotes key words and phrases from the Central Design Group's report throughout the paragraph but reserves the citation until the end to avoid too many intrusive parenthetical citations.

SAMPLE 12.17 continued

One of the few pages unbroken by a heading or graphic. Should something be added to break up the text?

of spin-off industries. The fields of electronics, transportation, communications, and medicine are a few of the areas that will provoke valuable industrial innovation.

The study of superconductivity holds perhaps the greatest potential of all. Superconducting cable is currently being tested for use in communications and high-speed trains. A breakthrough in the field of superconductivity could revolutionize communications technology and allow high-speed trains to travel at three hundred miles per hour by magnetic levitation (Mag Lev) on superconducting magnets. American companies making superconducting cable for the Super Collider have increased the electrical current-carrying capacity of superconducting wire by 50 percent. Because the Super Collider project requires such a large amount of superconducting cable, the project is creating an infrastructure for cable in America that will enhance energy storage and Mag Lev, reduce the cost of lifesaving magnetic resonance-imaging (MRI), and assist in proton-beam therapy for cancer. The SSC is making the United States a world leader in the development and manufacture of superconducting cable (SSC Central Design Group, 1992). "In terms of societal impact, [the technology of superconductivity] could well be the breakthrough of the 1980's in the sense that the transistor was the breakthrough of the 1950's," says Alan Schriensheim, director of Argonne National Laboratory (Lemonick, 1987, p. 74). The SSC will carry this technology into the 21st century.

Note that magnetic technology and MRI are mentioned here as well as elsewhere in the report. Should these references be consolidated? (See below.)

Note the use of brackets to clarify meaning in the direct quotation. The terms in brackets replace the word "this" in the original source.

Hundreds of thousands of microprocessors will be needed to control the Superconducting Super Collider. A new generation of supercomputers will be necessary to track the collisions at each experimental point. These computers, coupled with the advances of superconductivity, will lead to better electrical power, better load-leveling technology, and improved electric motors. Personal computer technology will also improve. Business computers will increase data-processing capabilities due to Super Collider breakthroughs. IBM, in conjunction with the SSC, is developing a system that will process the information contained on 10,000 floppy disks in one second (SSC Central Design Group, 1992).

The reporter uses the familiar (floppy disks) to explain the unfamiliar.

Superconducting technology is changing the face of modern medicine. Superconducting magnets are at the heart of MRI. New magnet technology will advance medicine even further. The technology used in some cancer treatments is the result of particle accelerator technology. Proton beam treatments are currently being used by doctors when a cancer is in an area where it is too dangerous to operate. The beam is directed at the tumor and is able to destroy the entire cancer with minimal damage to surrounding tissue. Further advances in proton beam treatments could eradicate cancer surgery (Lemonick, 1987).

Here's the other reference to MRI and magnetic technology.

SAMPLE 12.17 continued

Cost of Killing the Superconducting Super Collider

To scrap the SSC at this point would be an exercise in self-defeating economics. The Super Collider is not an item of consumption, as nuclear weapons were, but an item of investment that will generate wealth far into the next century. The investment is not only in long-term scientific progress, but in immediate and long-range economic development (Silber, 1992).

Termination of the Superconducting Super Collider at this point would require the United States to pay $180 million to cover and phase out 2,000 jobs, cancel contracts, and cover expenses. Over 6,000 technical and manufacturing jobs would be lost by the cancellation of the thousands of SSC contracts.

High-energy physics research in America would be devastated by project cancellation:

- Over 100 universities will lose millions of dollars that have been invested in research associated with the SSC.
- Over 100 universities would lose millions of dollars in research grants.
- Scientists and graduate students would have to go abroad to conduct high-energy physics research.
- Industry would lose the opportunity to develop an infrastructure for superconductivity in the United States.
- The U.S. would lose the competitive edge in the manufacture of superconducting wire, supercomputing, cryogenics, superconducting magnets, and other critical technologies.

Bulleted list provides emphasis, heightens readability, and breaks the text into smaller chunks.

Cancellation of the Super Collider would hurt our country's standing in the scientific world and reinforce the United States' increasing reputation as a country that lacks vision (SSC Central Design Group, 1992).

Note the use of "vision" for transition between sections.

National Vision

The Superconducting Super Collider will not just be the world's largest scientific research facility, but an investment in education, science, medicine, and product development. It is essential that America make this investment to ensure our ability to compete in the global marketplace (Silber, 1992).

Conclusion addresses "national vision" as a concession to the context of use (congressional decision-making).

Americans supporting the Super Collider are true visionaries. The responsibility of Congress is to legislate what is best for the United States today and in the future. The Superconducting Super Collider is right for America and will pay for itself in the growth of new industries and diversification of old ones. This project is essential for the United States to lead the rest of the technological world into the 21st century.

SAMPLE 12.17 continued

Reference list appears on separate page.

Note the use of "hanging indent."

See Chapter 9 for APA style.

References

Anonymous. (1990, October 28). What's a super collider and other questions. <u>Waxahachie Daily Light,</u> p. 1E.

Augustynowicz, S. (1992, April 16). Personal interview.

Jordan, N. B. (1992, April 16). Personal interview.

Lemonick, M. (1987, May 11). Superconductors! <u>Time,</u> pp. 65–75.

Silber, J. (1992, July 24). An Apollo project for the 90's. <u>The Wall Street Journal,</u> p. 12.

Superconducting Super Collider Central Design Group. (1997). <u>Superconducting super collider briefing book 1992.</u> Dallas: Universities Research Association.

United States Department of Energy. (n.d.). <u>Not for scientists only: Technology spin-offs from high-energy physics and the super collider.</u> Washington: Government Printing Office.

Watkins, J. (1992, July 21). Save the super collider. <u>The Washington Post,</u> p. 8.

Wylie, R. (n.d.). <u>News news news . . . super collider.</u> Dallas: Universities Research Association.

SAMPLE 12.17 continued

Biowaste Recycling
An Economic and Environmentally Sound Solution
To Municipal Waste Disposal Problems

by

Christopher J. Powers

Texas A&M University

October 26, 1994

Our Wastefulness = Our Problems

Many large cities across the United States currently face waste disposal problems which, if not addressed soon, will become critical public health hazards and economic nightmares. The question is an easy one to phrase: Where do we throw away our waste? However, the "away" of yesterday no longer exists (Hartman, 1989). Our growing nation faces space limitations as well as ethical limitations arising from society's keener awareness of how our careless practice of landfilling adversely affects the environment we have to live in.

The purpose of this report is to inform the mayor and city officials of San Diego about biowaste recycling because they have the power to prevent a waste disposal crisis while benefiting from the reuse of biowaste. Through the collection and treatment of biowaste, the city will not only be able to dramatically cut down on disposal costs paid to already near-capacity landfills, but it also has the potential to profit from a material that, in the past, was discarded as waste and forgotten.

The Past

For the past sixty years coastal cities like San Diego and New York relied on ocean dumping to dispose of most sludge and conveniently deposited the remainder of the waste in landfills (Goldstein, 1992). These methods of waste disposal are now proven to be environmentally hazardous and economically inefficient.

Ocean Dumping Is No Longer a Solution

Several new factors change the manner in which cities are dealing with their waste. Due to the Ocean Dumping Ban Act of 1988, implemented by the Environmental Protection Agency (EPA) as of June 1, 1991, cities no longer have the option of dumping biowaste in the ocean (Goldstein, 1992). Now that ocean dumping is illegal, cities such as San Diego have to decide how to dispose of ever-increasing amounts of biowaste in both an environmentally and economically sound way. Another problem exists.

SAMPLE 12.18   Student report on options (recommendation report)

In this short report, Chris avoids the overuse of headings but uses what he has effectively. He presents only one order of headings, omitting structural headings such as "Introduction," "Discussion," and "Conclusion," instead preferring informative headings, or *headlines*.

Landfilling Solutions—A Time Bomb

Despite the past acceptance of landfilling waste, many problems are now surfacing which always existed but were never discussed because of the "out of sight, out of mind" theory (Hartman, 1989). Not only are existing landfills closing, but many are near their capacity and are no longer of any use. The landfills are closing due to stricter EPA regulations, which require costly materials to line the landfill to prevent leachate from contaminating ground water supplies (U.S. EPA, 1989). Landfills which do remain operational and comply with EPA regulations now have to charge higher tipping fees to cover costs of the new technologies.

Even if tipping fees are made affordable, the environmental impact in the future will be disastrous. No landfill liner can be totally effective in keeping leachates from reaching the groundwater supply, contaminating and destroying our drinking water. Eventually these landfills will cause environmental problems and will have to be cleaned up, costing millions. Landfilling is neither economical nor environmentally safe.

Despite the fact that landfills will always be needed, the amount and type of materials deposited in landfills can be reduced substantially by recycling biowaste. If we decrease reliance on landfills, we can reduce the adverse impacts landfills have on the environment and the taxpayer's wallets.

Chris "chunks" the text, offering frequent breaks to enhance readability and structural integrity. Every page is broken at least twice by a new heading or a graphic. Lists appear frequently as well.

Biowaste: What Is It and Its Significance?

Biowaste is a material which is capable of being broken down by microorganisms into simple stable compounds. This includes most organic wastes, such as food wastes and paper, which are rich in nutrients. FIGURE 1 shows a percentage of how much of each type of waste is typically generated each year (Hartman, 1989).

Through the collection and treatment of biowaste, San Diego could not only dramatically cut down on disposal costs paid to already near-capacity landfills, but also profit from a material that, in the past, was discarded as waste and forgotten.

Biowaste Recycling Proven Effective in New York City

Several cities have shown that biowaste recycling is an effective alternative to landfilling. New York City, for instance, produces 355 dry tons of sludge per day, all of which, in the past, was dumped offshore. Landfill space is very limited and expensive to use. Therefore, city officials and engineers decided to try recycling the biowaste and creating a marketable product. The product is a fertilizer pellet, made from dried sludge, that can sell for up to $110 per ton. FIGURE 2 shows the processing steps the sludge goes through to make a liming agent and conditioner (fertilizer) (Purcell, 1992, 34).

This new product can be used for land application/reclamation, lawn care, gardening, and golf course/recreation cover.

SAMPLE 12.18 continued

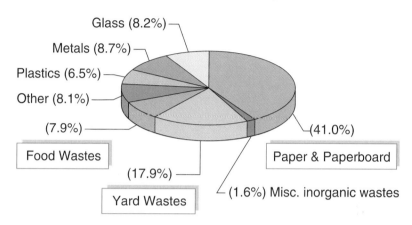

Typical Municipal Waste Stream
(Percent of Total)

Glass (8.2%)

Metals (8.7%)

Plastics (6.5%)

Other (8.1%)

(7.9%)

Food Wastes

(41.0%)

Paper & Paperboard

(17.9%)

(1.6%) Misc. inorganic wastes

Yard Wastes

66.8% is BIOWASTE and can be reused!!!

FIGURE 1. *The materials that are framed constitute biowaste.*

Public Concern

Some concern for the safe application of sludge-produced fertilizer has been voiced by the public. People are particularly concerned about areas such as parks in which children are exposed to the sludge fertilizer and about the use of pellets in agriculture, which produces consumable products. Many people believe that this product is not safe due to the pathogens that are present in the collected sludge. Pathogenic organisms do exist in raw sludge; however, through processing, the sludge is categorized according to pathogen reduction and metals content and applied to appropriate areas as seen in TABLE 1 (Purcell, 1992). As long as the classified sludge, A, B, or C, is applied for its designated use, there is no threat to public health.

Biowaste Recycling Proven Effective in Los Angeles

New York is not the only city to make advances in biowaste recycling. Los Angeles now reuses all of its 500 dry tons/day of raw sludge solids. This is accomplished by

The balance between graphical displays and complementary text is strong. Chris uses graphics to display, verbal text to interpret and recommend. He refers to his figures and tables in the text and takes a little time to explain each one. He doesn't use so many lists and graphics that the text becomes incoherent, and he continues to create effective transitions in the verbal text despite his practice of "chunking."

SAMPLE 12.18 continued

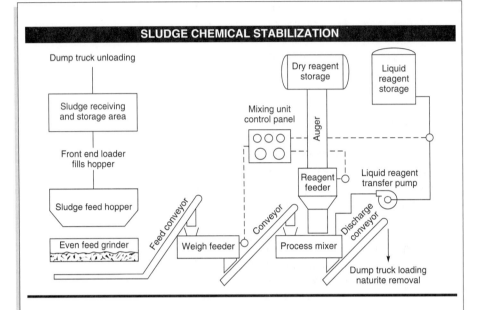

FIGURE 2. *This processing facility is located in Middlesex County, New Jersey. It produces the fertilizer Meadowlife from recycled sludge.*

combining on-site energy recovery processes and off-site reuse. Specifically, 400 **wet** tons per day of sludge are applied to a company's farm reclamation and crop enhancement program. Another contractor uses 150 wet tons per day for land application at sites within Los Angeles. In just four years, 1988–1991, the cost of managing the sludge dropped from an average of $43 per wet ton to $30 per wet ton in 1991 (Goldstein, 1992). In the long run, biowaste recycling pays off nicely.

TABLE 1. *Sludge Categorization*

SLUDGE CATEGORIZATION FOR SUITABLE APPLICATIONS AFTER PROCESSING

CLASS A   Sludge acceptable for land application at agricultural rates

CLASS B   Suitable for general land applications

CLASS C   Unsuitable for land application

SAMPLE 12.18 continued

Private Industry Can Profit From Biowaste Recycling

Private industry also has the ability to benefit from biowaste recycling. The Anheuser-Busch Brewery, located in Baldwinsville, New York, has demonstrated this. The brewery generates approximately 30,000 wet pounds of biowaste per day. This amounts to approximately 7,500 pounds of dry solids, which normally had to be land-filled. Instead of sending this waste to a landfill, though, the company now uses the waste to produce high-quality compost from its food processing biosolids. Results of the composting program are shown in TABLE 2 (Beers & Getz, 1992). Through biowaste recycling, the brewery reduces sludge output by 90 percent while generating 86,000 cubic yards of marketable compost. This waste disposal technique has saved 73,900 cubic yards in landfill space while saving the company more than 3.3 million dollars in dumping fees. It is obviously advantageous for industries that deal with bio-waste, such as food processing, to consider biowaste recycling, for it could save them millions in disposal costs (Beers & Getz, 1992).

Chris prefers paraphrase to direct quotation but judiciously uses information sources to add authority and specific details to his positions.

What Are Our Options?

San Diego can let waste disposal continue to cause problems, or the city can take advantage of the waste by recycling. The city can benefit from biowaste recycling in many ways. First, the city will save millions in landfill tipping fees. Second, landfill space will be saved for unusable material which has to be disposed of to keep the city clean. Finally, the environment will be saved from unnecessary landfills that clutter the countryside and produce leachate that contaminates our groundwater.

Paragraphs begin with strong sentences that tend to make the main point of the paragraph.

Numerous biowaste materials are available for reuse. By recycling these materials the waste disposal problems that haunt many cities will no longer cost the cities millions of dollars each year; rather they will produce an opportunity for profit.

Though he mixes in a good number of passive voice sentences (typical of engineers and scientists), Chris works hard to keep his style predominantly active.

How Does San Diego Implement Biowaste Recycling?

The only way this solution will be employed is if people are informed of the biowaste recycling technique. Education is the key to gaining support for such a pro-

TABLE 2. *Results of Composting Program*

	1989	1990	1991	Total
Wet tons sludge processed	9,170	17,800	17,900	44,870
Dry tons sludge processed	2,290	4,450	4,860	11,600
Cubic yards sludge processed	15,100	29,300	29,500	73,900
Cubic yards compost produced	13,000	40,000	33,300	86,300
Savings on landfill tip fees	$550,000	$1,157,000	$1,521,500	**$3,228,500**

SAMPLE 12.18 continued

gram. Once people know about biowaste recycling and its benefits to society and the environment, the option will be supported at a government level. Now that the legislators know that the people support biowaste recycling, legislation should be passed requiring:

- private industrial businesses to recycle biowaste
- appropriation of funding for biowaste recycling facilities, similar to what New York City has accomplished, to deal with biowaste generated by the public

San Diego not only can profit from the reuse of biowaste but will also save millions in landfill dumping costs, save the environment, and keep San Diego out of a waste disposal crisis.

SAMPLE 12.18

<u>References</u>

Beers, A. R, and Getz, T. J. (1992). Composting biosolids saves $3.3 million in landfill costs. *Biocycle 76,* 42, 76–77.

Goldstein, N. (1991). Beneficial reuse anchors shift from land to sea. *Biocycle 75,* 36–39.

Hartman, R. A. (1989). *Good riddance.* College Station, TX: privately printed.

Purcell, L. J., & Johnson, T. D. (1992). From sludge to brokered biosolids. *Civil Engineering 62,* 32–35.

United States Environmental Protection Agency. (1989). *Decision makers guide to solid waste management.* U.S. Government Printing Office: 1990, 257–977.

SAMPLE 12.18 continued

Experimental Reports and Scientific Papers

CHAPTER OUTLINE
▼▼▼▼▼▼▼▼▼▼▼▼▼▼▼▼▼▼▼▼▼▼▼▼▼▼▼

Step One: The Audience-Action Analysis for an Experimental Report or a Scientific Paper
Step Two: Writing the Five-Part Introduction for a Scientific Paper
Step Three: Writing the Scientific Paper
Reviewing Scientific Papers

CHAPTER OBJECTIVES
▼▼▼▼▼▼▼▼▼▼▼▼▼▼▼▼▼▼▼▼▼▼▼▼▼▼▼

After you have worked through this chapter, you should be able to do the following:

- Do an audience-action analysis for an experimental report or a scientific paper
- Discover the appropriate content and emphasis for your own writing task
- Plan and prepare a formal report or paper on experimental work in science, engineering, or the social sciences

Technical reports (discussed in Chapter 12) differ from reports on scientific analysis and experiments (discussed in this chapter) in one major way. Technical reporters usually address an audience of nontechnical decision makers (who are perhaps aided by technical reviewers), but the authors of scientific papers address an audience of their peers—other specialists in their own field. In other words, the context of production and the context of use overlap a great deal. The purpose of these documents is to advance a research agenda by contributing new information to a specialized field of scientific study.

Usually published in specialized journals, scientific papers are typically written by academic scientists, engineers, and social scientists. But industry and government employees may also produce scientific papers.

Although many of the instructions for good technical reporting apply to this genre as well, the shift in audience and purpose engenders a few key differences. We will discuss these differences below, but first it is important to realize that they give rise to a substantially different set of CORE steps. The sequence of steps for scientific papers is as follows (see also Figure 13.1):

1. Do an audience-action analysis for the paper that will serve as your planning notes.
2. Write a five-part introduction that follows the general outline of your paper. This is your core document. Share it with trial readers from your audience and revise it before you draft your final paper.

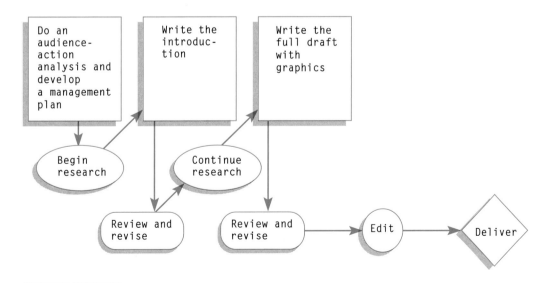

FIGURE 13.1

The CORE method adapted for producing scientific papers.

3. Develop the full scientific paper, following the IMRaD structure (to be explained in this chapter) and the principles in Chapter 6 for generating technical graphics.

You may also do an oral briefing between Steps Two and Three or a summary oral report after Step Three. If you do, you can follow the guidelines for oral reporting in Chapters 8 and 12. But this chapter will concentrate on the three steps listed above.

Step One: The Audience-Action Analysis for an Experimental Report or a Scientific Paper

The purpose of a scientific paper is to communicate an author's contribution to a research program in a specialized field for an audience of other specialists in the field.

A *contribution* to the field is defined as new information (a difference that makes a difference) in a special discipline such as nuclear physics, rural sociology, or mechanical engineering. The findings that you report may be theoretical, analytical, or experimental, but their importance depends on their relation to the work of other researchers in the field. They extend knowledge by affirming or challenging previous findings and then drawing a conclusion that has theoretical significance.

As an author, your objective is to influence the direction of the *research program*—to change the way readers carry out their experiments and think about their research. The *action* that you want to generate, then, is *experimental and discursive action*—research and writing—rather than action in the "real world" of politics, technology, and social service.

As we have said, the *author* and the *audience* for the scientific paper are experts trained in the special field of knowledge. Like members of any community, they have certain "badges of identity." In this case the badge is an ability to demonstrate mastery of key theoretical problems, special methods of investigation, a body of literature, and a set of stylistic and structural conventions.

As you did for a technical report (working through Chapter 12), you should construct a set of planning notes that identifies each of these components and determines how they will work together in your paper. Using Figure 13.2 as a point of reference, you should be able to answer the questions in Guidelines at a Glance 46 as a first step in developing your scientific paper.

Step Two: Writing the Five-Part Introduction for a Scientific Paper

The purpose of scientific writing shows in the conventions that authors and editors use for scientific papers. Though there are definitely differences in papers

GUIDELINES AT A GLANCE 46

Audience-action analysis for scientific papers

- Who is the audience for this work? What branch of research is your information most likely to affect? What kinds of writing or research will your audience do in response to your paper?
- What's new or theoretically significant about your contribution to the research program? How does it relate to familiar problems in your field?
- By what methods did you arrive at the information? Are these in any way different from ordinary approaches?
- In what ways are your methods or your findings likely to change the ideas or practices of your peers?

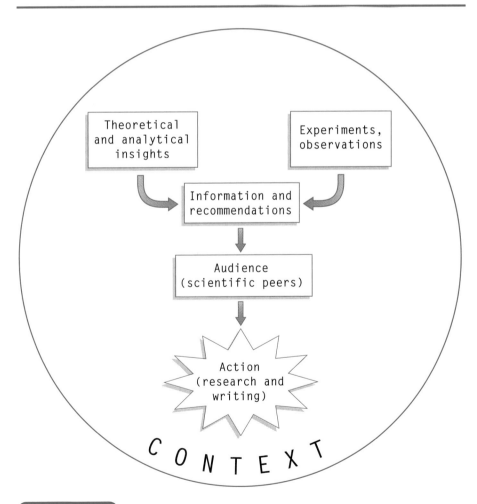

FIGURE 13.2

The pathway from information to action in a scientific paper.

written for different disciplines and different types of research (basic and applied, for example), the general style and structure are consistent: Scientific style is conventionally passive and objective, and the structure is fairly rigid.[1]

Using the IMRaD Structure

Many journals in science, social science, and engineering have stopped requiring authors to use the rather mechanical headings "Introduction," "Methods," "Results," and "Discussion" (or "Conclusions")—the first letters of which form the acronym IMRaD, a widely used reminder for the predominant structure. Nevertheless, most reports on experimental findings still follow the basic structure implied by these headings. That is, for better or worse, they move from a statement of a problem and its background (introduction) to a description of how to address the problem (methods and materials) to a presentation of experimental results obtained by using the method (results) and on to an explanation of the results in light of the major theoretical issues in the field of study (discussion).

When you write the introduction, you should follow a similar five-part structure that previews the rest of the paper. Proceed from the past to the present, stating first the problem you are investigating and its history, then briefly describing your own method, results, and conclusion. The introduction should establish your credibility with your audience by demonstrating your mastery of the key problems in the field and your understanding of how your findings relate to others reported in the literature.

Here are the steps toward writing an introduction as recommended by the scientific editor Robert A. Day:[2]

1. State the problem that your paper addresses.
2. Review the most pertinent literature on the topic.
3. Briefly state your method or approach to the problem, emphasizing differences between your approach and those reported in the literature.
4. Briefly summarize your results, again emphasizing differences between your findings and those reported in the literature.
5. State your most important conclusions.

As you can see, Steps 3, 4, and 5 give the reader a preview of your argument and provide a cognitive map for the rest of the paper. As Robert Day suggests, nobody likes suspense in a scientific paper.

---

1. See M. Jimmie Killingsworth and Michael Gilbertson, *Signs, Genres, and Communities in Technical Communication* (Amityville, NY: Baywood, 1992), Chapter 8. See also Charles Bazerman, *Shaping Written Knowledge: The Genre and Activity of the Experimental Article in Science* (Madison: University of Wisconsin Press, 1988); and Alan Gross, *The Rhetoric of Science* (Cambridge, MA: Harvard University Press, 1990).

2. Robert A. Day, *How to Write and Publish a Scientific Paper*, 3rd Edition (Phoenix, AZ: Oryx, 1988) 33.

Using an Objective Style and Scientific Terms

As you write your introduction, you'll immediately have to face some stylistic choices, some of which we covered in Chapter 9. In scientific papers, you'll re-member, the author tends to "disappear" in the style, using the passive voice sentence to emphasize the action rather than the actor:

> *The procedure was carried out . . .*

> *The burner was lit . . .*

rather than

> *We carried out the procedure . . .*

> *I lit the burner . . .*

Scientific writing tries to convey the impression that it is impersonal and re-producible. Anyone could have carried out the procedure or lit the burner; it is not important that the author did it.

While fully aware of this convention, however, many scientific authors have worked to reduce overuse of the passive voice in their writing. You would do well to follow these authors' lead. Passive constructions may project the tone of cool distance you are cultivating, but they can also hamper readability, as nu-merous studies have shown.[3] Certainly, you should avoid all passive construc-tions that obscure responsibility for actions. We saw in Chapter 9 how sentences such as "the village was bombed" create smokescreens for writers who are un-willing to accept or assign blame. In scientific writing, such constructions can also obscure understanding. For example, if you write, "the procedure was car-ried out" in a section that mingles literature review with a report of original re-search, the readers may lose track of who is performing the procedure—you or some other researcher whom they've read about. In short, any advantage that you gain by habitual use of passive constructions is more than offset by the dis-advantages. So work to increase the number of active constructions in your writing.

Scientific papers also use more scientific or special terminology than do technical reports. These special terms form a kind of shorthand that enables au-thors to avoid long descriptive phrases. It's simply more efficient to say "ben-thic organisms" than to say "plants and animals that spend their entire lives dwelling in formations attached to the ocean floor."

Such terms also allow scientists to provide precise descriptions of objects that familiar language might render ambiguous. There is only one scientific name for any given shrub, for instance, though there may be many common

3. See Thomas Warren, "The Passive Voice: An Annotated Bibliography," *Journal of Technical Writ-ing and Communication* 11 (1981): 271–286, 373–389.

names. Moreover, a single common name, such as "rabbitbush," may apply to several different varieties of shrubs.

Using scientific terminology to increase the precision of your writing is, of course, a sensible practice. Where you can go wrong is using technical terminology when simpler language will work just as well or piling up technical abstractions in such profusion that even specialists will wince. The best approach is to worry less about making yourself appear professional by donning stylistic badges of identity and worry more about how a busy reader will receive your writing. Use active sentences and ordinary language whenever you can without sacrificing precision. Your readers—even the most highly trained ones—will appreciate your effort.

Notice the biographical note, which gives the field and institutional affiliation of each coauthor.

"Synopsis" is simply another term for abstract.

Suicide in the State of Maryland, 1970–80

GARY POPOLI, MA
STEVEN SOBELMAN, PhD
NORMA FOX KANAREK, PhD

Mr. Popoli is a Research Statistician for the Local Health Administration in the Maryland Department of Health and Mental Hygiene. Dr. Sobelman is an Associate Professor of Psychology in the Psychology Department at the Loyola College of Baltimore. Dr. Kanarek is Chief of Chronic Disease and Epidemiology in the Maryland Department of Health and Mental Hygiene.

Tearsheet requests to Mr. Popoli, Maryland Department of Health and Mental Health, 201 W. Preston St., Rm. 306-C, Baltimore, MD 21201.

Synopsis..

A univariate and multivariate analysis of factors associated with suicide for residents of the State of Maryland was conducted. The investigation was statistically oriented in its approach, examining the relationships of age, race, sex, marital status, and month of death with suicide. Besides the usual death rates, percentages, and age-specific rates, a discriminant analysis was performed to test this approach.

Data were obtained on all suicides of Maryland residents, regardless of where the deaths occurred. Univariate analysis showed that the relationships between suicide and age, race, sex, and marital status are consistent with those in the literature. No significant relationship appeared to exist between the month of death and suicide. During multivariate analysis, the discriminant function correctly predicted 80 percent of all the deaths, 74 percent of all the suicides, and 80 percent of all other causes, in their respective categories.

SAMPLE 13.1   Introduction to a published scientific paper

There is abundant research which emphasizes the concrete, statistical relationships which exist between the demographic variables of age, race, sex, marital status, and specific month of suicidal deaths. The purpose of this investigation was twofold: first, to determine the universality of these relationships in the State of Maryland; and second, to determine exactly how reliable these relationships really are in actually differentiating suicidal deaths from all other causes of death statistically; that is, to predict whether a death is a suicide or not merely be examining these five demographic variables.

A brief summary of the literature revealed the significant relationships of age, race, sex, marital status, and specific month of death with suicide. Consistent with the results reported in these studies, one would expect the following relationships in the State of Maryland data:

1. The highest rate of increase in suicide over the past 11 years occurred to those under 30 years of age. (*1*).
2. Males have a higher frequency of suicides than females. (*2–5*).
3. Whites have a higher frequency of suicides than nonwhites. (*6*).
4. The rate of suicide is higher for persons who are widowed or divorced than for those married or single. (*3, 7*).
5. More suicidal deaths occur in the spring and fall months than the summer and winter months. (*8–10*).

There have been many studies that support these "universal" relationships between suicide and the specific demographic variables used in this investigation. This study is distinguished from others in the literature by the multivariate analysis used. Discriminate analysis incorporates the effects of five demographic variables—age, race, sex, marital status, and month of death—in order to determine how well the five variables predict suicide and all other causes of death.

The first paragraph of the introduction states the problem under investigation and establishes the context for the investigation.

The second paragraph reviews the relevant findings from related studies.

The third paragraph emphasizes what makes this paper original—its methodology.

Having adequately covered the first three parts of the introduction (problem statement, literature review, and methods statement), the authors should have gone on to preview the findings and conclusion.

SAMPLE 13.1 continued

Sample 13.1 is the first page of a typical professional scientific article entitled "Suicide in the State of Maryland, 1970–80."[4] The unheaded introduction follows the title, biographical note, and synopsis (abstract). It clearly states the problem under consideration, briefly notes the trends in the relevant literature, and shows how the methodology of these researchers differs from that of previous researchers. It does not, however, go as far as it might in previewing results and conclusions. This may be because the authors merely confirm what others have reported and concluded. The contribution of the paper is mainly a methodological innovation. Even so, the introduction should have forecasted this outcome more clearly for the reader.

4. Gary Popoli, Steven Sobelman, and Norma Fox Kanarek, "Suicide in the State of Maryland, 1970–80," *Public Health Reports* 104, no. 3 (May–June 1989): 298–301.

Step Three: Writing the Scientific Paper

Once you have tested your introduction out on trial readers and revised it in accordance with their feedback, you can begin drafting the rest of your paper. Follow the IMRaD structure, using your introduction as an expandable core.

The Methods Section

The methods section gives you a place to show your mastery of the special approaches to problems taken in your field of study. If the introduction tells *what* you did, the method section says *how* you did it.

The account should be short and fairly schematic. It should simply go through the steps: "We did A; then we did B." (Or "This was done; then that was done.") Sample 13.2 is the methods section of "Suicide in the State of Maryland, 1970–80."

The Results Section

The results section tells the outcome of your experiment or analysis. Following good technical reporting practices, you should display as many findings as possible in graphical form, using brief explanations in the text to point out the most significant data points and findings. Save major interpretations for the discussion.

Sample 13.3 shows the results section of "Suicide in the State of Maryland, 1970–80." Notice that the text compares the findings to those reported in the literature. It is typical for scientific papers to analyze results according to categories

---

Method

In order to examine the relationships between suicidal deaths and the five predictor variables more closely, cross tabulations, frequencies, and crude death rates were calculated for the different age groups, races, sexes, marital statuses, and months of deaths. To test statistically whether or not a suicidal death could be differentiated from all other causes of death, a discriminant analysis was performed using the "Statistical Package for the Social Sciences," (SPSS) (*11*).

Data for statistical analyses performed in this investigation were supplied by the Maryland Center for Health Statistics (MCHS). Researchers were given permission by MCHS to access a nonconfidential data set which contained deaths of Maryland residents for calendar years 1970 through 1980. It is important to note that the dates used in this study were those reported on the death certificate although this date and the actual date of death may differ (*12*).

SAMPLE 13.2   Methods section of a published scientific paper

Results

Data elements were extracted from the death certificates, including the underlying cause of death. There were 5,040 deaths that were coded as suicides according to the Eighth and Ninth Revisions of the International Classification of Diseases (*13*) from 1970 to 1980; and a total of 345,137 deaths during this period.

Because the International Classification of Diseases nomenclature changed from the Eighth to the Ninth Revision, differences in rates due to these changes may be observed. The suicide deaths' comparability ratio (Ninth compared with the Eighth) is 1.0032 (95 percent confidence interval, 0.9950-1.0114) (*14*). In this study, increases in rates were observed only in the youngest persons. An artifactual effect on these data would operate across all population groups, and thus it was not an important factor in this analysis of Maryland suicides.

The elements of interest were age, race, sex, marital status, and month of death. When the relationships between each independent variable and suicidal deaths were examined, the results of this investigation were consistent with those in the literature. The largest rate of increase in suicide in Maryland has occurred among those under 34 years (table 1); males committed suicide about three times as often as females in each of the 11 years (table 2); the suicide rate for whites was approximately twice that of the nonwhites (table 3); and divorced and widowed persons had higher rates than the married and single (table 4).

Table 1. Age-specific percent increases in suicide rates per 100,000 population, Maryland residents[1] and the United States,[2] 1970 and 1980

Age group (years)	1970 Mary-land	1970 United States	1980 Mary-land	1980 United States	Percent increase Mary-land	Percent increase United States
5–14	0.2	0.3	0.2	0.4	[3]	[3]
15–24	8.5	8.8	12.3	12.3	44.7	39.8
25–34	13.3	14.1	15.3	16.0	15.0	13.5
35–44	13.5	16.9	13.4	15.4	–0.7	–8.9
45–54	18.9	20.0	15.2	15.9	–19.6	–20.5
55–64	24.1	21.4	18.3	15.9	–24.1	–25.7
65–74	24.4	20.8	15.0	16.9	–38.5	–18.8
75–84	26.5	21.2	13.8	19.1	–47.9	–9.9
85 and older	5.1	19.0	24.5	19.2	[3]	1.1

[1]Data from reference 16.
[2]Data from reference 17.
[3]Due to small numbers, percent increases were not calculated.

continued

SAMPLE 13.3 Results section of a published scientific paper

Table 2. Suicide rates per 100,000 population, by sex for Maryland residents[1] and the United States,[2] 1970–80

| Year | Males | | Females | |
	Maryland	United States	Maryland	United States
1970	16.2	16.8	6.0	6.6
1971	15.5	16.7	6.4	6.8
1972	16.7	17.5	7.1	6.8
1973	14.9	17.7	6.7	6.5
1974	16.3	18.1	6.0	6.5
1975	17.5	18.9	6.2	6.8
1976	18.0	18.7	5.5	6.7
1977	19.9	20.1	5.9	6.8
1978	17.8	19.0	5.6	6.3
1979	16.9	18.9	6.3	6.1
1980	17.5	18.6	5.8	5.5

[1]Data from reference 16.
[2]Data from reference 17.

Table 3. Suicide rates per 100,000 population by race for Maryland residents[1] and the United States,[2] 1970–80

| Year | White | | Nonwhite | |
	Maryland	United States	Maryland	United States
1970	12.1	12.4	5.7	5.6
1971	12.1	12.5	5.2	5.8
1972	12.8	12.8	7.5	6.6
1973	11.9	12.8	6.1	6.4
1974	12.0	13.0	7.1	6.5
1975	12.8	13.6	7.7	6.8
1976	12.9	13.3	6.6	7.0
1977	14.2	14.2	7.4	7.3
1978	13.2	13.4	6.1	6.9
1979	12.5	13.1	8.3	7.5
1980	13.1	12.7	6.5	6.4

[1]Data from reference 16.
[2]Data from reference 17.

SAMPLE 13.3 continued

Table 4. Suicide rates per 100,000 population by marital status for Maryland residents,[1] 1970–80

Year	Married	Single	Widowed	Divorced
1970	14.5	12.8	17.9	28.3
1971	12.7	13.9	22.9	27.7
1972	14.3	15.5	19.8	41.0
1973	14.0	13.1	21.8	19.5
1974	12.5	15.0	20.2	38.3
1975	13.9	14.8	18.5	41.0
1976	12.2	17.8	22.8	30.6
1977	14.6	18.4	13.3	36.4
1978	12.4	17.5	19.3	24.7
1979	11.6	18.1	18.4	24.3
1980	12.7	14.1	16.7	32.3

[1]Data from reference 16.
Note: Rates for U.S. not available.

In terms of the specific month of death of Marylanders' suicides, there was a pattern of nonuniformity throughout the 11-year period. For the entire period, however, the most suicides occurred in the months of August through November, with September being the month with the largest proportion of these deaths.

A discriminant function was calculated using age, race, sex, marital status, and month of death to predict suicide or other cause of death. The resulting discriminant function (table 5) explained 29 percent of the variation (canonical correlation = 0.1712). Age, with a standardized coefficient of 1.02324, was the most potent predictor of suicides or other causes of death, with race next (table 6). Overall, the discriminant function correctly predicted 80 percent of the 345,137 deaths in the respective category.

Table 5. Results of the discriminate analysis

Actual group membership	Number	Predicted group membership	
		Suicides	All other causes
Suicides:			
Number	5,040	3,729	1,311
Percent	100.0	74.0	26.0
All other causes:			
Number	340,097	67,169	272,928
Percent	100.0	19.7	80.3

SAMPLE 13.3 continued

The discriminant function had a sensitivity (predicted positives—all true positives) of 74 percent and a specificity (predicted negatives—all true negatives) of 80 percent. The low sensitivity yields a low positive predictive validity, 5.3 percent.

Table 6. Standardized canonical discriminant function coefficients

Variable	Coefficient
1. Month of death	−0.01374
2. Sex	0.09753
3. Race	0.30328
4. Marital status	−0.09718
5. Age	1.02324

SAMPLE 13.3 continued

derived from literature in the discipline. The authors create a table for each of the variables reported in the literature as significant determinants of suicidal tendencies—age, sex, race, marital status, and so on. In this way they organize *data* as *information*. The interpretation (synthesis) of this information constitutes the specialized *knowledge* of a scientific field. We can conceptualize this relationship as a pyramid, the pinnacle of which is wisdom, the ultimate application of scientific knowledge in decision making and human action (Figure 13.3).[5]

The Discussion Section

The discussion section contains your conclusions. It places the findings in the overall context of knowledge in the field and *synthesizes* these findings with those of other researchers. It explains why the results should have been expected according to the best theoretical knowledge in the field, or it explains why the results were a surprise and argues for a revision of a theoretical model on the basis of these findings. Finally, the discussion may suggest directions for further research. Sample 13.4 on page 440 is the discussion section of "Suicide in the State of Maryland, 1970–80."

5. Scientific papers, which are knowledge-generating instruments, generally make no claims to wisdom. Unlike technical reports, they stop short of direct applications of knowledge to problems in the public realm. See M. Jimmie Killingsworth and Michael Gilbertson, *Signs, Genres, and Communities in Technical Communication* (Amityville, NY: Baywood, 1992), Chapter 8; see also Killingsworth and J. S. Palmer, *Ecospeak: Rhetoric and Environmental Politics in America* (Carbondale: Southern Illinois University Press, 1992), Chapter 3.

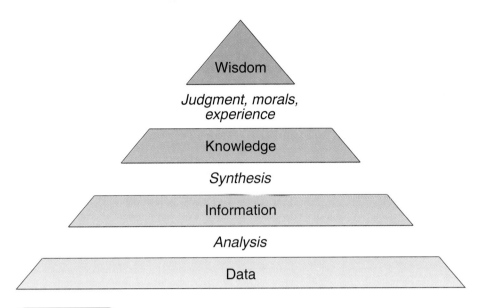

FIGURE 13.3

The information pyramid.
Source: Based on a diagram in G. Steven Tuthill and Susan T. Levy, *Knowledge-Based Systems: A Manager's Perspective* (Blue Ridge Summit, PA: Tab Books) 32.

Reviewing Scientific Papers

Once you have drafted your paper, you'll need to review and revise it carefully. You can use Guidelines at a Glance 47 for writing and reviewing scientific papers, both your own and those of others. Each guideline contains direction for writing ("Do such and such") along with criteria for evaluation (what each part of the paper should do to be successful). Exercise 13.3 provides practice in evaluation.

Do an audience-action analysis for a scientific paper.   Drawing on the research you are doing in one of your major laboratory or field courses, construct a set of notes to guide you as you write your paper. Use Figure 13.2 as a reference and answer the questions in Guidelines at a Glance 45. When you have answered the questions, discuss the notes with other students. Then expand or refine the notes to account for their suggestions and questions.

EXERCISE **13.1**

Discussion

The bases of this investigation were (a) to examine the "universal" relationships between age, race, sex, marital status, and month of death and suicide for Maryland residents and (b) to determine if suicidal deaths can be statistically distinguished from all other causes of death using these five variables as discriminating factors.

When the demographic variables of age, race, sex, and marital status were examined univariately for suicidal deaths of Maryland residents, this study did not reveal novel findings.

Older persons, whites, males, and divorced and widowed persons have the highest suicide rates (number of suicides per 100,000 population). All of these relationships are highly consistent with those published in the literature. Interactions of race and sex may show differential trends by age (*15*), but they were not examined in this investigation. No significant relationship appears to exist between suicidal deaths and seasonal variation. However, with the diversity of results already published on this variable, such findings in this investigation are neither consistent nor inconsistent with the literature (*15*).

The discriminant function performed does lend some useful information regarding the ability to predict possible suicides using age, race, sex, marital status, and specific month of death as discriminating variables. By use of a set of easily obtained variables, 74 percent of suicides and 80 percent of all other causes of death were correctly categorized. However, improvements could be made in the model to make the results more practical. For example, adding other variables, such as age-squared or an interaction term of calendar year by age, may improve prediction. This step would serve to raise the sensitivity of the predicted outcomes and the specificity. A way of raising the predictive validity of the model would be to increase the proportion of suicides in the deaths analyzed. Presently, suicides account for only 1.4 percent of all deaths. Eliminating deaths due to natural causes or nonsuicidal deaths, such as homicides and accidents, from the analysis would accomplish this.

SAMPLE 13.4   Discussion section of a published scientific paper

---

EXER**CISE**
13.2

Writing your introduction.   After reviewing your research notes and the notes you took during your audience-action analysis, you can draft your introduction. Use the five-part structure we've discussed, previewing each part of the whole paper. Share your draft with a group of your fellow students. Use their suggestions and answer their questions as you revise.

EXER**CISE**
13.3

Evaluating scientific papers.   At the end of this chapter you will find two papers for evaluation. One is a specialized professional paper; the other is a student lab report. Use the Guidelines at a Glance 46 to review each paper.

Evaluation A.   Sample 13.5 at the end of this chapter (pages 443–454) is a professional article in ecological science written by two biologists. The annotations

GUIDELINES AT A GLANCE

Writing scientific papers

- Create a title that works like a very short abstract, providing a specific description of the research topic. The title should be informative enough that when readers see it by itself in a reference list or bibliography, they can easily know what the paper is about.
- Write an abstract that extends the informativeness of the title, giving a brief overview of all parts of the report and stressing its significance in the research program of a specialized field of study.
- In writing the abstract and the paper itself you may use the "objective style," placing more emphasis on research action than on you and your fellow researchers. You may need passive sentences in places (as in "the measurements were taken"), but you should not overuse them. You can use the first-person plural "we" to avoid overuse of the passive (as in "we took the measurements").
- Technical terminology and specialized jargon are appropriate, but use these words only with great precision. Also, avoid great concentrations of abstract language, and do everything you can to enhance the fluency of the style (see Chapter 9).
- Use a five-part introduction, which states the problem, reviews the pertinent literature, briefly describes your method, summarizes your most significant findings, and forecasts your conclusions.
- Follow the IMRaD structure. After the *Introduction,* present sections on *Methods* (which describes your procedures, emphasizing how your approach differs from the usual way of doing things), *Results* (in which you say what happened in your research, organizing your data into categories based on well-known theories or previous research and using graphics to summarize as much of the data as possible), and *Discussion* (in which you state and explain your conclusions, saying *why* things turned out as they did and why these outcomes are important in your field).
- Document your information by using a standard form of citation for all references (see Chapter 10).
- Follow good design principles. Use graphics to display, and text to interpret, information. Create informative headings wherever convention allows. Use good forecasting and summarizing techniques. (See Chapter 7 for more techniques.)

in the margins give hints for review and mention some of the positive features. Note other instances of good writing that you find. Then make a list of things you would change if you were a contributing coauthor of this paper. When you have completed your list, share it with a group of fellow students, and come to an agreement on the good points and the changes needed. Discuss these with your instructor and with the class as a whole.

Evaluation B.   Sample 13.6 at the end of this chapter (pages 455–459) is a lab report from a course in mechanical engineering. The student authors follow the conventions of scientific style and structure fairly closely, but again I've pointed out a few features for special notice. As in Evaluation A, make a list of good points and possible changes, negotiate over these in your discussion group, and then discuss your consensus list with the class as a whole.

Recommendations for Further Reading

Day, Robert A. *How to Write and Publish a Scientific Paper.* 3rd Edition. Phoenix, AZ: Oryx, 1988.

Additional Reading for Advanced Research

Battalio, John T., ed. *Essays in the Study of Scientific Discourse: Methods, Practice, and Pedagogy.* Stamford, CT: Ablex, 1998.

Bazerman, Charles. *Shaping Written Knowledge: The Genre and Activity of the Experimental Article in Science.* Madison: University of Wisconsin Press, 1988.

Gross, Alan. *The Rhetoric of Science.* Cambridge, MA: Harvard University Press, 1990.

Killingsworth, M. Jimmie, and Michael Gilbertson. *Signs, Genres, and Communities in Technical Communication.* Amityville, NY: Baywood, 1992.

Killingsworth, M. Jimmie, and Jacqueline S. Palmer. *Ecospeak: Rhetoric and Environmental Politics.* Carbondale: Southern Illinois University Press, 1992.

Latour, Bruno. *Science in Action: How to Follow Scientists and Engineers through Society.* Cambridge, MA: Harvard University Press, 1987.

Prelli, Lawrence. *A Rhetoric of Science: Inventing Scientific Discourse.* Columbia: University of South Carolina Press, 1990.

Selzer, Jack, ed. *Understanding Scientific Prose.* Madison: University of Wisconsin Press, 1993.

Documents for Review

Fish predation in size-structured populations of treefrog tadpoles

Raymond D. Semlitsch[1] and J. Whitfield Gibbons[2]

[1]Meeman Biological Field Station and Department of Biology, Memphis State University, Memphis, TN 38152, USA

[2]University of Georgia's Savannah River Ecology Laboratory, P.O. Drawer E, Aiken, SC 29802, USA

Summary. The effects of tadpole body size, tadpole sibship, and fish body size on predation of gray treefrog tadpoles, *Hyla chrysoscelis,* were studied in laboratory and artificial pond experiments. Tadpole body size had a significantly positive effect on the survival of tadpoles in all experiments. The relationship between tadpole biomass eaten and biomass available suggested that fish were not satiated when consuming the largest tadpoles. Large tadpoles were probably better able to evade predators. A difference in survival among full sib families of tadpoles was only present in one family, suggesting that genetic differences in predator avoidance behavior or palatability were probably secondarily important to body size per se. Fish body size had a significantly negative effect on the survival of tadpoles. Larger fish consumed a larger number and proportion of tadpoles as well as greater biomass. These results indicate that environmental factors affecting the growth rate of tadpoles can dramatically alter their vulnerability to gape-limited predators.

Key words: Amphibian — Body size — Fish — Growth — *Hyla chrysoscelis* — *Lepomis macrochirus* — Predation — Sibship — Survival — Tadpole

Theories of life history evolution through demographic changes place a heavy emphasis on age-specific traits (Emlen 1970; King and Anderson 1971; Schaffer 1974; Thompson 1975; Michod 1979; Charlesworth 1980). Yet, some traits, such as body size, exhibit extreme variability within a given age and may be ecologically more important. In amphibians for example, body size is positively related to reproductive traits such as egg number or mating success (Salthe 1969; Kaplan and Salthe 1979; Howard 1980, 1983; Berven 1981, 1982; Verrell 1982). Predator avoidance and escape ability are also correlated with body size and developmental stage in amphibians (Heyer at al. 1975; Caldwell et al. 1980; Huey 1980; Smith 1983; Travis et al. 1985). Therefore, during the life history of an organism it is important to separate the effects of age from those of size and to measure their influence on traits such as survival or reproduction.

Most amphibians have complex life cycles with an aquatic larval stage and a terrestrial adult stage (Wilbur 1980). The larval stages of anurans are excellent subjects for studying predation in age- or size-structured populations. Many species of anurans are prolonged breeders, with females in the population ovipositing over several months or individual females producing multiple clutches in a year. This results in multiple age- and size-cohorts within the same breeding pond. Additionally, growth rates in even-aged cohorts are strongly affected by environmental variability as well as by genetic

The abstract for this paper is called a "summary." Read the third sentence. Could the authors have just said, "The fish could not get enough to eat when only large tadpoles were present"? What, if anything, do the authors gain by using their arcane vocabulary? Besides the style, what are the strengths and weaknesses of the summary?

SAMPLE 13.5 Professional paper in biology

Source: Raymond D. Semlitsch and J. Whitfield Gibbons, "Fish Predation in Size-Structured Populations of Treefrog Tadpoles, *Oecologia 75* (1988): 321–326.

Notice that the authors use the standard IMRaD headings but also include more informative sub-headings— "Natural history," "Laboratory experiments," and so on. How effective are these? Could the technique have been extended?

factors (Wilbur 1980; Travis 1980, 1981). These differences in growth rates can be further amplified by competition (Wilbur 1980). Consequently, individuals of the same age may vary dramatically in body size during larval development.

Although the body size of amphibian prey species has been shown to affect predation rates (Cooke 1974; Caldwell et al. 1980; Crump 1984; Travis et al. 1985a; Cronin and Travis 1986), it is not clear if this is due to body size per se or due to age-related or genetic differences in antipredatory behavior, chemical production, or social behavior (Liem 1961; Arnold and Wassersug 1978; Brodie et al. 1978; Formanowicz and Brodie 1982; Alford 1986). The purpose of this study was to test the effects of body size and parentage (i.e., full sib family) on the survival rates of even-aged cohorts of tadpoles exposed to a fish predator. The primary question we addressed was: Is large body size per se an effective antipredator defense for the survival of larval anurans?

Materials and methods

Natural history

The gray treefrog (*Hyla chrysoscelis:* Hylidae) is a prolonged summer-breeding anuran (May–August; Godwin and Roble 1983; Semlitsch, personal observation). Males chorus on warm rainy nights from perches surrounding the breeding sites. Females oviposit primarily in backwater pools along streams or ephemeral ponds but also can be found using more permanent ponds (e.g., farm ponds. Godwin and Roble 1983; Semlitsch, personal observation). The tadpoles are active, mid-water feeders and metamorphose in three weeks to four months in artificial ponds depending on growth rate (Wilbur and Alford 1985). Natural larval densities for other species of *Hyla* tadpoles range from 3–175 per m^2 (Turnipseed and Altig 1975; Caldwell et al. 1980).

The bluegill sunfish (*Lepomis macrochirus:* Centrarchidae) is the most widely distributed sunfish native to the eastern United States. It was chosen as a model predator because of its ubiquitous distribution, diverse diet, small body size, and adaptation to experimental manipulations (Keast and Welch 1968; Bauman and Kitchell 1974). Although unreported in the literature, this species opportunistically feeds on the eggs and larvae of amphibians (Semlitsch, personal observation). Densities of *L. macrochirus* in natural populations range from 0.1–2.2 fish per m^2 (Hall and Werner 1977).

Laboratory experiments

Effect of tadpole body size, fish body size, and tadpole sibship on the survival of *H. chrysoscelis* tadpoles was examined in a three-way factorial experimental design. The design included all combinations of three levels of tadpole body size (small, medium, large), two levels of fish body size (small, large), and five full sib families of tadpoles. The resulting 30-treatment combinations were replicated only once (Cochran and Cox 1957). Data were analysed by a three-way analysis of variance model for the three main effects and *a priori* selected two-way interaction effects. The remaining interaction terms were used as the estimate of error and the residual test term in each F-test (pp. 218–219, Cochran and Cox 1957). Percent survival of the tadpoles after 15 h was angularly transformed before analysis (Snedecor and Cochran 1980).

On 20 May 1986, five amplecting pairs of *H. chrysoscelis* were collected in Durham County, North Carolina, USA, from a breeding chorus. Pairs were placed in separate dishpans containing pond water and allowed to oviposit. The

SAMPLE 13.5 continued

full sib clutches were a sample from a selected portion of the breeding population on a single night. Each clutch was then divided and placed into three plastic wading pools (1.52 m diameter, 15 cm depth) at the Savannah River Ecology Laboratory in South Carolina, USA at three initial egg densities (low (X), medium (3X), high (6X)). The five sibships divided into 15 wading pools were reared for 20 days to generate three, even-aged, size cohorts of tadpoles (27 ± 1, $115 + 2$, 310 ± 24 mg) via food-limited growth. Tadpoles of all sizes were robust (i.e., round and not emaciated) and vigorous in appearance.

Lepomis macrochirus were collected from PAR Pond in Aiken County, South Carolina, USA by electro-shocking. Fish were placed in plastic wading pools for 48 h. On 8 June 1986 fish were measured and sorted into two size-classes (15 small, 2.2 ± 0.1 g; 15 large, 9.8 ± 0.3 g) and randomly assigned to 30 dishpans (18 cm × 34 cm × 47 cm) filled with 12 l of filtered pond water. These thirty dishpans were randomly assigned to positions on a large table, and 30 treatments were randomly assigned to the dishpans.

On 9 June 1986 tadpoles were counted and sorted into three non-overlapping size-classes. At 1500 h two cohorts of 20 tadpoles from each of the 15 sibship-size class combinations (30 total cohorts) were randomly assigned, within fish size-class, to the 30 dishpans. The 20 tadpoles were placed into an aluminum screen enclosure (bottom and topless cylinder, 10 cm high and 10 cm in diameter) positioned in the center of each dishpan. At 2100 h the screen enclosures were removed and the number of live and dead tadpoles were then censused at 3-h intervals until 1200 h on 10 June 1986.

Another laboratory experiment was performed to identify the body size that tadpoles must attain to escape death or mutilation by a gape-limited fish. Five *L.*

macrochirus of various body sizes (1.6–5.4 g) and not used in the previous experiment were randomly assigned to five dishpans (18 cm × 34 cm × 47 cm) each filled with 12 l of filtered pond water. After a 48-h acclimation period, tadpoles of eight size-classes were presented to the fish. Unused *H. chrysoscelis* tadpoles from one of the sibships used in the previous experiment were sorted into eight distinct size-classes (8–600 mg) of 10 individuals each. One tadpole from each of the size-classes was presented to each of the five fish in a randomly determined order at 3-h intervals. Tadpoles were classified as live, dead and mutilated, or eaten at the end of each 3-h interval. At the end of 24 h each fish had received one of each tadpole size. The remaining five tadpoles from each size-class and all of the fish were preserved to measure body size.

Pond experiment

Effect of tadpole body size and fish body size on the survival of *H. chrysoscelis* tadpoles was examined in artificial ponds. Two size-classes of *H. chrysoscelis* tadpoles were reared together in the presence of either small or large *L. macrochirus* or no fish. The three predator treatments were replicated three times.

Nine galvanized steel cattle-watering tanks (2.2 m diameter, 0.6 m deep, 2100 l volume), painted with epoxy enamel, were used for artificial ponds. Tanks were filled with water on 8 May 1986. Tanks were kept in an open field and exposed to natural temperature and photoperiod. Equal amounts (2.0 kg) of dried grass and leaf litter were added to each tank. Tanks were immediately covered with lids made of fiberglass window screening to prevent colonization by predatory aquatic insects. A mixture of zooplankton

SAMPLE 13.5 continued

Is this page well-designed? Can the reader go easily from text to graphics, and back again?

collected from several natural ponds was added in equal volumes to each tank several times during May.

Ten clutches of *H. chrysoscelis* eggs were obtained on 11 June 1986 in Durham County, North Carolina, USA. All eggs were mixed and tadpoles were reared at low (X) and high (10X) densities in plastic wading pools at the Savannah River Ecology Laboratory in South Carolina, USA. On 4 July 1986 nine cohorts of 100 small and 40 large *H. chrysoscelis* tadpoles were counted and sorted. One cohort of large and small tadpoles was randomly added to each of the nine artificial ponds. On 10 July 1986 two small *L. macrochirus* were each randomly added to three artificial ponds, and two large *L. macrochirus* were randomly added to three other ponds. The remaining three ponds had no fish as controls. The numbers of tadpoles and fish added to the ponds are within the range of natural densities (see above). After 72 h, the contents of all nine ponds were removed and the remaining tadpoles were counted and measured.

The results section arranges the experimental data in analytical categories derived from the literature review (with special emphasis on predator avoidance, escape ability or survivability, and growth rates).

The percentage of the initial cohort of tadpoles surviving among the three fish treatments was analysed by a one-way analysis of variance. Difference in the survival of large and small tadpoles was analysed by a *t*-test. Data were angularly transformed before analysis.

Results

Tadpole size

The body size of tadpoles had a dramatic effect on their survival in the presence of fish. In the laboratory experiment, tadpole size significantly affected ($P < 0.0001$) survival after 15 h (tables 1 and 2). A Scheffe's multiple range test indicated that all three size-classes had significantly different survival ($P < 0.05$). A higher percentage of the large tadpoles survived (85%) than the medium- (38%) or small- (1.5%) sized individuals. The effects of tadpole size were apparent at all time intervals during the 15 h but small tadpoles incurred most of the mortality in the first three hours (fig. 1). There was no significant two-way interaction effect of other factors with tadpole size (table 2).

Body size also had a significant effect on tadpole survival rate (table 3). Large tadpoles had significantly higher survival (89%) than small tadpoles (58%) in all ponds combined ($t = 4.71$, df = 16, $P = 0.0002$).

The effect of tadpole body size was also apparent when single tadpoles were fed to fish over a 24-h period. A distinct size boundary existed between those tadpoles eaten and those mutilated or escaping predation (fig. 2). Small fish (< 2 g) did not eat tadpoles above 150 mg body mass whereas large fish (> 5 g) consumed tadpoles > 500 mg in body mass (fig. 2).

Fish size

The size of fish predators significantly affected ($P = 0.0019$) the survival of tadpoles in the laboratory experiment (tables 1 and 2). At the end of 15 h, significantly more tadpoles of all sizes survived (51%) in the presence of small fish than in the presence of large fish. Large fish consumed more tadpoles of all sizes at all times intervals during the 15 h (fig. 1). There was no significant two-way interaction effect of other factors with fish size (table 2).

Fish size had no effect on the survival of tadpoles in artificial ponds (tables 3 and 4). Percent survival in the control ponds was higher ($\bar{X} = 95\%$ and 69%, respectively for large and small tadpoles) than in all the fish treatment ponds ($\bar{X} = 86\%$ and 52%, respectively for large

SAMPLE 13.5 continued

Table 1. Summary of the number and proportion of *Hyla chrysoscelis* tadpoles surviving in the presence of large and small *Lepomis macrochirus*. Values represent means ± one standard error.

Tadpoles			Fish			
			Large (9.8 ± 0.3 g)		Small (2.2 ± 0.1 g)	
			Number Surviving	Proportion Surviving	Number Surviving	Proportion Surviving
Small	Sibship	Mass				
	1	28	0	0.00	0	0.00
	2	23	0	0.00	3	0.15
	3	28	0	0.00	0	0.00
	4	30	0	0.00	0	0.00
	5	26	0	0.00	0	0.00
		$\bar{X} = 27 \pm 1$ mg	0	0.00	0.6 ± 0.6	0.03 ± 0.03
Medium	Sibship	Mass				
	1	115	4	0.20	16	0.80
	2	114	0	0.00	6	0.30
	3	112	4	0.20	10	0.50
	4	124	12	0.60	12	0.60
	5	112	2	0.10	10	0.50
		$\bar{X} = 115 \pm 2$ mg	4.4 ± 2.0	0.22 ± 0.10	10.8 ± 1.6	0.54 ± 0.08
Large	Sibship	Mass				
	1	351	14	0.70	20	1.00
	2	220	12	0.60	16	0.80
	3	313	13	0.65	20	1.00
	4	314	20	1.00	20	1.00
	5	351	16	0.80	19	0.95
		$\bar{X} = 310 \pm 24$ mg	15.0 ± 1.4	0.75 ± 0.07	19.0 ± 0.8	0.95 ± 0.04

Table 1 makes effective use of placement and white space, centering and framing the most telling findings, which appear in the column headed "Proportion Surviving." What are other strengths and weaknesses of the table on this page?

and small tadpoles; table 3). However, comparing just ponds with fish indicated there was no difference in percent survival of tadpoles ($\bar{X} = 65\%$ and 73%, respectively for large and small fish).

Sibship

The full sib family from which tadpoles came did not significantly affect ($P = 0.1208$) their survival when exposed to fish predators in the laboratory experiment (tables 1 and 2). The Scheffe's multiple range test did, however, detect a significant pairwise difference between sibship 4 and sibships 1, 2, 3, and 5 ($P < 0.05$). Tadpoles from sibship 4 had higher survival over all treatments (53%) than did individuals from the other sibships (30–45%). There was no significant two-way interaction effect of other factors with sibship (table 2).

SAMPLE 13.5 continued

Table 2. Summary of the three-way analysis of variance of the percentage of surviving *Hyla chrysoscelis* tadpoles. Data were angularly transformed by arcsine-square root before analysis.

Sources of variation	df	SS	MS	*F*-ratio	*P*-value
Main effects	7	8.420	1.203	29.9	<0.0001
Sibship	4	0.346	0.086	2.2	0.1208
Tadpole size	2	7.520	3.760	93.7	<0.0001
Fish size	1	0.554	0.554	13.8	0.0019
Interaction effects	6	0.313	0.052	1.3	0.3128
Sibship × fish	4	0.168	0.042	1.0	0.4131
Tadpole × fish	2	0.144	0.072	1.8	0.1973
Residual	16	0.642	0.040		
Total	29	9.375			

The authors write like scientists, maintaining the objective style throughout the article, but they also make a considerable effort to avoid *overusing* the passive voice. Notice the use of first-person plural ("our data") and the total avoidance of the passive in the first paragraph of the discussion.

Discussion

The most important result of this study is that large body size of tadpoles does increase their survival in the presence of a gape-limited vertebrate predator. Our data strongly indicate that large tadpoles are less likely to be eaten than small ones, regardless of additional effects of age or parentage. It is difficult, however, to distinguish among possible mechanisms accounting for this differential predation. At least two mechanisms may reduce predation on large tadpoles: satiation and capture or handling efficiency (Travis et al. 1985a; Formanowicz 1986). Larger tadpoles have more biomass and could satiate predators more quickly, resulting in fewer individuals being eaten relative to small tadpoles. Although our laboratory experiment was short-term and satiation was possible, it seemed unlikely. The amount of biomass ingested by fish across treatments was not constant with regard to the amount available (fig. 3). Both large and small fish consumed a greater biomass of medium tadpoles than that of small tadpoles, suggesting that fish in the small tadpole treatment may not have been fully satiated. However, when presented with even more biomass in the form of large tadpoles, biomass consumption decreased. In fact, for small fish the biomass consumed in large tadpoles was lower than the biomass consumed in small tadpoles (fig. 3). This latter result indicates that decreased handling efficiency or increased escape efficiency of large tadpoles led to the increased survival. Medium-sized tadpoles may have increased their survival by a combination of satiation and capture (or handling) efficiency. Handling efficiency may also explain the relationship between fish size and tadpole size in fig. 2, where satiation was probably not important. Caldwell et al. (1980) also presented evidence that salamanders feeding on tadpoles were gape-limited.

Positive effects of tadpole body size were also present in the artificial pond experiment. It is unlikely that satiation was a factor over this 3-day period, with relatively few tadpoles being eaten. More likely, tadpoles were able behaviorally to avoid fish by hiding in the leaf litter on

SAMPLE 13.5 continued

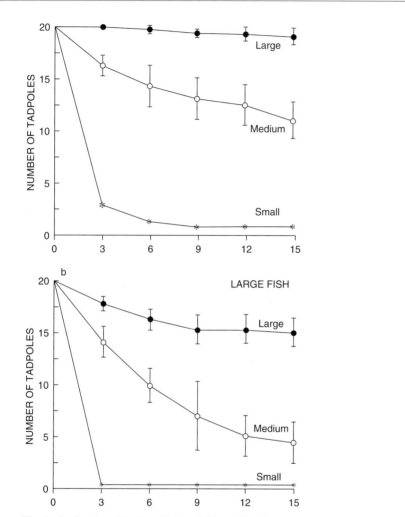

Fig. 1a, b. Survivorship of small (*asterisks*), medium (*open circles*), and large (*solid circles*) tadpoles exposed to large and small fish predators for 15 h. Data points represent means ± one standard error.

the bottom of the ponds. Tadpoles can detect and avoid predators such as newts (Morin 1986). Formanowicz (1986) found that large tadpoles were more effective than small tadpoles at evading small predators (i.e., more strikes and fewer first attempt captures by beetle larvae). Additionally, spring velocity in tad-

SAMPLE 13.5 continued

Table 3. Summary of the number and proportion of surviving *Hyla chrysoscelis* tadpoles reared in artificial ponds. Values represent means ± one standard error.

Fish	Replicate	Large (622 ± 67.2 mg) N = 40		Small (30 ± 2.9 mg) N = 100	
		Number Surviving	Proportion Surviving	Number Surviving	Proportion Surviving
Large (6.74 ± 0.11 g)	I	32	0.80	57	0.57
n = 5[a]	II	22	0.55	29	0.29
	III[a]	40	1.00	70	0.70
Small (2.68 ± 0.20 g)	I	37	0.92	61	0.61
n = 6	II	39	0.98	63	0.63
	III	37	0.92	33	0.33
No fish (control)	I	39	0.98	62	0.62
	II	38	0.95	71	0.71
	III	37	0.92	74	0.74

[a]One fish died during the experiment.

poles increases with developmental stage and body size (Huey 1980). Perhaps the fish seldom had more than one chance to capture a tadpole in the complex habitat of the artificial ponds. Predator avoidance behavior was evident from the noted absence of tadpoles in the open water and higher proportion of surviving tadpoles in the artificial ponds relative to the laboratory dishpans.

Genetic (i.e., full sib family) differences were not apparent in our predation experiment. There is, however, evidence for genetic variation in growth rate of larval anurans (Travis 1980, 1981, 1983). Such genetic variation has been demonstrated to lead to differential competitive ability as well as vulnerability to predation (Travis et al. 1985b; Alford 1986). Although Alford (1986) indicated that the differences in vulnerability to newt predation were due largely to body size, other factors such as genetic differences in behavior or palatability, might account for some of the variation. Body-size differences were apparent among our sibships reared in wading pools, but we deliberately selected similar size-classes of tadpoles for the experiments. Yet, there was a difference in the overall survival of sibship 4 that was not concordant with body size differences (table 1). The lack of strong sibship differences may likely be explained by the efficiency of the predator we used. Fish such as *L. macrochirus* are agile and relentless predators relative to other vertebrae predators such as newts and salamanders (Semlitsch, personal observation). Such predator efficiency would probably ob-

SAMPLE 13.5 continued

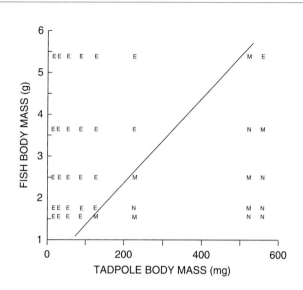

Fig. 2. Relationship between predator body size and tadpole body size. Each letter represents a single observation of whether a tadpole was eaten (*E*), mutilated (*M*), or not eaten (*N*). The line separating the tadpoles eaten from those mutilated or not eaten was drawn by eye.

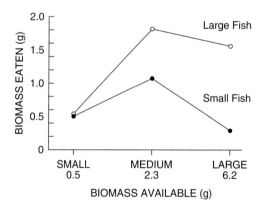

Fig. 3. Relationship between the total amount of tadpole biomass available and the total amount eaten by large (*open circles*) and small fish (*solid circles*) predators. Points represent means calculated from the five sibships.

SAMPLE 13.5 continued

Table 4. Summary of the one-way analysis of variance of the percentage of large and small surviving *Hyla chrysoscelis* tadpoles reared in artificial ponds. Data were angularly transformed by arcsine-square root before analysis.

Sources of variation	df	SS	MS	F-ratio	P-value
Large tadpoles					
Fish effects	2	0.058	0.029	0.60	0.5789
Residual	6	0.293	0.049		
Total	8	0.351			
Small tadpoles					
Fish effects	2	0.061	0.030	1.14	0.3794
Residual	6	0.160	0.027		
Total	8	0.221			

scure subtle differences in avoidance behavior in the confinement of laboratory containers. More complex experimental systems such as the artificial ponds may be more conducive to elucidating the subtle genetic effects from sibships for predator avoidance. Nevertheless genetic differences in palatability (i.e., chemical production) would not depend on predator efficiency. Palatability differences were not apparent among the sibships we used, also suggesting that genetic differences were secondarily important to environmentally induced body size differences.

> The discussion synthesizes the findings in theoretical terms, showing the significance of this experiment for the theory of evolution.

These results indicate that environmental factors affecting the growth rate of tadpoles can dramatically alter their vulnerability to gape-limited predators. Tadpoles (large and small) are restricted to the shallow water and thick vegetation along the edge of ponds inhabited by fish (Turnipseed and Altig 1975; Semlitsch, unpublished work). Small juvenile *L. macrochirus* are also restricted to the same microhabitat because of potential predation by larger fish (Werner et al. 1983). These juvenile fish represent a continuous but varying risk to tadpoles during their entire larval period. Tadpoles escaping predation through rapid growth and large body size will leave greater number of offspring than smaller, more vulnerable tadpoles, provided there are no compensatory changes later. In this sense differential predation on tadpole size can act as an agent for natural selection to increase growth rate. It is now important to learn whether other selective mechanisms to increase body size (e.g., competitive ability of larvae, dessication resistance of terrestrial juveniles, fecundity, mating success) positively interact with predator vulnerability or represent tradeoffs to balance other processes and life history traits (e.g., maintenance metabolism, maneuverability).

Acknowledgements. We thank J. D. Congdon, R. Fischer, J. L. Greene, T. Lamb, J. Wallin, and especially C. A. West for their help in counting, measuring, and sorting tadpoles and fish. The manuscript benefited from the comments of W. Gutzke, G. Harkey, R. Harris, T. Lamb, J. Van Buskirk, L. Wheat, and C. West. This research was supported by U.S. Department of Energy Contract DE-AC09-76SR00819 with the University of Georgia. Data analysis and manuscript preparation were

SAMPLE 13.5 continued

supported by the Department of Biology at Memphis State University.

References

Alford RD (1986) Effects of parentage on competitive ability and vulnerability to predation in *Hyla chrysoscelis* tadpoles. Oecologia (Berlin) 68:199–204

Arnold SJ, Wassersug RJ (1978) Differential predation on metamorphic anurans by garter snake (*Thamnophis*): Social behavior as a possible defense. Ecology 59:1014–1022

Baumann PC, Kitchell JF (1974) Diel patterns of distribution and feeding of bluegill *Lepomis macrochirus* in Lake Wingra, Wisconsin. Trans Amer Fish Soc 103:255–260

Berven KA (1981) Mate choice in the wood frog, *Rana sylvatica*. Evolution 35:707–722

Berven KA (1982) The genetic basis of altitudinal variation in the wood frog, *Rana sylvatica*. I. An experimental analysis of life history traits. Evolution 36:962–983

Brodie ED Jr., Formanowicz DR Jr., Brodie ED III (1978) The development of noxiousness of *Bufo americanus* tadpoles to aquatic insect predators. Herpetologica 34:302–306

Caldwell JP, Thorp JH, Jervey TO (1980) Predator-prey relationships among larval dragonflies, salamanders, and frogs. Oecologia (Berlin) 46:285–289

Charlesworth B (1980) Evolution in age structured populations. Cambridge University Press, Cambridge

Cochran WG, Cox GM (1957) Experimental designs. 2nd ed. John Wiley and Sons, New York, New York, USA

Cooke AS (1974) Differential predation by newts on anuran tadpoles. British J Herpetol 5:386–390

Cronin JT, Travis J (1986) Size-limited predation on larval *Rana areolata* (Anura: Ranidae) by two species of backswimmer (Insecta: Hemiptera: Notonectidae). Herpetologica 42:171–174

Crump ML (1984) Ontogenetic changes in vulnerability to predation in tadpoles of *Hyla pseudopuma*. Herpetologica 40:265–271

Emlen JM (1970) Age specificity and ecological theory. Ecology 51:588–601

Formanowicz DR Jr (1986) Anuran tadpole/aquatic insect predator-prey interactions: tadpole size and predator capture success. Herpetologica 42:367–373

Formanowicz DR Jr, Brodie ED Jr (1982) Relative palatabilities of members of a larval amphibian community. Copeia 1982:91–97

Godwin GJ, Roble SM (1983) Mating success in male treefrogs, *Hyla chrysoscelis* (Anura: Hylidae). Herpetologica 39:141–146

Hall DJ, Werner EE (1977) Seasonal distribution and abundance of fishes in the littoral zone of a Michigan lake. Tans Am Fish Soc 106:545–555

Heyer WR, McDiarmid RW, Weigmann DL (1975) Tadpoles, predation and pond habitats in the tropics. Biotropica 7:100–111

Howard RD (1980) Mating behavior and mating success in woodfrogs, *Rana sylvatica*. Anim Behav 28:705–716

Howard RD (1983) Sexual selection and variation in reproductive success in a long-lived organism. Am Nat 122:301–325

Huey RB (1980) Spring velocity of tadpoles (*Bufo boreas*) through metamorphosis. Copeia 1980:537–540

Kaplan RH, Salthe SN (1979) The allometry of reproduction: an empirical view in salamanders. Am Nat 113:671–689

Keast A, Welch L (1968) Daily feeding periodicities, food uptake rates, and some dietary changes with hours of the day in some lake fishes. J Fish Res Board Can 25:1113–1114

King CE, Anderson WW (1971) Age specific selection. II. The interaction between r- and K- during population growth. Am Nat 105:137–156

Liem KF (1961) On the taxonomic status and the granular patches of the Javanese frog *Rana chalconota*. Herpetologica 17:69–71

Michod RE (1979) Evolution of life histories in response to age-specific mortality factors. Am Nat 113:531–550

Morin PJ (1986) Interactions between intraspecific competition and predation in an amphibian predator-prey system. Ecology 67:713–720

Salthe SN (1969) Reproductive modes and the number and sizes of ova in the urodeles. Am Midl Nat 81:467–490

Schaffer WM (1974) Selection for optimal life histories: the effects of age structure. Ecology 55:291–303

SAMPLE 13.5 continued

Smith DC (1983) Factors controlling tadpole populations of the chorus frog (Pseudacris triseriata) on Isle Royale, Michigan. Ecology 64:501–510

Snedecor GW, Cochran WG (1980) Statistical methods. 7th ed. Iowa State University Press, Ames, Iowa, USA

Thompson DJ (1975) Towards a predator-prey model incorporating age structure: the effects of predator and prey size on the predation of *Daphnia magna* by *Ischnura elegans.* J Anim Ecol 44:907–916

Travis J (1980) Genetic variation for larval specific growth rate in the frog *Hyla gratiosa.* Growth 44:167–181

Travis J (1981) The control of larval growth variation in a population of *Pseudacris triseriata* (Anura: Hylidae). Evolution 35:423–432

Travis J (1983) Variation in growth and survival of *Hyla gratiosa* larvae in experimental enclosures. Copeia 1983:232–237

Travis J, Keen WH, Juiliana J (1985a) The role of relative body size in a predator-prey re-lationship between dragonfly naiads and larval anurans. Oikos 45:59–65

Travis J, Keen WH, Juiliana J (1985b) The effect of multiple factors on viability selection in *Hyla gratiosa* tadpoles. Evolution 39:1087–1099

Turnipseed G, Altig R (1975) Population density and age structure of three species of hylid tadpoles. J. Herpetol 9:287–291

Verrell PA (1982) Male newts prefer large females as mates. Anim Behav 30:1254–1255

Werner EF, Gilliam JF, Hall DJ, Mittelbach GG (1983) An experimental test of the effects of predation risk on habitat use in fish. Ecology 64:1540–1548

Wilbur HM (1980) Complex life cycles. Ann Rev Ecol Syst 11:67–93

Wilbur HM, Alford RA (1985) Priority effects in experimental communities: response of *Hyla* to *Bufo* and *Rana.* Ecology 66:1106–1114

Received August 3, 1987

SAMPLE 13.5 continued

Lab #14 Modulus of Rupture: Brittle Materials

J. R. Smith, M. Jones, and K. R. Singh

December 9, 1991

Introduction:

   In today's industry as in the past, the need for hard
materials with little or no ductility is prevalent. From
glass structures of the past to ceramic engine blocks of the
not-so-far-away future, brittle materials fill this need.
The strengths and other properties of brittle materials are
very important in their design for any application. Because
brittle materials are susceptible to the mechanical gripping
incorporated in the tensile testing of metals and plastics,
beam-bending (no grips required) is incorporated to overcome
this problem. The Modulus of Rupture (MOR) is a measure of
the tensile stress required to cause failure in the bending
of a beam of a brittle material. ASTM Standard C 158, Flex-
ure Testing of Glass (Determination of Modulus of Rupture),
is referenced for the test procedure. The variables associated with the test procedure are rate of stressing, test
environment, and the area of the specimen subjected to the
stress. The method for determination of the MOR is discussed
later in the Calculation section of this report.

Theory:

   Ductile materials display an engineering stress-strain
curve that goes through a maximum at the tensile strength.
In more brittle materials, the maximum load tensile strength
occurs at the point of failure. In extremely brittle materials, such as ceramics, the yield strength, tensile strength,
and breaking strength are all the same.

   In many brittle materials, particularly ceramics and certain composite materials, the normal tensile test cannot be
easily performed due to the presence of flaws at the surface. Often, placing the material in the grips of the
tensile testing machine will cause these flaws to promote
cracking, invalidating the test. Preparation of tensile
specimens of brittle materials may also be expensive. One
approach used to minimize these problems is the bend test.
By applying the load at three points and causing bending,
a tensile force acts on the material opposite the middle
point. Fracture begins at this location.

   A. A. Griffith conducted experiments to determine the
fracture strength of glass containing small flaws. Sharp
notches stay sharp in a perfectly brittle material. The
radius at the tip of the crack is at the atomic level and
atomic bonds must be ruptured for fracture to occur.

The authors use
a slight variation
of the usual IMRaD
headings, follow-
ing guidelines for
the lab issued by
their professor.
At one time or
another, every
technical commu-
nicator has to
work within the
confines of a
"house style," so
compliance with
demands of dif-
ferent professors
is more realistic
than you may
think.

The introduction
refers to key the-
oretical problems
in the field of
materials engi-
neering and
mentions the lit-
erature (the work
of A. A. Griffith
and ASTM Stan-
dards). What's
missing in these
citations?

SAMPLE 13.6   Student report from a lab in mechanical engineering

In the procedure (methods) section the authors give the summary of procedural steps in imperative sentences, suggesting commands or instructions rather than reported events. This style suggests, in a somewhat unusual way, the reproducibility of the experiment. Most scientific writers would prefer a reporting style, saying "The plate glass specimens were obtained" rather than "Obtain the plate glass specimens." Here we have another instance of "house style," we suspect.

The authors seem to prefer the passive style of traditional scientific writing ("Four glass specimens were tested"), but occasionally vary this practice by beginning using the subject "we" in active voice sentences ("In Figure 1, we presented a plot"). Having thus broken the hold of the passive, should they use the "we" pattern more frequently?

   Since cracks and flaws tend to remain closed in compression, brittle materials are often designed so that only compressive stresses are acting on the part. Often, we find that brittle materials fail at much higher compressive stresses than tensile stresses, although ductile materials such as metals may have tensile and compressive stresses that are nearly equal.

Procedure:
1. Obtain the plate glass specimens, two scratched and two non-scratched specimens.
2. Measure specimen width and thickness.
3. Set the span length on the MOR testing jig to 4″ for one scratched and one non-scratched specimen and 5″ for the other two specimens.
4. Select strain rate of 0.002 inch/minute.
5. Operate test according to ASTM standard.
6. Examine fracture surface and determine failure mode.

For the scratched specimens we also measured the scratch half length (a) and related the fracture stress to the scratch half length.

Equipment Required:
   Four glass specimens were tested using an Instron Testing machine. Two specimens had a scratch on one side while the other two were scratch free. Calipers were used for dimension measurements of the specimens. To measure the precise crack length of the scratched specimens, a Vickers Hardness Tester was used.

Calculations Required:
The MOR is calculated using the following:

$$S = \frac{3PL}{2b(h)^2} \qquad (1)$$

S -- Rupture stress (psi)
P -- Load at fracture (lb)
L -- Length of span (in.)
b -- Specimen width (in.)
h -- Specimen thickness (in.)

For the scratched specimens of glass:

$$Sf = \frac{C}{(PIa)^{.5}} \qquad (2)$$

Sf -- Fracture stress (psi)
C -- Constant for glass
       $(28.14 \text{ psi in.}^{.6})$
a -- Half crack length (in.)

SAMPLE 13.6 continued

Results:

In Table 1 we entered the dimensions, failure load, failure stress, and half crack length (if applicable) for the four specimens. In Figure 1, we plotted fracture stress against the half crack length. We also included the two plots from the testing computer portraying load against time. One plot was for the scratched and non-scratched specimens tested at a 4" span length, and the other plot was for the 5" span length.

Analysis:

All four of the glass specimens failed under brittle fracture. The 4-inch cracked specimen had one very distinct crack at its failure point. The other three specimens, however, had many smaller cracks emanating from the point of failure. These failures are the result of the crack initiations on each of the samples. On the 4-inch cracked specimen, the initiated crack was large enough to cause failure to occur before any others were able to be initiated. On the two smooth specimens, several cracks had to be initiated before the entire specimen failed along one of these initiated cracks. On the 5-inch cracked glass, the previously initiated crack was not enough to cause failure immediately. Several other cracks had to be initiated to cause stress concentrations before the glass specimen could fail.

The length of the tested pieces played an important part in how much load they could take before failure. The longer pieces displayed more deformation before rupture than the small pieces. The 4-inch cracked specimen, displayed 0.011 inch deformation before failure while the 5-inch cracked specimen, displayed 0.016 inches deformation before failure. The 5-inch unscratched specimen had 0.02 inches deformation

Table 1

Sample	Width (in.)	Thickness (in.)	Failure Load (lbs)	Failure Stress (psi)	Half Crack Length (in.) Meas./Calc.
4 in. cracked	0.9705	0.2230	49.3	6129.0	0.003/6.7E-6
4 in. no crack	0.9700	0.2231	53.3	6626.3	N/A
5 in. cracked	0.9704	0.2231	44.8	6962.0	0.003/5.2E-6
5 in. no crack	0.9705	0.2230	56.5	8780.0	N/A

SAMPLE 13.6 continued

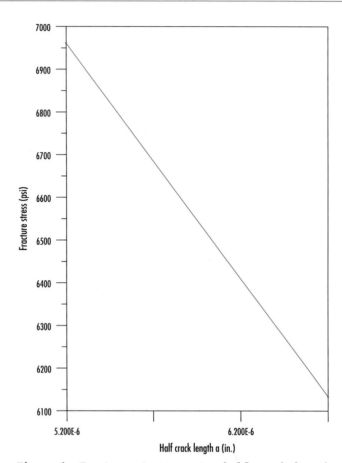

Figure 1. Fracture stress versus half crack length

at fracture and the 4-inch specimen showed 0.0106 inch at fracture. This result shows that the higher modulus of elasticity was displayed by the shorter samples.

   The values obtained for the half crack length when the calculated values for the fracture stress in Griffith's crack theory equation were used were much different than the measured half crack length. The measured half crack length was 0.003 inches. The calculated values for the half crack length for the 4- and 5-inch specimens were 6.7×10^{-6} and 5.2×10^{-6} inches, respectively. Even though these values seem to conflict with each other, the discrepancy can be easily explained. Griffith's equation deals with the radius

SAMPLE 13.6 continued

of a crack on the atomic level, since that is where bonds must be broken for failure to occur. The measured values were not taken on the atomic level; they were taken instead on the macroscopic level. The effect of these large cracks was still seen at the atomic level where the cracks were really initiated.

On the 5-inch specimen, the half crack length was calculated at 5.2×10^{-6} inches and the failure stress was 6962 psi. On the 4-inch sample, the half crack length was calculated to be 6.7×10^{-6} inches and the failure stress was 6129 psi. This follows Griffith's crack theory. The larger the initial crack, the lower the stress required before failure. Since the 4-inch specimen had the largest crack, it was the one to fail at the lower stress.

Conclusion:

Much was gained through the execution of this experiment. An initiated crack on the surface of a brittle material will cause the material to fail at a much lower stress level than expected. Not only does the presence of a crack cause premature failure, but the width, depth, and orientation of the crack also play an important part in determining just how much weaker the material will be made. The length of a brittle specimen under a load also affects its mechanical properties. The short specimens were much more rigid than the longer specimens. It was also seen that cracks initiate at the atomic level of a material. All in all, brittle materials exhibit a sensitivity to their surface conditions as well as their physical dimensions.

SAMPLE 13.6 continued

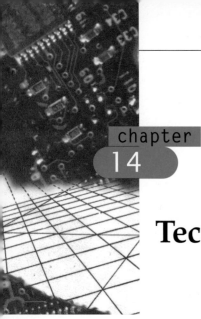

Technical Proposals

CHAPTER OUTLINE
▼▼▼▼▼▼▼▼▼▼▼▼▼▼▼▼▼▼▼▼▼▼▼▼▼▼▼

CHAPTER OBJECTIVES
▼▼▼▼▼▼▼▼▼▼▼▼▼▼▼▼▼▼▼▼▼▼▼▼▼▼▼

After you have worked through this chapter, you should be able to do the following:

- Recognize the generic purposes and typical audiences of technical proposals
- Use the CORE method to plan and prepare a technical proposal, beginning with a short concept paper as your core document
- Participate in peer reviews of concept papers and technical proposals
- Define and analyze the kinds of problems that technical researchers face in writing for audiences of their peers

T he proposal grew up with democracy. In the most famous communication manual in ancient times, the *Rhetoric,* Aristotle pointed to proposals, which he called "deliberative speeches," as one of the key means of deciding on the future actions of the democratically organized Athenian city-state. More than 2000 years later, the proposal once again became a major form of problem-solving discourse when, following World War II, the United States government made money available for research and development on a large scale. Proposals began linking the public and private sectors of the information economy.

The human situation that encompasses proposals has not changed that much since Aristotle. A problem surfaces. There are many ways in which it could be solved, many technical paths. Which one should the community take? From the context of production, proposers come forward with plans for solving the problem; and, representing the context of use, a formally or informally assembled team of evaluators considers the costs and benefits of the plans and their advantages and disadvantages. The ultimate decision for or against the proposal is based on the proposer's understanding of the problem, the ingenuity of the solution, the overall costs, the benefits to the public, and the evaluators' confidence in the proposer's ability to deliver what is promised.

In this chapter we will consider how you can take your place as a successful proposal writer in this process of public decision making. To face the challenge of writing effective proposals, you will learn to adapt the CORE method by following five basic steps (see Figure 14.1):

1. *Do an audience-action analysis for a technical proposal* that will serve as your planning notes.
2. *Write a concept paper* of no more than two pages. Share it with trial readers to get a sense of the kind of information your audience will need. The more readers, the better. The closer the readers are to your actual audience of evaluators, the better.
3. *Develop graphics* to support the main points of your concept paper. Cover as much information as you can in graphical form. Adapt some of these into transparencies for an oral presentation.
4. *Present the proposal orally* to a trial or actual audience.
5. *Develop the full formal proposal,* amplifying the concept paper and using the graphics that you designed for an audience of evaluative readers.

These five steps are the same whether you are developing informal technical proposals or more formal research or contract proposals. Differences between types of proposals within the genre result in modifications to the audience-action analysis and to the outline. Formal research and contract proposals generally require more components, as specified by potential funding agencies. We will address these differences in the appropriate sections.

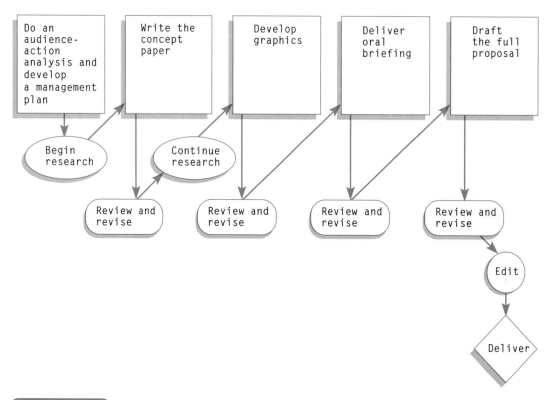

FIGURE 14.1

Applying the CORE method to the development of technical proposals.

Step One: The Audience-Action Analysis for a Technical Proposal

The purpose of a proposal is to put forward a plan that will convince an audience of evaluators that an author can solve a problem by taking a future action.

The *plan* takes the form of a narrative for future action. This is the heart of the proposal—the story of what the proposer will do. The challenge for all proposal writers is to make the plan credible by bringing the past to bear on the future. The past is factual and information-rich, but the future is nebulous, nonfactual, and a little scary for most audiences. To be a successful proposal writer, you must put your the audience at ease by inspiring confidence and making your plan seem doable. You should support your plan in three ways:

1. Marshall *information* to show that you (or someone like you) has done such work before.

2. Give good *reasons* for your proposed course of action, showing that it will produce tangible outcomes.
3. Demonstrate that your plan is better than other alternatives—especially the alternative of doing nothing.

The *audience* for a proposal is a group of decision makers, a panel of judges who have the power to say whether the plan can be carried out or not. Like the audience for technical reports (discussed in Chapter 12), proposal evaluators may include managers, government officials, and technically trained reviewers. Often, they are gathered specifically to review proposals on a certain issue.

Research and contract proposals assume that the author and the audience are on equal footing as technical peers. Typically, the author is a researcher in a university or research organization. The typical reviewer is a member of an expert review committee in an agency like the National Science Foundation or is part of a team of technical reviewers working for a company that needs additional expertise or subcontractors. The author and the audience have a mutual interest in advancing the knowledge and research program of a specialized field.

If you go to graduate school in engineering, in science, or in one of the social sciences, you will almost certainly contribute to a proposal or write one of your own. If you go to work in industry—especially in the defense industry or in other fields that compete for federal money—you will also contribute to research or contract proposals. Occasionally, undergraduates are also involved in this kind of work. Even if you aren't, you will find that your professors' (or employers') understanding of the proposal genre has been shaped by work with research proposals directed to federal agencies.

As indicated in Figure 14.2 (reproduced from Chapter 5), proposals may be *solicited* or *unsolicited*. The audience for unsolicited proposals is practically

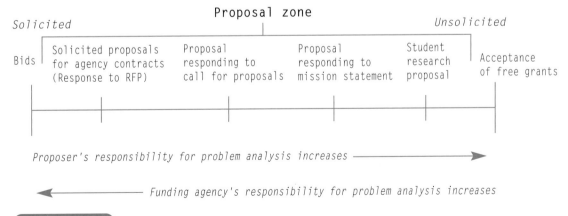

FIGURE 14.2

The proposal zone.

identical to the audience for reports on options (recommendation reports and feasibility studies)—a group of decision makers who need your views about what problems exist and how to solve them. By contrast, the audience for solicited proposals is quite aware of the problem and knows that it needs your help in solving it. The key difference between a report on options and a proposal is the author's perspective. Whereas the author of a report may or may not be directly involved in the project he or she is writing about, the author of a proposal is usually a key player in the plan, most often the project director or principal investigator. The proposal writer has a bigger stake in the outcome of the proposal, and the audience of judges is therefore more sensitive to the qualities and qualifications the author displays in the arguments, analyses, and plans put forward in the proposal.

The prevailing attitude among managers and evaluators of proposals is to leave things as they are—"If it ain't broke, don't fix it"—so your job as the author of a proposal is to show that something *is* broken and that you can fix it. This means convincing evaluators of three things:

1. There is a problem (or need).
2. You understand the problem.
3. You know how to solve the problem in the best way possible.

The *problem* must be a demonstrable and pressing *need* among a group of people for whom both the proposal writer and the audience have an abiding concern. It may be an obstacle that is impeding daily operations in a company, causing a group of clients to suffer, or standing in the way of scientific knowledge. The problem isn't always a roadblock, though. It can also be an exploration or expansion opportunity with potential to enhance the company's profits, reputation, or status when things are already going well.

A proposal usually centers on one of the following:

- *The development of a product design or technological innovation* (a computer program for student filing, a smoke alarm for the hearing-impaired, a new parking garage, a better mouse trap)
- *The implementation of a process or improved procedure* (online student registration, twenty-four-hour banking services, a new technique for environmental assessment, a management plan for a company, a reorganized curriculum in a special discipline of study, a pest control system)
- *A combination of technology and process* (using computers to manage a dairy farm, biofeedback as a way of dealing with learning-disabled children, electronic tracking systems for parolees, combining chemical and mechanical devices in pest control)

The tone for a proposal should be positive, hopeful. The proposal writer should communicate a faith in progress, in the ability to solve problems to make the world a better place. To borrow the words of Shakespeare, the proposal writer is striving to give shape and form to the "airy nothing" of the future.

In technical proposals, however, it is not poetic language that accomplishes this utopian purpose; it is the ability to match an analysis of a problem with a feasible *solution*. Unlike reports, which look to the future but focus on what has been done in the past, the proposal provides a detailed *plan of action* keyed to a specific set of *objectives* based on projected outcomes. If the objectives can be accomplished, argues the proposal writer, then the problem will be solved. And if the problem is solved, the proposal writer, the audience, and a broad constituency of their clients or dependents will benefit. (See Figure 14.3.)

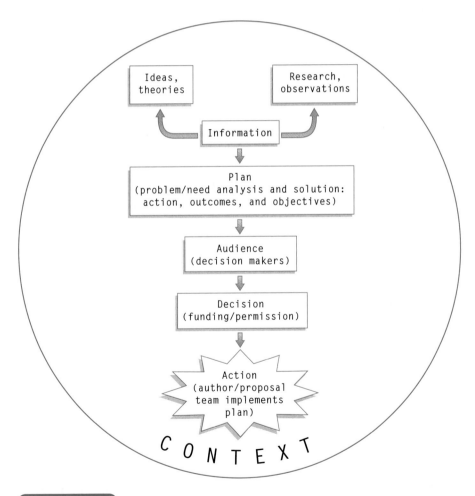

FIGURE 14.3

Pathway from information to action in a technical proposal.

Step Two: Writing the Concept Paper

Just as reports often start their lives as position papers (see Chapter 12), so proposals often begin as *concept papers,* short documents that give an overview of the problem, the proposed solution, and the benefits of the project to be proposed. In industry, *internal proposals* or *informal proposals* are almost identical to concept papers presented in the memo format (see Chapter 11 on correspondence). If management needs more information and justification for the proposal, they will request a fuller document or perhaps an oral presentation.

Most funding agencies require two stages for research proposals. For the first stage, proposers submit "preproposals" or concept papers. A panel of evaluators screens these short papers and invites the authors of the best ones to submit longer proposals.

The second step in the CORE method, then, is quite practical. Writing a concept paper allows you to present key ideas for the proposal to trial audiences without making great demands on either their reading time or your own preparation time.

Using the Standard Proposal Structure to Develop Your Concept Paper

The concept paper, as the core of the proposal in progress, follows roughly what will become the structure of the full document. In nearly every proposal you will ever write, you'll have to give, as a bare minimum, the following information:

- an analysis of the problem (or need)
- an analysis of possible solutions for the problem (or ways of fulfilling the need)
- a plan for the best solution (as you envision it)
- an analysis of the costs and benefits (advantages and disadvantages) of the proposed solution
- assurances that you are capable of carrying out the plan you offer

To meet these demands, you can use the minimal outline presented in Chapter 5, Guidelines at a Glance 8 (see page 130).

Applying Guidelines at a Glance 8

To demonstrate how this structure translates into steps toward a concept paper, let's examine the procedure that one student—Keith Bell, a biology major at Texas A&M—followed in developing a proposal for surveying aquatic ecological systems. At this point, you should be able to match Keith step for step in writing your own concept paper.

I. Introduction

    A. *Begin with a sentence that briefly states the problem that the proposal addresses.* Keith wrote the following:

        Current techniques of surveying marine life are expensive, time-consuming, and inaccurate.

    B. *Say how you plan to solve the problem.* Here's how Keith proceeded:

        The method proposed in this paper solves the problem by using techniques based on video technology and computer-aided image analysis.

    C. *Finish the first paragraph with a sentence or two explaining the purpose of the proposal in terms of the intended audience's interests.* Keith decided to come right out and claim that his project would benefit the funding agency (Apple Computers) directly:

        This proposal describes the project and requests support for the project from Apple Corporation. By donating or lending computer hardware, Apple will potentially open up new markets of computer users in oceanographic biology and environmental impact assessment.

II. Background on the problem

    Give the background in the second paragraph. Analyze the problem by describing the old way of doing things. Keith wrote the following:

        The common method of surveying marine life is a time-consuming manual counting process. A team of divers inventories the organisms by hand. The process is slow, expensive, dangerous, and open to inaccuracies and inconsistencies.

III. Project Plans and Requirements

    A. *In the third paragraph, describe how your project will solve the problem.* Note that Keith worked in an ecological benefit—a potential PR opportunity for Apple—when he wrote this:

        The new system uses a VHS video recorder in conjunction with a Macintosh IIsi microcomputer equipped with a video capture card to record, store, and analyze data. With the new system, scuba divers do not have to waste time manually counting and recording specimens. They simply take video images of the sea bottom. Once the images are digitized, the computer can automatically analyze the data. The method is faster, less expensive, more reliable, and less likely to disturb marine organisms and environments.

GUIDELINES AT A GLANCE 48
Reviewing concept papers

- Does the author clearly identify a problem or need? Are you convinced that the problem is worth addressing?
- Does the author provide enough information for you to understand the problem and the solution? What more would you need in a complete concept paper? What would be enough to make you want to commission a full proposal?
- Does the solution seem better than other possible solutions? Are there alternatives that the author overlooks?
- Do the benefits of the concept appear to justify the costs?
- Are the benefits widely distributed, or do they mainly accrue to the proposer?
- Are you satisfied that the proposer is capable of carrying the project to completion? What would you need in the concept paper to assure you about his or her credentials?
- Is the writing clear and free of difficult jargon?
- Is the tone appropriate? Does the author project enthusiasm, commitment, and intellectual energy?

B. *Next, say what the project needs to continue.* Keith wrote:

> To continue, the project needs more powerful hardware. Thus we are requesting the donation or loan of a Quadra 950 microcomputer.

IV. Benefits

> *Finish strong, giving the benefits of the project for all parties involved.* Keith wrote:

> The image analysis system is likely to become the method of choice. Mobil Oil Company has shown great interest in the project. The company has welcomed us to use its oil platforms as a testing ground and plans to use the information gathered to determine the environmental impact of oil spills. By investing in the project, Apple can contribute to a methodological breakthrough that not only benefits science and industry, but also opens the way for new applications of high technology in the booming field of applied ecology.

Sample 14.1 is the full concept paper. The annotations point out a few features that make the proposal particularly persuasive. Sample 14.2 is another student concept paper. As you develop your own concept paper, use Guidelines at a Glance 48 to check for effectiveness and completeness. You can practice doing this in Exercise 14.2 (at the end of the chapter).

Private Funding of Image-Analysis
Techniques for Ocean Floor Study
M. K. Bell

Current techniques of surveying marine life are expensive, time-consuming, and inaccurate. A method now under development at Texas A&M University promises to revolutionize the field by introducing techniques based on video technology and computer-aided image analysis. This proposal describes the project and requests support for the project from Apple Corporation. By donating or lending computer hardware, Apple will potentially open up new markets of computer users in oceanographic biology and environmental impact assessment.

The only method of surveying marine life now widely practiced is a time-consuming manual counting process. A team of divers lays a grid over the portion of the sea floor under investigation and then, by hand, takes an inventory of the organisms present within each section of the grid. The process is obviously slow, expensive, occasionally dangerous, and often open to inaccuracies and inconsistencies.

The system now under investigation uses a VHS video recorder in conjunction with a Macintosh IIsi microcomputer equipped with a video capture card to record, store, and statistically analyze data. Using the new system, scuba divers do not have to spend long hours beneath the surface manually counting and recording specimens. They simply take video images of the sea bottom. Once the images are digitized, the computer saves even more time and energy by automatically analyzing the data. The method is faster, less expensive, more reliable, and less likely to disturb marine organisms and environments.

To refine the technique, we need more powerful hardware. At present, we have exhausted the capability of the Mac IIsi. Thus we are requesting the donation or loan of a Macintosh Quadra 950 microcomputer. With this additional power, we believe we can perfect the method and promote its wide use in the field.

The image-analysis approach is likely to become the method of choice in several areas of scientific and technological investigation. Its promise for environmental impact analysis is particularly strong. A measure of its promise is the interest that Mobil Oil Company has shown in the project. The company has welcomed us to use its oil platforms as a testing ground and plans to use the information gathered to determine the environmental impact of oil spills. By investing in the project, Apple can make a major contribution to a methodological breakthrough that not only benefits science and industry, but also opens the way for new applications of high technology in the booming field of applied ecology.

Keith avoids technical jargon from the field of ocean studies. The original draft of the paper used the term "benthic" to describe bottom-anchored organisms. Keith replaced the term out of consideration for the audience, whose special expertise lies in another area (computer technology).

Keith analyzes the key features of the problem by breaking it into components—the old method is expensive, slow, unreliable, and potentially dangerous—and shows how the solution addresses each of these components of the problem.

The paper attempts to show how the benefits extend beyond the author's own field of study (ocean ecology).

It ties the project to a critical issue in the political economy at large—the problem of ecological degradation.

SAMPLE 14.1   Concept paper for a student proposal

Catering Care Packages
A Business Plan by Dan Eason

This concept paper outlines a plan for a company specializing in mail-order care packages for students at Texas A&M. There are over 40,000 students at A&M, most of them from out of town. Right now, if parents want to provide a gift for students, they have to order it from their hometown or bring it to College Station themselves. I am proposing an alternative--a service that allows parents to call and have a special package of healthy snacks hand-delivered to their son or daughter.

A small investment in such a business could mean high returns. On a small scale, I've already tested the concept, with excellent results. In the Fall semester of 1991, I started a care package business. It catered to the needs of both students and parents. I designed and printed 4,000 brochures and sent them out in three mailings throughout the semester. I got approximately a 4% response. Research on mail order businesses shows that 1.5% is a typical response and 2% a good one for a start-up business. I was excited to get the response I did, but being short on capital I was forced to put the business on hold after one semester. Even without further advertising, I have received several calls every semester on my 1-800 line.

There are several reasons why an investor should share my excitement. First, in a matter of only four months I achieved a successful response, a response that will no doubt grow when the company builds up a data bank of regular customers. Second, other than mother's clubs and A&M Food Services, there is little competition. Third, the profit margin of the packages is relatively high and would only get higher as the number of brochures printed and mailed increases. And finally, because I have already worked out many of the bugs that plague start-up companies, I can concentrate on improving the rate of return right from the start.

For these reasons, I strongly encourage investors to study the full business plan and to consider the tremendous potential of a care package company situated in the virtually untapped market of College Station, Texas.

SAMPLE 14.2   Concept paper for a student business plan

Step Three: Developing Graphics for the Proposal

The third step in the CORE method is to plan and design graphics to support the argument outlined in your concept paper. As you work on your graphics, you will be working simultaneously on Step Four of the CORE method. You can use the graphics you produce in combination with your concept paper as the basis for an oral proposal.

To develop graphics for a technical proposal, concentrate on the following points, which are summarized in Guidelines at a Glance 49 and covered at greater length in Chapter 6:

- *Use graphics and hybrid displays to preview and review your main points.* For the oral presentation, outline your main topics and subtopics and present bulleted lists of key points. For the written proposal, convert the topic outline to a table of contents, and use the lists to forecast points at the beginning or end of each section. You can use flowcharts and tables as forecasters and summarizing devices in both oral and written presentations.

- *Use graphics to orient the audience in space and time, showing readers where and when actions take place.* For both written and oral proposals, the best orientational graphics include maps, flowcharts, and diagrams. Line charts and other devices that project future trends are especially needed in proposals. With such devices you can draw on the solid experience of the past to convince an audience of future needs and directions.

- *Use high-density graphics to interpret findings and support claims.* Use complex tables, line graphs, and other data displays to condense and classify your research. In written proposals, high-density graphics are excellent for supporting arguments, and they help you to establish your credibility, showing that you have "done your homework" in great detail. In oral proposals, though, be careful not to overwhelm the reader with too much complexity.

- *Use low-density graphics for emphasis and impact.* Strong contrasts come through especially well in pie charts, bar graphs, and short tables. These are excellent for oral presentation because your audience can comprehend them at a glance. Impact is crucial in proposals because it may be the effect that distinguishes your presentation from others with which you are competing.

- *Use drawings and photographs for vividness and impact.* To create a strong visual image of your subject—an especially useful technique when you are discussing products and places—use line drawings, sketches, and photographs. Vivid renderings of prototypes and projects that are already under way can reassure an audience of evaluators about your ability to carry out the plans you are making.

GUIDELINES AT A GLANCE 49
Graphics for technical proposals

To preview and review your main points, use outlines, lists, tables, and flowcharts.

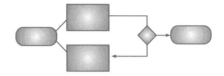

For a discussion of flowcharts and other forecasting and summary devices, see Chapter 6, pages 140–142.

For detailed summaries and to show trends over time, use line graphs and other high-density graphics.

For a discussion of line graphs and other high-density graphics, see Chapter 6, pages 149–156.

To orient the audience in space and time, use maps, flowcharts, and diagrams.

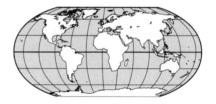

For a discussion of maps and other orientational graphics, see Chapter 6, pages 142–147.

To interpret findings and support claims, use tables, line graphs, and other data displays.

1993	1994	1995	Total

For a discussion of tables and graphical interpretation, see Chapter 6, pages 149–156.

For emphasis and contrast, use pie charts, bar graphs, and short tables.

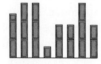

For a discussion of bar charts and other impact graphics, see Chapter 6, pages 156–163.

For vivid images, use line drawings, sketches, and photographs.

For a discussion of photographs and other high-impact graphics, see Chapter 6, pages 163–171.

Step Four: Planning and Presenting Oral Proposals

The fourth step in the CORE method is to plan and present a short oral version of your technical proposal, integrating the material developed in your concept paper and graphics plan. You should use the guidelines for effective oral presentation (see Chapter 8) and follow the same procedures that are used in preparing briefings for oral reports (see Chapter 12).

Begin by planning a set of transparencies. Make sure your transparencies have the following characteristics:

- They are clear, easy to read, and not overcrowded.
- They use primarily low-density graphics to make major points.
- They provide effective forecasts and summaries.
- They give enough information so that if the audience were not able to hear your oral presentation, they would still get the gist of your report.

With your transparencies completed, you are ready to prepare and rehearse your presentation. Keep in mind the following guidelines:

- Review your concept paper and research notes until you have a strong command of your subject matter.
- Practice looking forward, making eye contact with the audience, and maintaining a fluent delivery and an expressive variation of pitch and tone for emphasis.
- Time your report carefully, staying within your assigned limits.

In developing your oral proposal, remember that a strong persona is even more important for this genre than it is for reports. As a proposer, you or one of your coauthors will likely be in charge of managing the project you describe. So it's crucial to give a good account of yourself as a committed, knowledgeable, and credible project director.

Step Five: Developing the Proposal

With your notes from the audience-action analysis, your research, concept paper, and graphics in hand, and with an oral proposal behind you, you should be in good shape to put the proposal in its final form.

Informal Proposals

If you are drafting an informal technical proposal, you may have already based your concept paper on the outline suggested in Chapter 5. You can simply expand this outline for the full proposal under three conditions:

1. The proposed project has not proceeded far enough to have settled the specific questions of who will do what.
2. The proposal is an in-house proposal—from one department to another or from technical staff to management—so everyone already knows who will do what.
3. The proposal is a student proposal in a technical communication class. Often, teachers are more interested in how you handle the analysis of the problem and the argument for your plan than in how well you can dream up management schemes and analyze budgets that don't really exist. The need for this kind of work depends on your project. Keith Bell's project (Sample 14.1), for example, required very little attention to management and budget, but Dan Eason's (Sample 14.2) required a great deal more because it involved a business plan. To be convincing, it had to show that the proposer had good money sense and management skills.

If you use the abbreviated outline from Chapter 5, you will need to be certain that your conclusion projects the general costs of the project and argues that the benefits will justify the costs. The most convincing way to make this argument is to show that the alternatives either cost more or deliver fewer benefits than the proposed solution. The conclusion should also provide assurances on personnel and facilities—essentially, a shortened management plan.

Sample 14.3 shows how Keith Bell's proposal could be structured, using a version of the abbreviated outline that he adapted to suit his content.

Amplifying the Parts of the Proposal   In developing your outline into a full draft, you will usually follow a few (or several) standard *lines of argument.* Some arguments that are widely used in proposals are summarized in Guidelines at a Glance 50.

Writing the Problem Analysis   In analyzing his problem, Keith Bell uses the first line of argument in Guidelines at a Glance 50: The old method should be replaced totally by the new one. So after he gives the purpose of the proposal (outline section 1.1) and simply states the problem (outline section 1.2), he must analyze the old method of solving the problem by focusing on its negative effects. Then he can show how the new method answers each of these complaints, matching advantages against disadvantages. Here is his sketch of problems with the old method (outline section 1.3):

- The work can be **dangerous,** since it demands that the divers stay below the surface for a long time and exposes them to potentially dangerous plant and animal life living in and on the rocks.
- The work is obviously **time-consuming,** requiring many hours of time both at sea and in the lab.
- Because of the need to use highly trained personnel for long periods of time, the work is **expensive.** Projects usually require fairly large research ships to

GUIDELINES AT A GLANCE 50
Lines of argument used in amplifying proposals

1. The old way of doing things is totally wrong. I am offering a new way.
2. The old way is incomplete. My way fills in the gaps.
3. The problem has been handled in many ways before. I will combine several of the best ways.
4. The old way is good but has never been tried in this area. I will apply it in a new way.
5. There are X number of solutions, but only mine will work.
6. There are X number of solutions, but only mine is feasible (economically, socially, environmentally, etc.).

accommodate several people to lay the grids and also facilities for specimen storage. In addition, they often require special crews and equipment, such as winches and dredges.

- The work can be **environmentally harmful,** especially when it involves scraping and collecting.
- Worst of all, from a scientific viewpoint, the work is often **inaccurate.**

Proposal for Research in Image Analysis of Marine Ecosystems

1. Introduction
  1.1. Purpose: To request Apple Computer Corporation to donate or loan machines for research in image analysis technology
  1.2. The Problem: The difficulties of analyzing underwater ecosystems
  1.3. Current Method: Analyzing sea floors "by hand"
    - dangerous
    - time-consuming
    - expensive
    - often inaccurate
    - environmentally harmful
  1.4. The Solution: Analysis using videotaping, digital scanning, and computer analysis

2. Narrative
  2.1. Progress to Date
  2.2. Objectives
  2.3. Project Needs

3. Conclusion: Large initial capital outlay justified by overall savings and enhanced accuracy, safety, and environmental quality

Because marine biologists currently rely on only one method, Keith has only one option for solving the problem to analyze, not a whole range of alternatives.

In this part of the paper, Keith will tell the "story" of his project, bringing the reader up to date and saying what remains to be done—and what is required in the way of funding to get it done.

The abbreviated treatment of the budget goes here.

SAMPLE 14.3   Outline of a student proposal using the abbreviated form

When he describes his own solution (outline section 1.4), he uses the same organization to show the advantages of the new method, as the following draft shows:

- The method is **safer.** The divers spend less time below the surface and have no need to get as close to the rocks as they did when they were collecting or marking specimens.
- The method is **faster.** Researchers spend less time at sea and in the lab.
- The method is **less expensive.** Projects can use smaller crews, smaller ships, and less heavy equipment.
- The method is **less likely to disturb marine organisms.** It involves no stake-driving, no scraping, no collecting. The divers do not even have to touch the sea bottom.
- The method produces results that are **more reliable.** Automating the counting and analysis processes cuts down on the chance of error.

Writing the Narrative   The *narrative* is a story, a set of actions plotted on a time line. Stories do more than just provide entertainment. Because they encode cultural values, stories also motivate decisions. In our technological society, proposals that can convey such concepts as *progress, evolution,* and *Yankee ingenuity* may have a subliminal influence on reviewers. These are the informing myths of our culture. Our heroes are the inventor, the entrepreneur, the problem solver.[1]

Your job is to translate these "stories of value" into appropriate actions. The values that underlie your proposal—a desire to contribute to overall technological progress, a drive to create more efficient systems of production, a belief in universal education—should be inherent in your statement of the problem you are trying to solve. Anything worth dealing with in such a way has an important value in a society or a field of research. If the value is not clear, you must establish it as you analyze the problem.

The proposal tells how, on the basis of your past successes, you are going to remedy problems that exist in the present to create a better future.

Develop your proposal's narrative in three parts:

1. *Tell what you have done so far toward the project.* The most successful proposals are anchored in solid actions in the past and present. Good news about your accomplishments will make the future seem a bit less daunting to the evaluator.
2. *Give your objectives for extending the project.* An objective is a statement about a proposed action with a clear outcome. One of Keith's objectives, for example, is "To improve the speed and power of the analytical tools." The pro-

---

1. See Walter R. Fisher, *Human Communication as Narration* (Columbia: University of South Carolina Press, 1987).

posed action involves computer programming and testing. The outcome is an improved system.

3. *Say what you need to continue the project.* Again, base your request on past trials, expenditures, and accomplishments. Show that you have used what you have had responsibly.

Of course, not all projects are as far along as the one Keith narrates. If you are just starting, you can begin your narration by assuring the audience that what you are proposing has been tried elsewhere with some success or that you have done similar projects successfully. If you are not able to make these claims, then you'll have to write some mighty convincing objectives!

Writing the Conclusion   The conclusion can be fairly brief, but it should tie up the ends of your proposal with a strong impact. It should focus on the future, projecting what the overall benefits of the project will be in relation to the costs. Here you should justify your budget requests with reference to specific objectives and outcomes. Also, try to show the widest possible distribution of benefits.

Formal Research and Contract Proposals

If you are developing a more formal research or contract proposal, your outline will usually be provided by the funding agency in its mission statement or Request for Proposals (RFP). You will probably need all of the information you researched and used to develop your concept paper, and more. Specifically, you will need to greatly expand your problem analysis so that it becomes a section in itself. Also, you will probably need to include one or more of the following: a literature review, a rationale for the project, a list of objectives, descriptions and vitae for project management and personnel, a detailed timeline or Gantt chart, a list of products expected to result from the project and a description of how they will be disseminated, a project budget, a cost–benefit analysis, a description of a plan to evaluate the effectiveness of the project, and some guarantees or assurances that you or your team can do what you are proposing.

Problem Analysis   As Figure 14.2 indicates, solicited proposals often provide detailed specifications of the problem to be analyzed; you must find acceptable ways to address those specifications. Unsolicited proposals may target only a general area or field of study, so it will be up to you to identify and analyze a problem, then to describe how you or your team proposes to solve that problem. The extended problem analysis will vary in structure and content depending on whether the proposal is unsolicited or solicited. In unsolicited research proposals the problem analysis involves a rationale for the proposed study that you set up by reviewing the relevant literature in the field. In solicited proposals the problem analysis must be tailored to reflect the funding agency's requests.

Analyzing the Problem in a Solicited Proposal   The problem that you tackle in a solicited proposal will be at least partially outlined for you. Your job is to show that you understand the problem and that you have a compelling and feasible plan for a solution.

Sometimes the project will be fairly clear-cut. For example, this solicitation from the Air Force appeared in the January 29, 1991, issue of *The Commerce Business Daily:*

> The object of this research effort is to evaluate and recommend design modification of state-of-the-art liquid lubrication system component materials to attain weight reductions to meet Integrated High Performance Turbine Engine Technology (IHPTET) goals.

Despite the jaw-breaking language, this is a fairly straightforward engineering project. It needs very specific work from organizations or individuals with ongoing interests in this kind of technology.

The RFP might be even more explicit than this, going so far as to suggest an actual outline for analyzing the problem. The following solicitation from NASA, calling for proposals to develop a Laboratory for Atmospheres Support Services, breaks the problem into a definite set of smaller tasks:

> The contractor shall provide Spacecraft/Space Shuttle Interface Engineering and Project Administration support for the GFSC Laboratory for Atmospheres (LAS) Total Ozone Monitors (TOMS) and Shuttle Solar Backscatter Ultraviolet (SSBUV) projects. This requirement encompasses ultraviolet and infrared radiometric calibration and testing; design of spacecraft and space shuttle command power, data and mechanical interfaces, hardware and software development for MICROVAX, PC and PDP computer interfaces for experiment and payload group support equipment; spacecraft and space shuttle integrated systems testing and flight support analysis; preparation of mission requirements documentation, specifications, flight planning configuration management and project administration.

In RFPs like this, such details are not only helpful for analyzing the problem; they are a real requirement for your proposal. A responsive proposal should deal with each listed task, fleshing out each one with appropriate activities, omitting nothing that is mentioned in the RFP. You could even think of the topics listed—ultraviolet and infrared radiometric calibration and testing, design of spacecraft and space shuttle command power, data and mechanical interfaces, and so on—as possible chapter headings in your proposal. You will have to devote a section to explaining how you will handle each of these requirements.

At other times, the work requested will be more open to interpretation and variation. Here is part of an RFP from the Department of Energy for such a project:

> Research and Development for "Advanced Research in Direct Liquefaction." Solicitation No. DE-RA22-91PC91035. The purpose of this procurement is to solicit proposals for advanced research of new ideas to increase the basic un-

derstanding of coal conversion to liquid fuels. Proposed research should increase fundamental scientific and technical understanding of coal liquefaction and explore new and evolving concepts for the conversion of coal to liquid fuels. The intent is for the performance of laboratory research involving the most promising approaches to new technology or the significant extension of the understanding of previously identified concepts. The Department of Energy anticipates receiving research proposals in the following areas of program interest: 1) Role of the Solvent Properties in Catalytic Coal Liquefaction; and 2) Upgrading of Coal Liquids to Transportation Fuels. It is anticipated that multiple awards may result from this solicitation. All responsible individuals, corporations, non-profit organizations, educational institutions, and state or local governments may submit proposals for consideration.

The passage contains a number of key words that experienced readers will recognize as codes for the kind of research projects the DOE is seeking. Terms such as "basic" and "fundamental" suggest that the proposal should deal with scientific and advanced engineering concepts and should not, for example, simply propose designs for a new kind of coal liquefying machine. Successful proposals in this project will almost certainly involve heavy doses of experimentation in chemistry and chemical engineering. The researchers will develop the basic principles for new technological designs.

Notice also the number of hints about the proposal writer's need to be familiar with previous research in the field, especially the reference to "the most promising approaches to new technology or the significant extension of the understanding of previously identified concepts." This suggests that the contract will be awarded to researchers who present new approaches to old ways of doing things.

This mingling of conservative and progressive attitudes toward projects is a key to effective appeals in both solicited and unsolicited proposals. The idea needs to be new and different, but not so new and different that it cannot be easily understood by major authorities in the field (some of whom will no doubt evaluate your proposal).

Analyzing the Problem in an Unsolicited Proposal   In an unsolicited research proposal you are completely responsible for defining both the specific problem and the solution. You should begin by making sure your project fits the mission statement or program announcement of the funding agency.

Usually, a funding agency limits its objectives by stating a clear preference for proposals submitted in particular fields by particular kinds of researchers. Consider, for example, this statement of objectives from the Earhart Foundation:

> The sponsor provides support to individuals and organizations in the social sciences and humanities disciplines, including economics, international affairs, political science, and philosophy.

Obviously, this foundation accepts proposals from a wide variety of applicants, and this is typical of agencies that fund unsolicited research proposals.

However, the agency may in fact fund projects dealing with a range of topics much more limited in scope than this statement suggests. It is always wise to call a program officer and request an annual report to see what projects the foundation has funded recently and to discuss its interest in your field.

The problem that you analyze in an unsolicited proposal, however, will be generated not by the agency, but by the context of your research—your academic or professional discipline. Say, for example, that you are a polymer engineer and that, having tried to locate information about the durability of a particular plastic, you found nothing—*a gap in the knowledge base*. In your lab you have been working extensively with similar materials, and with proper funding for new equipment and personnel you could run a series of tests that would establish once and for all the strength of this experimental plastic. You could fill the gap in the knowledge base. A proposal for funding is in order.

Your first step would be to review the literature to show that, though people have had good results with similar plastics, either no one has paid much attention to your plastic or no one has found a good method for testing its strength. You would then demonstrate the importance of finding out the information, show how you would do it, and prove that you are capable of doing it. The result (with a good dose of luck) would be a winning proposal.

In an industrial setting, you might hit on a similar situation based on hints you receive about your competition. If your competitors have gone further in developing the material than you have, you might compose a proposal asking management to let you spend more time and money on this material in order to close the gap between your company and its competitors.

Once you have recognized a gap, either a gap in the general literature in your field or a gap between you and your competition, you are on your way to discovering the nature of the problem.

Refining the Problem

One way to advance your analysis is to create a well-formed problem out of an ill-formed problem. An *ill-formed problem* is stated at a fairly high level of generality, without giving the intermediate tasks that must be completed before the problem can be solved. For example, "No one really knows how strong this plastic really is," a statement of an ill-formed problem, might be the jumping off point for your analysis. A *well-formed problem* divides the problem into smaller tasks, then subdivides the task into steps, as in the NASA RFP on page 478. Ideally, with a well-formed problem you could solve the problem by carrying out all the steps in each task, as shown in Figure 14.4.

Ill-formed problems often appear in the last sentences of articles reporting research. They will say something like this: "Whereas the composition of durathene is well known, its range of functions must yet be established." This assertion paves the way for future research by stating the problem at a general level.

To make a good proposal based on such sentences (which often turn up in literature reviews), you would have to ask questions such as "How could the

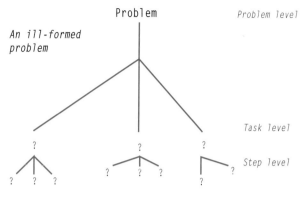

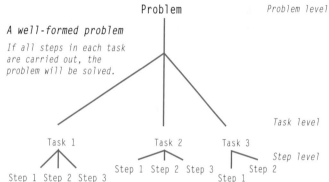

FIGURE 14.4

An ill-formed problem becomes a well-formed problem.

range of functions be established?" "What methods should be used?" "What steps should be followed?" In other words, you would need to develop the task level and the step level of a well-formed problem. The more specific you can make your analysis, the more convincing your proposal will be.

The tasks you designate at the second level of problem formation will become the basis for your objectives when you write your plan of action. They can also furnish you with the information you need to criticize alternatives to your plan. You can say, for example, that a certain alternative plan may satisfy Task 1 but not Task 2 and Task 3.

Here are some additional questions to guide you in analyzing the problem once you have identified it:

1. Why does the problem exist?
2. Why has no one noticed it? If they have, what did they say?

3. Is the problem something that the field or the company can live with? Knowledge is never complete, so why must we fill in this gap right now?
4. Who will benefit from the new knowledge?
5. How will the new knowledge improve the current research program in your field or your organization?

Few research projects worth undertaking can be rendered as perfectly well-formed problems. That luxury is reserved for simple problems such as getting the car washed. Even such easy chores can never be analyzed exhaustively. (Which soap to use? Which direction to rub the cloth on each pass over the car? Which muscle group to relax and which to flex?) So in every proposal you must aim for a technical analysis that makes the statement of the problem *relatively* well formed, or less ill formed than it was before you wrote the proposal.[2]

The Narrative of the Proposed Plan The proposal narrative is the heart of your proposal, even if it turns out not to be as long as your analysis of the problem. It says in as much detail as you can muster (in the space allowed) precisely what you plan to do. Once you have analyzed your problem, frame your solution as a story that will take place in the future. Since no hard facts exist in the future, you will have to imagine many of the actions in your text, drawing as much as possible on hard data from the past and present. It may be helpful to keep in mind a series of questions that has proved useful to storytellers for generations: *What? How? Who? When and where?*

How you approach the narrative may vary slightly in solicited and unsolicited proposals.

The Statement of Work in Solicited Proposals In solicited proposals the narrative is sometimes called a *Statement of Work*, or SOW. Because the reasons for the work are given in the RFP, there is no need for you to formulate extensive objectives; you need only give the details of what you plan to do. A list or a table with a timeline may suffice.

But don't overlook the importance of careful project planning. The SOW will form the basis for the contractual agreement between you and the funding agency, so be very careful in listing proposed activities.

Developing Objectives and Activities in Unsolicited Proposals In unsolicited proposals the story cannot be a mere narration of events. It must also supply motives and reasons for the actions you have planned. You should follow up the answers to each of your planning questions by supplying a response to another question: *Why?*

---

2. In *Writing and Technique* (Urbana, IL: NCTE, 1989), David Dobrin gives a lively analysis of the concept of "background knowledge," the kind of knowledge we must assume that any possible audience brings to the task of reading. He asks, for example, Should the instructions for a coffee mill advise the reader not to use the device to grind armadillo armor?

- *How* do you plan to solve the problem? *Why* this way rather than some other?
- *Who* will perform the actions, supply the funding and other support, evaluate the project? *Why* these people rather than others?
- *Who* will benefit from the project? *Why* and *how*?
- *When* and *where* will the activities take place? *Why* there rather than someplace else?

Complete answers to these questions should provide the subject matter you need for your narrative. Once you have answered the questions, ask them again. If you come up short on any of the answers, you have probably not done enough research or have failed to think through the problem carefully.

In your narrative, tie each of the activities you propose directly to a major objective. An *objective* is a statement of a specific outcome that you envision for your project if it is funded and carried out as planned. For example, an objective for a course in technical writing might be stated this way: "By the end of the course, students will be able to write and edit research proposals suitable for presentation to thesis committees in their own disciplines." Notice a couple of things:

1. *The objective deals with a fairly specific task:* the writing of proposals. It does not try to encompass the entire purpose of the course (to improve the student's ability as a communicator).
2. *The objective focuses on changes in performance, not changes in knowledge.* It does not say, "Students *will know* how to write proposals," but rather asserts that they *will be able* to write them. This may seem to be going out on the limb a bit, but that's precisely the kind of risk that proposal writers are expected to take. If the student can't write proposals by the end of the course, then the student (not to mention the instructor) has failed to meet a major course objective.

You should always show that objectives arise from documented needs and lead to appropriate activities. Clarify the needs in your analysis of the problem; the objectives and appropriate activities should emerge in your consideration of options and alternatives. The objective just mentioned—that students will be able to write proposals by the end of the technical writing course—could have arisen from a need expressed by the teacher's colleagues: Students in science and engineering should be able to write good research proposals. Or perhaps the writing instructor has read a survey showing that professional engineers and scientists spend up to half of their working lives writing proposals.

Most objectives can be accomplished through a number of activities. To accomplish the one stated above, for example, the technical writing teacher might plan lectures on proposal writing, make reading assignments in textbooks, require students to research types of proposals used in particular fields, and offer workshops in planning, drafting, illustrating, and editing proposals. Again, the activities you choose should correspond directly to your stated needs.

The table in Sample 14.4 (discussed in Chapter 6) summarizes a project in science education. It shows how to key objectives and activities directly to needs. A table like this can be very important in the planning stages of the project, especially if multiple authors are involved. The contributors can survey it fairly quickly and, by reading a single page, get a fair idea of the whole project. They can see quickly how needs, objectives, and activities relate to each other. For the same reasons, such a table can be effective in the proposal itself as a tool to summarize the project for busy evaluators.

Planning Outcomes, Products, and Dissemination   How specific you can be about the outcomes of your project will, of course, depend on the nature of the project. Be as specific as you can. How have similar projects turned out? What results can you predict accurately?

Although you may not be able to predict clearly the outcomes of experiments and scholarly activities, you must be able to make a strong case that the work will be important enough in the field to be useful and publishable. If you analyze the problem and reveal a gap that should be filled, the need for a published article or book will be clear enough. Your narrative can then simply state that the manuscript of this work will be submitted for publication at the end of the project.

What are the products of a training grant? Usually, they relate to *dissemination,* spreading the word. If you are writing a proposal to hold a series of training workshops, don't say that the project will be finished after the workshops are held. Make plans to publish an educational model for this kind of training or develop a set of videotapes based on expert panels or demonstrations delivered at the workshop. The greater the number of people who will benefit directly from your project, the greater the chances that the project will be funded.

Always complete your narrative, then, with a clear plan for delivering specific products and extending the range of the project through publication or other means of dissemination.

You'll be surprised at how easy it is to write the rest of your narrative after you have developed effective summaries of your objectives and activities. Use your prose text not to add more detail to every activity listed in your tables, lists, and charts but rather to show the chronological relation among the parts of the project and to emphasize the activities that you think are especially innovative or interesting.

Developing the Management Plan   The management plan says who will do what and how the work will be evaluated. You must show that the people you choose to handle each task are qualified and experienced (see "Providing Assurances" below). It is also important to show how the task of each person or group relates to the rest of the project.

Often, diagrams provide the best tool for demonstrating this network of relationships. Have a look at the diagram in Sample 14.5, for instance. Like the table that accompanies the same project (Sample 14.4), such a chart can be a powerful planning device for the project, as well as a good graphical display in the text. Its effectiveness as a summarizing device is considerable.

Science Futures Alliance Project

Needs and Objectives	Activities	Evaluation	Dissemination
Need: To enhance communication and cooperation among science teachers, researchers in science and education, and leaders in business and industry **Objective:** To organize and operate a community-based Science Education Alliance Network	1. Assemble an advisory board of community representatives who are interested in science education. 2. Develop a Science Teacher Corps and present in-service training sessions for cooperating school systems, involving recognized lead teachers from local schools as well as faculty researchers and teachers. 3. Organize and implement a Research Partnership Program for science teachers and student researchers. 4. Provide mini-grants to assist science teachers who are interested in participating in research and professional activities. 5. Develop an equipment grant/loan program for science teachers in community schools. 6. Develop and provide access to a database of science information (text and website) for all educational levels. 7. Produce a regional newsletter to provide a forum for project announcements, descriptions of activities, and the presentation of opinions and essays on scientific topics to stimulate ideas, participation in science, and communication among interested parties.	An ethnographic study will analyze and evaluate the effectiveness with which information is exchanged and cooperation is enhanced in the project's effort to overcome barriers in its promotion of scientific education, research, and communication. Documentation techniques will include audiotaping and videotaping of project activities, interviews with participants, and surveys taken through the Science Futures Alliance newsletter.	Public meetings, on-site in-service training sessions, newsletter, reports at professional meetings (such as NSTA)
Need: To increase the number of science majors and the number and quality of science teachers to ensure the continuation of the scientific enterprise **Objective:** To organize and implement a Science Teacher Recruitment Program	1. Organize an interuniversity/college planning committee to encourage and assist students in selecting science teaching careers. 2. Establish a Cadet Science Teacher Program that would train high school students in their respective schools and provide an opportunity for them to participate in pre-teacher preparation activities.	Ethnographic study	Brochures, newsletter, reports in teaching journals and at professional meetings
Need: To understand the efficacy of networking procedures in promoting scientific activity and in recruiting science teachers **Objective:** To produce an ethnographic study of a pilot networking project: the Science Futures Alliance Network	1. Document project, using audiotaping and videotaping, interviews with participants, and surveys attached to quarterly newsletters. 2. Analyze the data, taking into consideration the institutional, social, cultural, and personal goals that attach to each communicative interchange among the various participants. 3. Evaluate the effectiveness of the activities according to how well the objectives have been accomplished. 4. On the basis of the evaluation, make recommendations and present a model for science educators who are interested in developing similar networks.	Peer review for publication in established journals and/or scholarly presses	Scholarly articles and/or books

SAMPLE 14.4   Table summarizing a proposed project in science education

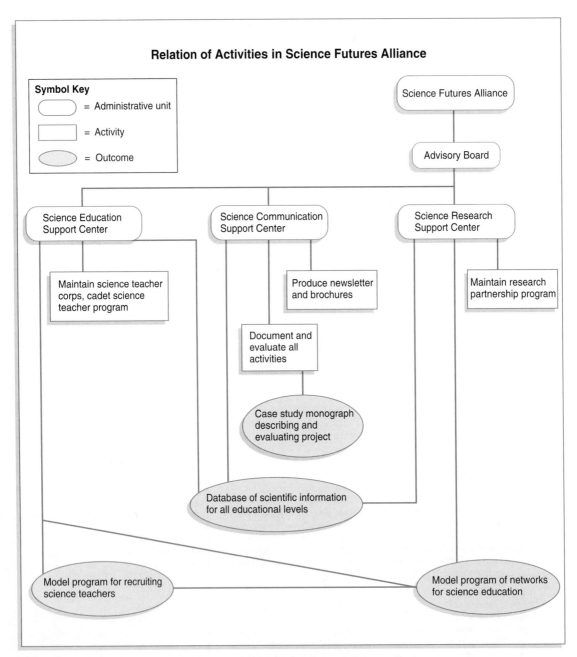

SAMPLE 14.5   Chart summarizing the same project in science education

Analyzing Costs and Benefits   In solicited proposals the cost analysis section consists primarily of an elaborate accounting plan. Unsolicited proposals also require a budget, in which you must justify each expenditure. For complex proposals, rely on professional help in developing your budget. Mention every budget item directly in the narrative. If you list "consultant fees" in the cost section, for example, explain in the narrative how you will use consultants to carry out the work and who you hope to employ in this role.

Unlike the analysis of costs, which can take special expertise in accounting, showing benefits can be a matter of simple arithmetic. In proposals devoted to training, for example, which are common in the field of education, you can just give, as accurately as possible, actual numbers for how many people will directly and indirectly benefit from your project. The more actual data you can give, the better.

There are many ways of multiplying the ultimate reach of the project. You can plan to train teams of trainers, for example, who, for a relatively small percentage of the project budget, can offer workshops at their home institutions or in geographical areas away from the initial training site. Those who attend these regional workshops will put what they learn into effect in their own classrooms or places of business, so others benefit indirectly by your work. If you want to claim a direct influence at those far reaches of the project, you can offer videotapes, literature, follow-up workshops, or an information hotline as further support.

The analysis of benefits from research proposals, especially in basic science or humanities, is more subtle. In these fields your main goal should be to show that you are advancing a recognized research program in your own specialty by breaking new ground in the field. If you can show that your work could have benefits in other fields, too, all the better.

The benefits of research will vary from discipline to discipline. In the humanities (fields devoted to the study of language, philosophy, and history), you should show that your project does one of the following:

- uncovers new information about an old problem (new poems by a well-known author, new biographical data on a former president, an unpublished work by a famous philosopher)
- presents new information in a neglected area of study (the works of an author whom history has overlooked)
- brings old information up to date with new interpretations

In theoretical science the aim is to solve puzzles posed by the leading theories in your field—the theory of evolution, for example, or relativity theory. The results of your research should do one of two things:

- confirm the work of other researchers on a major research problem
- place previous work in question

In applied science and engineering the aim is to solve problems of human existence through scientific experimentation and engineering design. The benefits of your research should be presented as one of two things:

- an actual solution to a major problem (traffic in Houston, pesticide runoff in the Mississippi delta)
- a step toward the solution of a major problem (a cure for AIDS, nuclear fusion)

In business (which, of course, overlaps with applied science and engineering and increasingly with the other fields of university research), the aim is toward one of two objectives:

- enhanced efficiency ("getting the most bang for the buck," that is, the highest-quality product for the least investment in time and money)
- enhanced effectiveness, especially in matters involving the public image of the organization you represent

In the human services field (education, social work, nonprofit organizations) the greatest benefits are realized when you do the following:

- apply new methods to old problems (use computers to motivate student writers)
- bring an underprivileged group or region up to a national standard (develop a method for motivating people in rural communities to take advantage of health care programs)
- *always do the greatest good for the greatest number*[3]

In presenting the projected benefits of your work, you must be honest yet enthusiastic about the range of your work. If you find, in writing the proposal, that your scope of influence is too small, you might consider ways to extend the breadth of the benefits. Or you can argue that considerations of *quality* are more important than considerations of *quantity* in your case. Not every proposal can offer a cure for cancer, but your work should be valuable within the limits approved in your field of study.

Providing Assurances   Once you have surveyed the literature, made an effective argument about the value and feasibility of your plan compared to others, described and justified your objectives, keyed the objectives to a reasonable and specific set of activities, stated your plan for managing the actors and actions included in the proposal, explained the costs, and demonstrated the ben-

---

3. This is the key goal of *the utilitarian ethic*, which drives technical communication and technological innovation. It is discussed in some detail in Chapter 2 on the ethics and politics of information.

efits of your plan to other people, you have done much to convince the reader of your qualifications to direct a project.

Nevertheless, most agencies require you to submit *resumes* for all participants (see Chapter 11 for suggestions on creating resumes). You will usually need to write at least a short section on personnel. Some questions to answer include the following:

- Do the people in your plan qualify for the project because of their previous experience, their education or training, or some other form of special knowledge?
- What special skills or abilities do your people possess that distinguish them from others?
- If you were going to get the best people in the country to do this work, who would they be? Why didn't you get them?

You may also need to say something about your *facilities:*

- Do you have equipment, or access to outside equipment, that makes your organization special?
- Do you have an appropriate climate for the work (literally and figuratively)?
- Does the population whose needs your proposal addresses have a special prominence in your region? Is the problem that you address especially important to the people in your location?
- Are you close to the support facilities that you will need?
- In general, what makes your place unique for a project like yours?

In sections on facilities, proposal authors are often tempted to use "boilerplate," or standard material in the files on their organization. Proposal evaluators often feel cheated by this kind of information. They want everything in the proposal to be specifically designed for the proposed project. So don't just toss in a few old brochures about your organization or download something from an old proposal on a computer disk. Write your facilities section with an eye to how the place is specifically appropriate for the work you are proposing. This extra bit of attention to your proposal can make a critical difference because it shows your commitment to the project.

In addition to what you can say about your organization's capabilities, you may need to provide letters of support from others who will attest to your qualifications. Try always to select well-known people in the field who are familiar enough with your work to give a fairly detailed account. Contact these people early, and give them plenty of lead time.

Finally, government agencies also require assurances about affirmative action, use of animals or human subjects in experimental projects, and other such information to show that your organization does not violate law or ethical research practices. Always take such requests seriously, and be sure you obtain the required forms and signatures well before the submission deadline.

Ethical Issues in Proposal Evaluation

The ethical integrity of the proposal rests largely on the authors' claims to knowledge, objectivity, and good judgment. All the issues covered in Guidelines at a Glance 43 (Chapter 12) also apply to proposals. But there is an additional set of concerns for proposals: Can the proposers deliver on their promises? Can they produce the results they say they can within the time frame and within the budget they propose?

We must remember that proposals are usually accepted in a competitive environment. A proposal may beat out a competitor because it proposes to deliver higher quality in a shorter time for less money. If the product turns out to be of lower quality than promised, or if the proposers eventually require more time and money, then they have not only broken a promise—they have competed unfairly with other proposers. Realistic promises concerning quality control, scheduling, and budget are not only a matter of good business; they are also an ethical requirement of good proposal writing.

These ethical requirements should be in the forefront of your mind as you develop the lines of argument in your amplified draft, as you plan your budget, and as you review and revise the final document.

Revising and Framing Proposals

Once you've got a rough draft of your proposal, assemble a set of critical reviews and then begin to fine-tune it. Direct your revisions toward making effective rhetorical appeals: Balance the technical quality of the content with an awareness of your audience's interests and knowledge, and remain sensitive to your context.

Always remember that a proposal is more than just a clear and thorough statement of a plan. It should not focus exclusively on you and your idea; it must make an *appeal* for the reader to join you in an action partnership. An appealing proposal must be *responsive, authoritative, credible, specific, vivid,* and *organized.*

Making It Responsive

The funding agency's mission statement or RFP is the initial guide to the needs and expectations of your audience. Do not try to persuade your readers that they are mistaken about their own needs or that their expectations are misguided. These are the marks of an unresponsive proposal.[4] You should study RFPs carefully and respond point by point. In addition, you should contact a

4. The best treatment of the concept of *responsiveness* in proposal writing is found in Herman Holtz and Terry Schmidt, *The Winning Proposal* (New York: McGraw-Hill, 1981).

program officer at your chosen agency directly to get a good sense of what's expected of you and your ability to meet those expectations. Many foundations and agencies supply copies of successful proposals on request. You can study these carefully to see what design and style conventions, as well as what kinds of projects, the organization prefers.

This does not mean that you should pander to the audience. Evaluators do not respond well to flattery or trickery. They follow specific criteria for making decisions, noting matters such as the quality of the research design, the importance of the work to the field, the credentials of the proposal writer and the project staff, the range and importance of the benefits, the cost efficiency of the project, and so on. Try to find out what the evaluators in your funding agency consider to be the most important criteria. Be aware that the relative importance of the criteria will change from agency to agency. Agencies that fund education projects are likely to be concerned with benefits, whereas agencies that fund scientific research may focus more on research design and significance.

Though the audience has the power of funding or permission, you hold the power of knowledge and action. Show the readers what you can do with their money or help. Your theme should be "Working together, we can do something important in this project."

Making It Authoritative

Responsiveness, which is an audience-directed goal, is balanced in a good proposal with an authoritative tone, the achievement of which reflects well on the author. Again, you are the expert. Demonstrate your expertise by showing that you have a thorough command of the information needed to understand the problem and carry out the project. Never try to deceive your reader, and never strain to impress anyone with your *person.* Aim instead to give the fullest and clearest account of *your work,* and let your plan of action speak for your personal knowledge and competence.

Making It Credible

Though proposals are stories about the future, they are not science fiction or fantasy stories. They maintain fidelity to past and current trends. Your most effective argument is "This is how things have been going, so they should continue along the same lines." The truth of such statements is always highly tentative—a catastrophe could change the course of history tomorrow. But arguments built on verifiable precedents still carry great weight with proposal evaluators.

This way of arguing is essentially narrative in structure. It involves the audience in a story that has already begun and is headed toward a predictable conclusion. But unlike a good detective story, the more predictable the outcome of a proposal, the better the proposal is.

Making It Specific

You also boost your credibility by working out as many details as possible. Instead of saying, "The project will be evaluated by a team of experts from industry," the best proposals will outline the precise methods of evaluation, show how these methods have worked for similar projects, explain why it is better to use industry evaluators rather than people from other sectors, and actually name the evaluators and comment on their personal qualifications.

Again, ground your claims about the future in sound information about past and present actions. Conventional wisdom says that the proposal most likely to win a contract or grant is the one whose project is most nearly finished. Indeed, if you have already done a great deal of the work you are projecting, the outcomes that you project will seem more like a sure thing. And every proposal evaluator wants to put money on a sure thing.

Making It Vivid

If you load too much specific information into too many high-density graphics and tables, you may overwhelm your readers. Instead, use high-impact graphics such as photographs and line drawings, and use a page design that emphasizes the graphics and breaks up the text.

A word of caution here, however: Some agencies, notably the military, frown upon presentations that are overly slick. Some even specifically prohibit "brochuremanship." Determine your limits through a careful analysis of your audience, and stay well within those bounds.

Pay careful attention to your writing style. Since your aim is to get the reader involved in an ongoing story of human action, write in sentences that have a narrative structure. Use active sentences with agents as subjects and actions stated in verbs. Avoid a style overloaded with nouns and abstractions. (Review Chapter 9 for help in revising along these lines.)

Vividness is never a substitute for authoritative, intelligent information. But it can seem like a breath of fresh air for the weary evaluator hunched over a stack of proposals written by your dullest competitors.

Making It Organized

Like a technical report, a formal proposal may include a number of accessory elements in its final frame. Most of these framing devices are discussed at the end of Chapter 12. A fully framed proposal might include the following:

Letter of Transmittal
Title Page
Executive Summary
Abstract
Table of Contents

GUIDELINES AT A GLANCE 51
Evaluating technical proposals

- Is the proposal *authoritative*? That is, does the author (or the team of authors) seem to have mastered the content and methods necessary to carry out the project?
- Is it *persuasive*? Does the author analyze the problem carefully and make a convincing case that the proposed solution to the problem is better than the competing alternatives?
- Is it *credible*? That is, does the project seem doable? Do you believe that the objectives can be accomplished?
- Is it *specific*? Are there enough details so that you get a clear sense of what the project is all about?
- Is it *vivid*? When you read the proposal, do clear images of actors, actions, and outcomes form in your mind?
- Is it *ethical*? Can the proposers live up to their claims of knowledge, objectivity, and good judgment, and can they deliver on their promises? (Also refer to the questions in the Guidelines at a Glance 31 in Chapter 10.)

   Proposal Proper
      Introduction
      Narration
      Management and Personnel
      Budget
   Appendices

Reviewing and Evaluating Proposals

Technical proposals are usually reviewed extensively both in-house and by evaluators from funding agencies. Contributing authors and all personnel involved in the project read the document closely for accuracy and rhetorical effectiveness before it is submitted for its final evaluation. So to be a contributing member of a proposal-writing team, you need to develop your skills as a critical reader and analyst of projects—both your own and those of others. Guidelines at a Glance 51 provides criteria that you can use in your critical reviews of proposals. If you are developing a research or contract proposal, you should also apply Guidelines at a Glance 52.

Selecting a topic and doing an audience-action analysis.   As you read this chapter, you will probably be working with a topic that you have devised yourself or that someone else (your boss or one of your instructors) has assigned.

EXERCISE
14.1

GUIDELINES AT A GLANCE 52
Evaluating research proposals

- Does the proposal clearly identify and analyze a problem? Are you convinced that the problem is worth addressing? Is the problem well formed? Does the author use a review of the literature to identify the problem?
- Does the author provide enough information for you to understand the proposed research? Is the narrative or statement of work complete and clear? What more would you need? Is the information likely to be sufficient for an audience of the author's peers?
- Does the research clearly advance the research program in the author's field? What information in the proposal leads you to think it does (or does not)?
- Do the benefits of the research appear to justify the costs? Are the benefits widely distributed, or do they mainly accrue to the proposer?
- Are you satisfied that the proposer is capable of carrying the project to completion? What would you need in the concept paper to assure you about the research team's credentials?
- What about the style and document design? Is the presentation vivid and specific enough?
- Is the tone appropriate? Does the author project enthusiasm, commitment, objectivity, or intellectual energy? What needs to be added or changed to project the best tone?

Once you have a clear idea of your topic, you can begin your research (following the guidelines in Chapter 4). But early on, you should create a set of notes that identifies the following elements:

- the problem (need)
- the audience
- the kinds of information you will need to make your case
- the solution (the proposed action, projected outcomes, objectives, benefits)

Once you have completed your notes, discuss your project with a small group of your fellow students and with your instructor. Listen carefully, and adjust your notes to accommodate the questions and advice you receive. During research, use this audience-action analysis to decide which information is most useful. Review it carefully before you write your concept paper.

EXERCISE
14.2

Part 1. Evaluating concept papers. In your discussion group, evaluate the concept paper in Sample 14.2 on page 470. Written by a major in business administration, it represents a different kind of proposal—a business plan, an appeal to investors or bankers to provide support for a project that promises a strong return over time.

To some extent, all proposals are sales documents. But business plans differ from other technical proposals in their emphasis on markets and profit margins. Also, note that in contrast to proposals for engineering and scientific projects, which muffle their enthusiasm in the tone of confident, but cautious expertise, business plans allow the principal's excitement to show through. Dan openly relishes his 4% return, for example, and refers to the potential of the business as "tremendous."

Use Guidelines at a Glance 48 to guide your evaluation. When you have completed your review, discuss it with the class as a whole.

Part 2. Writing a concept paper.   Following the outline and procedures described, draft a concept paper on the topic for which you did an audience-action analysis in Exercise 14.1. Bring your draft to class and discuss it in your group, applying the questions in Guidelines at a Glance 48 to evaluate its potential effectiveness. Use the feedback you get to revise the draft, then submit the paper to your instructor for evaluation.

Part 1.   In your discussion group, return to the concept papers in Samples 14.1 and 14.2 (pages 469 and 470). Make a list of the graphics you would need in a full proposal based on these concept papers. Which would be most appropriate for the oral presentation and which should be saved for the formal written proposal? Compare your list with other groups' lists.

Part 2.   Now consider your own concept paper (developed in the second part of Exercise 14.2). Make a list of graphics that you will use first in your oral presentation and then in your full written proposal. Share the list with your discussion group, and get further ideas. Complete the list, and submit it to your instructor for further advice.

EXER**CISE**
14.3

Preparing and delivering an oral proposal.   Use the following steps:

1. On the basis of the concept paper and graphics plan you developed in Exercises 14.2 and 14.3, develop a plan for a short oral presentation. Draft a set of transparencies for the presentation.
2. Show your tentative transparencies to your discussion group. For every group member's transparencies, ask the following review questions:
   - Who is the audience?
   - How is the intended audience likely to respond to each transparency?
   - Do the transparencies clearly convey the author's intended position?
   - Would the audience be able to give a short account of the topic just from reading the transparencies?

EXER**CISE**
14.4

3. After the discussion, revise your transparencies and rehearse a presentation based on the transparencies and your concept paper.

4. Take turns delivering your presentations to the class or to a small group from the class. In an informal group discussion, critique each presenter, using Guidelines at a Glance 13 in Chapter 8 and the special considerations discussed above.

EXERCISE 14.5

Using your concept paper and the materials you developed in previous exercises, draft an outline based upon the model in Sample 14.3 (page 475). Feel free to adapt the parts to fit your topic. Take the completed outline to class, and compare it to the outlines of others in your discussion group. As you review each outline, focus especially on the adaptations. Are these justified in light of the project? Why or why not? Use the feedback you get in the discussion to revise and refine your outline.

EXERCISE 14.6

Evaluating technical proposals. This exercise helps you to develop your skills as an evaluator. Use the questions in Guidelines at a Glance 51 to direct your evaluation of the sample proposals given below.

Evaluation A. Your first proposal to evaluate is Sample 14.6 at the end of this chapter (pages 499–503), the full final proposal submitted to Apple Computer by Keith Bell. Make a list of things you would change, omit, or add if you were the coauthor of this proposal. In your discussion group, come to an agreement about your list of revisions, then share the consensus list with the whole class.

Evaluation B. Sample 14.7 at the end of this chapter (pages 504–510) is another student example based on the abbreviated outline: a planning proposal or preproposal that represents a single step beyond the genre we called report on options or recommendation report in Chapter 12. Reports on options merely point the way to a better option, but proposals ask for permission and funding to begin a definite, if limited, plan for action.

This sample is a classic engineering proposal, providing a way to "build a better mousetrap." Like Sample 14.6, it solves a general social problem, but it takes a more technological approach. It begins by analyzing the problem, in this case a water treatment plant's mechanized system. Reviewing the literature briefly, it shows what's wrong with the system and then goes on to show how a better one would work to deliver clean water more efficiently and, from an environmental perspective, more responsibly.

In this case the proposal was successful in a very practical way. It began as a student project in a technical writing class but ended by accomplishing at least part of its actual stated purpose. After reading about ozonation, the student author, Michael Glass, visited the local water treatment plant and talked with the manager about switching systems. The manager was very interested and, after hearing Michael's ideas, offered him an internship.

Even with this success, the proposal could be improved. Again, make a list of changes you would make if you were Michael's coauthor. Discuss the list in your group, then make a consensus list. Compare your group's list with those of other groups.

Evaluation C.   Sample 14.8 at the end of this chapter (pages 511–519) is a professional proposal that more or less follows the same abbreviated outline as the student samples. It analyzes a problem, develops a solution, and narrates a plan for implementing the solution. But there are some differences. Make note of these in your review. Especially notice the following features:

- The analysis of the problem concerns a much broader population of potential beneficiaries. The proposal aims to solve a big problem in public health and therefore to provide benefits beyond the scientific community. It addresses a social need rather than just a methodological problem or an extension of knowledge and technology. The need is established by referring to specific statistics in the introduction.
- The objectives are stated as much more specific outcomes, and clear measures of those outcomes are built into the mechanism for evaluation. This approach is typical of good proposals in the social sciences and in education.
- The narration of the plan (Methodology and Project Implementation) is extremely specific.
- The work is planned down to the last detail.

Despite its merits, however, this proposal could be improved. How, for example, could you apply some of the principles from Chapter 9 to make the style more readable? Again, make a list of potential revisions, discuss these with your group, and decide on a final list. Compare the consensus list with other groups' lists.

Writing your own proposal.   You should be ready at this point to draft your own full proposal. Use the following steps:

EXERCISE **14.7**

1. Either as an individual author or as a member of a proposal team, draft a proposal based on the concept paper, graphics, oral presentation, and outline that you developed in the previous exercises in this chapter.
2. Share your draft paper with your discussion group. Use the questions in Guidelines at a Glance 51 to determine the potential effectiveness of your proposal and to locate areas that need serious revision or further development.
3. After the discussion, revise your proposal, and submit it to your instructor for further evaluation.

Recommendations for Further Reading

Aristotle. *On Rhetoric.* Trans. George Kennedy. New York: Oxford University Press, 1991.

Bowman, Joel P., and Bernadine P. Bradchaw. *How to Write Proposals That Produce.* Phoenix: Oryx, 1992.

Sanders, Scott P. "How Can Technical Writing Be Persuasive?" *Solving Problems in Technical Writing.* Lynn Beene and Peter White, eds. New York: Oxford University Press, 1988: 55–78.

Zimmerman, Muriel, and Hugh Marsh. "Storyboarding an Industrial Proposal: A Case Study of Teaching and Producing Writing." *Worlds of Writing: Teaching and Learning in Discourse Communities at Work.* Carolyn B. Matalene, ed. New York: Random House, 1989: 203–221.

Additional Reading
for Advanced Research

Fisher, Walter R. *Human Communication as Narration.* Columbia: University of South Carolina Press, 1987.

Killingsworth, M. Jimmie. "A Bibliography on Proposal Writing." *IEEE Transactions on Professional Communications* PC-26 (1983): 79–83.

Documents for Review

**Image-Analysis Techniques for Ocean Floor Study:
A Proposal**

Submitted to Apple Computer Corporation

by
M. K. Bell
Texas A&M University

October 22, 1992

Abstract

    Current techniques of surveying marine life are expensive, time-consuming, and inaccurate. A method now under development at Texas A&M University promises to revolutionize the field by introducing techniques based on video technology and computer-aided image analysis. This proposal describes the project and requests support for the project from Apple Corporation. By donating or lending computer hardware, Apple will make a solid investment, opening up new markets of computer users in oceanographic biology and environmental impact assessment.

Introduction

    Researchers in science and industry frequently have a need to take surveys of segments of the ocean floor to determine the condition of organisms living in hard substrates, such as coral, rocks, and jetties. Marine biologists, oceanographers, and environmental impact assessors are only a few people who must carry out such studies beneath the sea. Their work has proven difficult, expensive, and often unreliable.

    This proposal presents a method to make their work easier, more efficient, and more accurate. Using recent advances in video and computer technology, researchers at Texas A&M have developed a new method for sea-floor surveys. With additional funding from Apple Computers, we can refine the method and create new applications for existing hardware and software, thereby opening new markets for Apple products in the burgeoning field of environmental studies.

The Problem with Current Methods

    Though photographic technology is now being applied experimentally (Miller & Carefoot, 1989), the only method of surveying marine life now widely practiced is a time-consuming manual counting process (Foster, Harrold, and Hardin, 1991). A team of divers lays a grid over the portion of the sea floor under investigation and then, by hand, takes an inventory of the organisms present within each section of the grid. The divers create the grid by driving steel spikes into the rock or coral and then stretching nylon line between each stake, forming a system of squares or quadrants.

    To carry out the inventory, the divers usually follow one of two methods. Either they lay a clear plastic sheet over the grid, mark each organism on the plastic, and then carry the plastic back to the lab to record and analyze the information gathered (Brace and Quicke, 1986). Or they chisel and scrape organisms off the rock, store the

---

The title and abstract give only enough information to identify the author and project. What else could have been included? What additional information might be given in a letter of transmittal?

The paper uses both structural headings (introduction, narrative, etc.) and informative subheadings. Structural headings may be required by funding agencies.

In the problem analysis the author reviews important literature. Is this necessary? Why or why not?

SAMPLE 14.6   Student proposal based on an abbreviated outline

material, then (back at the lab) sieve, count, and record the specimens (Bellan, Desrosiers, and Willsie, 1988).

Both methods cause problems:

- The work can be **dangerous,** since it demands that the divers stay below the surface for a long time and exposes them to potentially dangerous plant and animal life living in and on the rocks. One of our crew, for example, was injured when exposed to an outcropping of fire coral.
- The work is obviously **time-consuming,** requiring many hours of time both at sea and in the lab.
- Because of the need to use highly trained personnel for long periods of time, the work is **expensive.** The projects usually require fairly large research ships to accommodate the several people to lay the grids and the facilities for specimen storage. In addition, the need for special crews and equipment, such as winches and dredges, often arises.
- The work can be **environmentally harmful,** especially when scraping and collecting occur.
- Worst of all, from a scientific viewpoint, the work is often **inaccurate.** Even the existing photographic techniques consistently undersample plots (Whorf and Griffing, 1992). Often the underwater researchers are forced to estimate the number and composition of organisms and formations (Foster et al., 1991).

**The Solution: Videotaping, Digital
Scanning, and Computer Analysis**

The system now under investigation uses a VHS video recorder in conjunction with a Macintosh IIsi microcomputer equipped with a video capture card to record, store, and statistically analyze data (Whorff and Griffing, 1992). Using the new system, scuba divers do not have to spend long hours beneath the surface manually counting and recording specimens. They simply take video images of the sea bottom. Then, using the ship's on-board computers, they digitize and analyze the images.

The method is better in several ways:

- It is **safer.** The divers spend less time below the surface and have no need to get as close to the rocks as they did when they were collecting or marking specimens.
- The method is **faster.** Researchers spend less time at sea and in the lab. In experimental tests, researchers recorded sixty quadrants (30.25 cm^2 each) in five minutes and had analyzed an astonishing 11,000 quadrants by the time the project was completed (Whorf and Griffing 1992).
- The method is **less expensive.** Projects can use smaller crews, smaller ships, and less heavy equipment.
- The method is **less likely to disturb marine organisms.** Our current experiments still use the conventional grids. However, our research team is developing a software program which creates computer-generated quadrants which can be superimposed over the digitized images. The technique would then involve no stake-driving, no scraping, no collecting. The divers would not even have to touch the sea floor.
- The method produces results that are **more reliable.** Automating the counting and analysis processes cuts down on the chance of error.

What does the author achieve by using bulleted lists with bold key words? Is this a good example of document design? Why or why not?

SAMPLE 14.6 continued

Narrative

Progress to Date

The heart of the system has been developed and field tested. A test on an experimental jetty demonstrated the potential value of visual imaging as a technique for analyzing data. Figure 1 shows the results of biological disturbance as captured by this technique. The first column shows an undisturbed segment of ocean reef. The second column shows the effects of a single predator. The species coded green begins to disappear, while the species coded yellow begins to proliferate. The third column shows a reef subjected to numerous predators (Whorf and Griffing, 1992).

The results so far are encouraging. The Texas Advanced Technology program has provided $129,513 in funding, and the National Science Foundation has offered grants totaling $2,240,513. Further support has come from Mobil Exploration and Producing U.S. Inc.

Objectives

As we continue to develop the technology for studying submarine environments, we hope to accomplish the following objectives:

- To improve the speed and power of the analytical tools
- To enable the system to handle greater amounts of data
- To enhance the system's capability for determining size and color difference in specimens under analysis
- To increase storage capability

Project Needs

To accomplish these objectives, we need more powerful hardware. The new programs require an enormous amount of memory. At present, we have exhausted the capability of the Mac IIsi.

Thus we are requesting the donation or loan of a Macintosh Quadra 950 microcomputer. With this additional power, we believe we can perfect the method and promote its wide use in the field.

Conclusion

The image-analysis approach is likely to become the method of choice in several areas of scientific and technological investigation. Its promise for environmental impact analysis is particularly strong. A measure of its promise is the interest that Mobil Oil Company has shown in the project. The company has welcomed us to use its oil platforms as a testing ground and plans to use the information gathered to determine the environmental impact of oil spills.

By investing in the project, Apple can make a major contribution to a methodological breakthrough that not only benefits science and industry, but also opens the way for new applications of high technology in the booming field of applied ecology.

What are the possible gains and the possible risks involved in mentioning other sources of funding? Is there an ethical issue involved here?

Notice the slight shift in tone in the conclusion. Words like "breakthrough" and "booming" convey enthusiasm. Is it a good idea to conclude on this high note? Why or why not?

SAMPLE 14.6 continued

Figure 1 is the only graphic in the proposal. Should there be more? What additional graphics could have been added under ideal conditions of production?

Figure 1. Color Coded Analysis of Reef Organisms

SAMPLE 14.6 continued

References

Bellan, G., Desrosiers, G., and Willsie, A. (1988). Use of annelid pollution index for monitoring a moderately polluted littoral zone. *Marine Pollution Bulletin 19,* 12: 662–65.

Brace, R. C., and Quicke, D. L. J. (1986). Dynamics of colonization by the beadlet anemone. *Journal of Marine Biology Association* U.K. 66: 21–47.

Foster, M. S., Harrold, C., Hardin, D. D. (1991). Point vs. photo quadrant estimates of the cover of sessile marine organisms. *Journal of Experimental Marine Biology and Ecology* 146: 192–203.

Miller, K. M., and Carefoot, T. H. (1989). The role of spatial and size refuges in their interaction between juvenile barnacles and grazing limpets. *Journal of Experimental Marine Biology and Ecology* 134: 157–74.

Whorff, J. S., and Griffing, L. (1992). A video recording and analysis system used to sample intertidal communities. *Journal of Experimental Marine Biology and Ecology* 148: 1–12.

SAMPLE 14.6 continued

**Ozonation as a Primary Water Treatment Process:
A Proposal**

submitted to

The City of Bryan Water Treatment Facilities

by
Michael N. Glass
Department of Mechanical Engineering
Texas A&M University

November 2, 1993

ABSTRACT

Current techniques used by the city of Bryan in the treatment of drinking water involve the use of large amounts of extremely hazardous and environmentally unsafe chemicals. An optional method of treating the community's water must be implemented at the city water treatment facilities in order to meet new federal regulations and to create trends toward better environmental standards for the future. This proposal introduces an improved process called ozonation. Ozone used as a source of water treatment is a safe and effective means of complying with regulations and ensuring safe water resources for the future. This proposal asks for funding to generate a technical and cost proposal to be submitted to the Environmental Protection Agency as a first step toward implementing an ozone-based treatment system for the city of Bryan.

This proposal avoids the standard structural headings (introduction, narrative, etc.), using only informative headings. It begins with an unlabeled introduction.

**Ozonation as a Primary Water Treatment Process:
A Proposal**

One of the major provisions of the National Safe Drinking Water Act is the initiation of research on medical, economic, and technological problems related to drinking water supplies (Bryant, Fulton, and Budd, 1992). This legislation calls attention to environmentally unsafe processes currently used in water treatment facilities, many of which are also quite expensive.

Chlorination is one such process. Though widely used for many years, research and experience have increasingly pointed to chlorine as a hazardous and environmentally unsafe chemical.

Because of dependence upon chlorination processes, the city of Bryan's water treatment facilities have begun to appear dated and potentially dangerous. The city may soon face forced compliance with the new regulations.

This proposal presents an optional method of water treatment, called ozonation, which has no hazardous by-products, is cost effective over the long term, and is also environmentally friendly. Ozonation outperforms the current technique of chlorination while meeting and exceeding the demands of all current regulations. With the implementation of the ozonation process, the city of Bryan water treatment division can open doors to improved environmental and health standards for the future.

SAMPLE 14.7   Student-authored engineering proposal

The Problem: Chlorine Is Hazardous to Human Health and the Environment

The three water treatment plants located in the city of Bryan use chlorine in drinking water treatment (Drews, 1993). Until recent studies, chlorine was considered to be a good treatment because of its high levels of residual and anti-bacterial treatment capabilities. But several problems are now known to arise with the use of chlorine:

- **Chlorine is hazardous to human health.** There is an increased concern over the formation of chlorinated organics in drinking water along with the discovery of newly recognized waterborn disease-causing organisms not effectively controlled by the disinfection process of chlorination. The hazardous by-products represent the biggest problem. They are the result of reactions between standard chlorine and naturally occurring elements found in many water resources. The result of these chemical interactions is the formation of trihalomethanes (THM's), which have been linked to cancer in animals and humans. Since chlorine is the chemical responsible for these chloro-organics, we are only compounding the problem every time we use chlorine to treat drinking water. As research continues, there is concern that many of today's health defects may be traced back to chlorine (Jacangleo, 1984).

- **Chlorine is environmentally unsafe.** Over 10.5 million tons of chlorine are transported on our major roadways yearly. The transport of this highly reactive chemical creates serious concern about the high levels of impact it could have on the environment. In 1989, 37 major chemical spills or leaks were recorded. There is no telling how many occurred that were not reported (Singer, 1990). In addition, since it is not feasible to produce chlorine on site, the chlorine must be transported by large diesel trucks, which ultimately increase the level of air pollutants (LePage, 1988). Air pollution, in turn, directly affects future water resources.

- **Chlorine's cost is on the rise.** The Environmental Protection Agency (EPA) is enforcing the new regulations from the Safe Drinking Water Act with heavy fines and possible termination of operations for those not cooperating (Lepage, 1988). Water facilities that depend on chlorine as the major water treatment will see an increased cost of running their facilities due to these fines.

To solve these problems now and avoid future fines, Bryan water treatment facilities must find a cost-effective way to eliminate chlorination as the primary treatment process.

The Solution: Ozonation

One method of solving problems involves the implementation of a safe and effective water treatment program known as ozonation. This ozone will not protect you from harmful ultra-violet rays, but it will give you a more effective and safer method of water treatment than the current method of chlorination.

Ozone is the strongest oxidant of the common disinfecting agents. Its safety is ensured by its characteristic odor, which can be detected by most humans at very low concentrations, far below the levels of acute toxicity (Bryant, Fulton, and Budd, 1992).

As in Sample 14.6, the problem analysis focuses on a single alternative—the most widely used method.

SAMPLE 14.7 continued

The city of Bryan could benefit in many ways by replacing chlorine with ozone:

- **Ozonation is effective.** Ozone provides 100% disinfection of drinking water and is a much better viricide than chlorine. Unlike chlorine, when ozone is applied to water, there are no impurities produced in your drinking water such as chloro-organics. Ozone is generally the most effective oxidizing reagent for the control of taste and odor in drinking water treatment (Rice, Bollylky, and Lacy, 1986).

- **Ozonation is safe for the community and environment.** Ozonation treatment leaves no hazardous by-products to harm people, animals, or the environment. When applied correctly, the only by-product formed from ozone is oxygen. Also, any excess ozone produced can be readily destroyed through a thermal catalytic breakdown process (Figure 1). The excess ozone passes through a pressure vessel containing a thermal catalyst. Here there is a breakdown to 0.1% yield. A heat recovery unit then brings the gas to a stable temperature, and the non-harmful gas is vented into the atmosphere (Langlais, Relkaun, and Brink, 1991).

- **Generation and application of ozone are simple.** Electric discharge is the most common generation process of ozone (Figure 2). Here ozone is formed by the interaction between an electric discharge and oxygen molecules in a gas stream that passes through the generator. This is a fairly simple process, making on-site generation of ozone possible and thus eliminating the hazards of transporting chemicals. The application of ozone is just as simple. After the ozone is produced, it is applied to the water by a two-staged "Contacting System Method" (Figure 3) (Langlais, Relkaun, and Brink, 1991).

This sample has more graphics than Sample 14.6. Are the graphics effective and well integrated with the written text?

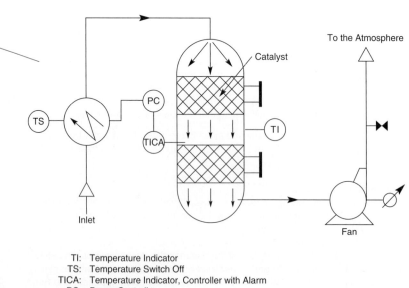

TI: Temperature Indicator
TS: Temperature Switch Off
TICA: Temperature Indicator, Controller with Alarm
PC: Power Controller

Figure 1. Destruction of Excess Ozone

SAMPLE 14.7 continued

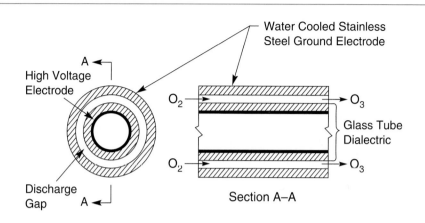

Figure 2. Generation of Ozone

- **Ozonation is cost effective.** The initial high cost of implementing the ozone system is made up in the low long-term cost of maintenance, cutback in operator hours, reduction in transportation, and lower annual chemical costs for the plant (Wang and Hoven, 1986).
- **Ozonation meets and exceeds new regulations.** With ozone, city water facilities will create healthier drinking water, provide a safer and cleaner

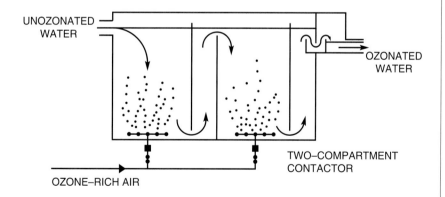

Figure 3. Application of Ozone (Contacting Method)

Source: Figs. 1, 2, and 3 from Brink, D.R., Langlais, B., and Reckhow, D.A., *Introduction,* I-1, 5, in *Ozone in Water Treatment,* Langlais, B., Reckhow, D.A., and Brink, D.R., Eds., AWWA Research Foundation and Lewis Publishers, 1991, an imprint of CRC Press, Boca Raton, Florida. With permission.

SAMPLE 14.7 continued

environment, and solve many problems associated with the use of chlorine as a drinking water treatment.

NARRATIVE

Ozonation's Credibility

Ozone may seem new as a water treatment, but in fact, it has been used on the municipal scale since 1906. Today, more than 2200 water treatment plants in Europe use ozonation as their primary disinfection method. As of 1992, there were 50 ozone plants in the United States, with the third largest in the world residing in Los Angeles, California (Figure 4). Over 50% of the plants in the United States use ozone in the treatment of taste, color, and disinfection of drinking water (Figure 5), while the others use ozone to treat such things as waste water or pool water (Langais, Relkaun, and Brink, 1991).

Project Needs: Funding for an Implementation Study

We thus believe it is a realistic goal to replace chlorination with ozonation in the city of Bryan. We must think of ozonation not as an alternative water treatment but as a new *primary treatment,* and we must be willing to take the initiative to implement the necessary changes.

Having determined the technical effectiveness of the treatment by review of the literature, we must now develop an implementation plan for our facilities. The EPA has recently designated funds for such projects, and there is every indication that the city of Bryan would be eligible to receive at least partial funding for implementation.

We thus recommend that the Water Treatment Facility request a planning grant from the city to cover the cost of developing a proposal to the EPA. The exact

Figure 4. Location of U.S. Water Treatment Plants Using Ozone

SAMPLE 14.7 continued

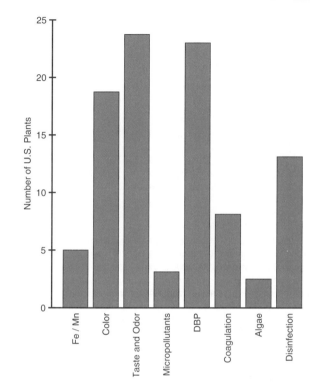

Figure 5. Applications of Ozone in U.S. Drinking Water Plants

Source: From Bablon, G., et al., *Practical Application of Ozone: Principles and Case Studies,* III-1, 136, in *Ozone in Water Treatment,* Langlais, B., Reckhow, D.A., Brink. D.R., Eds., AWWA Research Foundation and Lewis Publishers, 1991, an imprint of CRC Press, Boca Raton, Florida. With permission.

costs of this project should be determined by the facility's accounting department and should be sufficient to cover the following items:

1. **Summer salary for an intern from Texas A&M's Department of Mechanical Engineering.** The intern will provide technical expertise and will write the technical proposal. Because ozonation is a well-tested procedure, no original research is required, so an advanced undergraduate student should be able to handle the assignment.
2. **Cost study.** The grant should cover the time the accounting department will need to analyze the costs of implementation.
3. **Travel.** The study team may benefit by a visit to Dallas, the site of the only ozonation plant now operating in Texas. Some travel funding would thus be useful.
4. **Overhead.** Some charges for secretarial time, printing, copies, and computer time may also be appropriate.

The project has obviously not proceeded very far. What would you add to convince the audience that the project is worth undertaking?

SAMPLE 14.7 continued

CONCLUSION

With implementation of ozonation as a primary water treatment method, we hope to achieve goals of

- Providing the public with cleaner and much healthier water
- Improving future environmental conditions by removing harmful by-products from water
- Improving long-term cost of water facilities
- Meeting and exceeding all government regulations of the Safe Drinking Water Act and its amendments

Ozone is the future. The city of Bryan must seize the opportunity to be a leader among Texas cities in setting forth new standards for future water facilities.

References

Bryant, E., Fulton, G., and Budd, G. (1992). *Disinfection alternatives for safe drinking water.* New York: Van Nostrand Reinhold.

Drews, R. (1993, October 29). Personal Interview. Bryan, Texas.

Jacangleo, J. G. (1984, August). Ozonation: Its role in the formation and control of disinfection by-products. *American Water Works Association Journal,* pp. 74–84.

Jolley, R. L. (1978). *Water chlorination: Environmental impact and health effects.* Ann Arbor, Michigan: Ann Arbor Science.

Langlais, B., Relkaun, D., and Brink, D. (1991). *Ozone in water treatment: Application and engineering.* Chelsea, Michigan: Lewis Publishers.

LePage, W. (1988, August). A treatment plant operator assesses ozonation, *American Water Works Association Journal,* pp. 44–48.

Rice, G., Bollylky, J., and Lacy, W. (1986). *Analytical aspects of ozone treatment of water and waste water.* Chelsea, Michigan: Lewis Publishers.

Singer, P. (1990, October). Assessing ozonation research needs in water treatment. *American Water Works Association Journal,* pp. 78–88.

Wang, C. and Hoven, D. (1986, November). Ozonation: An economic choice for water treatment. *American Water Works Association Journal,* pp. 83–89.

SAMPLE 14.7 continued

A Proposal for Strengthening Medical School Training in STD Prevention Techniques

JANE K. STEINBERG, MPH
JOANNE WELLMAN, MPH
JOAN MELROD, MPH

Synopsis ....

Despite increases in national rates of sexually transmitted disease (STD), surveys indicate that medical students generally lack programs to train them in STD prevention techniques and in counseling patients about STD. The authors of this proposal investigated STD education for medical students at the University of California at Los Angeles and propose a project to involve third-year medical students in STD prevention techniques during their STD-clinic rotation. The long-term goal is to decrease the incidence and prevalence of STD. The immediate aims are to increase medical students' knowledge of STD prevention and to help them develop the communication skills necessary to effectively counsel patients about STD.

Interventions would consist of a series of lectures and workshops using audiovisual aids, small group discussions, and role playing, and would be conducted by health educators and guest lecturers. A quasi-experimental research design would be used in testing the effectiveness of the project in two experimental and two control groups involving a total of 80 third-year medical students. The first intervention would be a 1-week lecture series. Written tests would be given before and after the first intervention to measure the baseline of the students' knowledge of STD prevention methods. The second intervention would be a 1-week workshop series. Students' interviews with patients would be videotaped before and after the second intervention so that the interviewer's communication skills may be assessed and compared. Sets of interventions would be scheduled for the fall of 1990 and the spring of 1991. Six months after the completion of the project, a follow-up questionnaire would be given to evaluate the project's overall effectiveness.

ONE OF THE MAJOR PUBLIC HEALTH objectives is the prevention and control of sexually transmitted disease (STD) (*1, 2*). However, despite the attention and resources directed to the problem, the incidence and prevalence of STD are steadily increasing in the United States, spreading rapidly in cities and rural areas, and they show no signs of slackening. STD falls hardest on the young and the poor, especially urban teenagers and young adults, who engage in sex earlier and with more partners than in the past (*3*).

The magnitude of the problem in Los Angeles County, CA, is evidenced by the fact that reported cases of all stages of syphilis increased from 24 per 100,000 persons in 1985 to an estimated 55 per 100,000 in 1989. By 1994, an estimated 100,000 women will become infertile because of untreated syphilis infection, and thousands of infants will become infected

SAMPLE 14.8   Professional proposal structured on the abbreviated outline

Source: Jane K. Steinberg, Joanne Wellman, Joan Melrod, "A Proposal for Strengthening Medical School Training in STD Prevention Techniques," *Public Health Reports* 106, no. 2 (March–April 1991): 196–202.

through birth (*4*). The incidence of 140 cases of penicillinase-producing *Neisseria gonorrhoeae* (PPNG) strains resistant to penicillin in 1982 increased to 2,200 in 1988 (*4*). Untreated PPNG often results in pelvic inflammatory disease and sterility. The number of physician visits for genital warts has increased sevenfold since 1966. About 750,000 cases are diagnosed annually (*4*). Untreated genital warts are known to increase a woman's risk for cervical cancer. The rates of STD in Los Angeles County indicate that a growing number of infants, youth, and women are contributing to overall STD morbidity.

Nationwide, about 12 million persons are likely to develop new cases of STD within a year, all preventable, as a result of the disease not having been detected or simply for lack of treatment (*5*).

Nearly half of all physicians will encounter STD-infected patients in the course of their practice, most likely in the fields of dermatology, emergency medicine, family medicine, general medicine, infectious diseases, obstetrics and gynecology, adolescent pediatrics, proctology, and urology (*6*).

According to one Los Angeles County physician, programs to train medical students and physicians in STD prevention and counseling are crucial to reducing the incidence of STD among their patients (personal communication, Deborah Cohen, MD, Los Angeles County-University of Southern California Hospital, December 20, 1990). Nearly 44 percent of all medical schools in the United States have no clinical training program in STD prevention and counseling (*7*), however. Of 127 medical schools surveyed in 1982, 54 percent had no hospital or health department-based STD clinic for training students, and 70 percent offered no STD clinical training, leaving 62,000 students without such specialized clinical training.

Where available, clinical training programs involved a minority of the class, a mean of 30 students, and tended to be brief, a mean of 6 hours duration (*8*).

Medical school instruction in venereology has declined during the last 15 years largely because of decreases in the incidence of syphilis during the 1970s (*8*). When an increase in syphilis occurred during the late 1980s, many physicians were inadequately trained in effective diagnosis, treatment, and counseling of STD patients (personal communication, Deborah Cohen, MD, Los Angeles County-University of Southern California Hospital, December 20, 1990).

At the same time, Federal and State funds for STD control have been shifted to projects involving HIV infection and the acquired immunodeficiency syndrome (AIDS) epidemic. Many health workers and researchers who were trained in and working in the area of STD were redirected to the AIDS effort (*4*), resulting in increasing numbers of cases of STD going undetected or untreated (personal communication, Deborah Cohen, MD, Los Angeles County-University of Southern California Hospital, December 20, 1990).

A major deficiency in medical education involving STD is a lack of training in communication skills necessary for physicians to teach prevention techniques to patients and counsel them on problems arising from infection (*9*). Medical students commonly demonstrate in their interviewing practices a reluctance to inquire into relevant psychosocial aspects of the patient's condition (*9*). The physician treating the STD patient needs to be able to communicate effectively to elicit responses that enable him or her to determine the patient's disease risk factors. Effective communication skills are important in providing information about prevention methods to reduce the spread

SAMPLE 14.8 continued

of disease. Korsch and coworkers discussed how a patient's perception of the physician as having good communication skills was consistently associated with increased satisfaction and compliance with prescribed regimens (*10*).

Objectives

The project is designed to enhance the STD prevention and counseling techniques of medical students in their third year at the University of California at Los Angeles (UCLA) School of Medicine. Immediate objectives are for

• 90 percent of program participants to achieve a 75 percent or higher score on an assessment of their communication skills during a videotaped interview held after the 1-week workshop series.

• 90 percent of the program participants to achieve a 75 percent or higher score on a posttest questionnaire given 3 days after the final intervention. The questionnaire would measure their knowledge of STD preventive measures.

Methodology

Interventions would be held at the Los Angeles Free Clinic during the fall and spring semesters. The interventions are to consist of a 1-week lecture series and a 1-week workshop series, each with two health educators as facilitators. The workshop series would be 2 hours in length each morning. The planned sequence of activities is

Lecture 1: *Overview of STDs.* Incidence, prevalence, modes of transmission, symptoms, complications, diagnosis, and treatment of 13 diseases, briefly discussed

Lecture 2: *Prevention of STD through safer sex practices.* In-depth discussion of safer sex methods, including proper condom use, communication with partner, and reduced number of partners

> 'A major deficiency in medical education involving STD is a lack of training in communication skills necessary for physicians to teach prevention techniques to patients and counsel them on problems arising from infection.'

Lecture 3: *Everything you want to know about condoms.* Discussion of advantages and disadvantages, seven steps to encourage a partner to use a condom, demonstration of how to teach patients to use a condom properly, and an informational videotape on principles of condom use.

Lecture 4: *Basic communication variables and facts to consider when interviewing a patient.* Consideration of 16 communication variables (*11*) in the clinical interview, such as seating arrangements, eye contact, and use of silence. Two guest speakers from the Los Angeles County Health Department discuss methods for communicating about sexuality with patients from different cultures. Facts to consider when taking a sexual history are discussed, such as level of sexual activity, history of STD, patient's knowledge of STD and how the disease was acquired, risky behaviors, and knowledge about condom use.

Lecture 5: *The patient–physician relationship in the management of STD.* Discussion of patient's sexual preferences and behaviors, failure of treatment, case reporting, confidentiality issues, counseling partners, and review of the previous four lectures.

Workshop 1: *Values clarification exercise.* Focuses on identifying students' feelings about their own sexuality and identifying discrepancies between intellectual and subjective reactions that may impact upon effective communication with patients. Students are placed in

SAMPLE 14.8 continued

Test Administered Before and After
the Medical Student's STD Clinic Rotation

1. Name four safer sex measures that sexually active persons can engage in.
2. List three advantages of condoms.
3. Name four features to look for when purchasing a condom.
4. Why is it better to use a lubricated condom?
5. Where should condoms not be stored and why?
6. What is the function of nonoxynol-9?
7. True or false: It is essential that patients attending STD clinics understand that the information they offer will be kept confidential.
8. List four steps you would take to encourage a patient to use a condom.
9. If a patient tells you that he or she has anonymous contacts, one night stands, group sex, or multiple partners with inconsistent use of condoms, how often should he or she be screened for STD, (*a*) every 1 to 3 months, (*b*) every 3 to 6 months, or (*c*) every 6 to 12 months?
10. Which of the following condom features are the more effective in preventing STD, (*a*) latex, (*b*) lubricated, (*c*) lambskin, and (*d*) nonoxynol-9?
11. Which of the following are low-risk sex practices, (*a*) dry kissing without having open sores, (*b*) masturbation alone or with partner, (*c*) caressing, (*d*) massage, or (*e*) all of the above?
12. Studies have found that when counseling a patient certain words or phrases are more appropriate to use than others. Which words and phrases are most appropriate, (*a*) are you homosexual, (*b*) are you bisexual, (*c*) do you engage in sexual activity with men, (*d*) do you engage in sexual activity with women, (*e*) do you engage in sexual activity with both men and women, (*f*) are you gay, (*g*) do you have a sexual

contact, (*h*) partner, (*i*) boyfriend, (*j*) girlfriend, (*k*) husband, (*l*) wife?
13. List eight sexually transmitted diseases and give their symptoms for men and for women.

Correct answers:

1. Partner communication, reduce number of partners, low-risk sex (mutual masturbation, caressing, massage, dry kissing without open sores), avoid exchange of body fluids, regular checkups, inspect partner's genitals, wash genitals before and after sexual contact, use spermicide, condom usage
2. Available without prescription, no harmful side effects, tidy because semen is contained
3. Lubricated (as opposed to nonlubricated), reservoir tip, latex rather than lambskin, contains spermicide
4. Less of chance of breaking the condom when it is lubricated
5. Condoms should not be stored in hot places because they deteriorate faster
6. Nonoxynol-9 kills the AIDS virus
7. True
8. Desensitize the patient, offer limited amount of specific information, explicitly endorse the use of condoms, rehearse specific scenarios with the patient, provide behavioral scripts, suggest gradual practice, identify alternatives to unprotected intercourse.
9. A
10. A, B, D
11. E
12. C, D, E, G, H
13. Syphilis, gonorrhea, AIDS, herpes simplex II, chlamydia, genital warts, trichomoniasis, gardnerella vaginalis, moniliasis, nongonococcal urethritis, viral hepatitis, pubic lice.

---

groups and instructed to react to a set of topics related to sexuality, such as homosexuality and abortion. Each person is asked to react or pass.

Workshop 2: *Brainstorming exercise.* The exercise tries to make students comfortable with sex-related language that is initially foreign or uncomfortable that

SAMPLE 14.8 continued

they may encounter when working with STD patients. Students are divided into groups and with the assistance of the facilitators discuss a list of terms related to sex, such as street language terms referring to sexual activity or anatomy.

Workshop 3: *Videotaped session and critique.* Students watch a videotape of a physician counseling a patient with STD using an appropriate communication style and discussing preventive measures. After viewing the tape, each student appears in a video in which one role-plays the patient and another the physician. The student acting the part of the physician uses the communication skills discussed in lecture 4 in taking a STD history, discussing a diagnosis, and explaining preventive measures. The students exchange roles. Students view the videotapes, which are critiqued by other students and the facilitator for feedback and reinforcement.

Project Implementation

Because of the high incidence and prevalence of STD in Los Angeles County, we investigated the extent of STD education at the UCLA School of Medicine to determine whether communication or prevention skills were taught and to examine students' communication skills and their ability to discuss prevention methods with patients. All first-year medical students take a behavioral science course focusing on issues of communication about sexuality as it relates to sexual dysfunctions. Communication with patients about STD was not emphasized in the curriculum (personal observation, 1989). Microbiology, pathology and immunology are required courses for all second year students. Diagnosis and treatment aspects of STD are emphasized, with little instruction on the prevention or communication skills used in

---

Evaluator's Scale for Rating 16 Communication Variables on a Scale of 1 to 4

1. Beginning of interview: poor to positive and smooth
2. Seating arrangement: closed to open
3. Body posture: bad to good
4. Eye contact: inappropriate or excessive to appropriate
5. Interruptions: frequent to none
6. Use of facilitation: did not use to frequent use
7. Maintaining relevance: none to kept patient to relevant matter
8. Psychosocial and cultural concerns: did not cover to covered concerns
9. Empathy: no empathic statements to frequent use of empathic statements
10. Use of silence: inappropriate to appropriate use of silence
11. Personal and social: avoided personal issues to did not avoid personal issues
12. Verbal and nonverbal leads: failed to pick up leads to picked up leads
13. Warmth: did not accept patient as a person to accepted patient as a person
14. Question style: inappropriate question style to good question style
15. Clarity: lack of clarification to good clarification
16. End of interview: abrupt or imprecise to smooth and definite

---

counseling STD patients (personal communication, Lydia Lopez, Clinic Coordinator and UCLA liaison, STD Clinic, Los Angeles Free Clinic, November 1989).

During their third year, about half of the students take an elective medical course that can, depending on clinic rotations that involve STD, such as dermatology and obstetrics and gynecology, cover STD diagnosis and treatment (personal communication, Lydia Lopez, Clinic Coordinator and UCLA liaison, STD Clinic, Los Angeles Free Clinic, November 1989). An STD clinic elective at the Los

SAMPLE 14.8 continued

Angeles Free Clinic gave students an opportunity to work exclusively with the clinic's STD patients.

Our project would be directed to 40 third-year students. The intervention would be a pilot project implemented and evaluated at the STD clinic at the Los Angeles Free Clinic, which serves predominantly a low-income, multi-ethnic population residing in Los Angeles County. The clinic is an appropriate site because health clinics are often the only place where those at highest risk for STD can get sex education (12).

The clinic environment is important because the students can immediately apply the communication skills that they have learned. Depending on the number of clinic patients and constraints on students' time, each student will be involved in STD counseling with about 50 patients. We have permission from a medical school administrator and the director of the free clinic to implement the project on a 1-year trial basis as a component of the STD clinic elective. A 12-month project budget of $80,130 is shown.

Category	Cost
Full time principal investigator	$32,000
Health educators, 2 part time at $16,000	32,000
Consultants and support: 1 graduate student (video), 1 clerical, 2 evaluators, 1 biostatistician	7,000
Equipment and supplies:	
Computer and software	3,500
Video supplies: VCR, video camera, tripod, tapes, monitor	1,775
File cabinet	200
Offices supplies	300
Photocopying, syllabus, evaluation forms, tests	500
Rental, sex education films	30
Program supplies	75
Postage	50
Incentives for patients completing evaluations	2,000
Telephone	700
Total	$80,130

Summary of Evaluation Methods

A pretest–posttest, nonequivalent control group design will be used. The experimental group would receive written and video pretests, 1 week of lectures, a written posttest, 1 week of workshops, a video posttest, and 2 weeks with patients and laboratory work. The control group would receive laboratory and pharmacology rotations in place of the STD lectures and workshops.

A written pretest and videotape will be issued to each of 80 students on the first day of the rotation to assess the baseline of their knowledge of STD prevention methods and communication skills. The written test to be given to both the experimental and the control groups during their respective orientations is shown in the box. After completing the pretest, the students will be videotaped for 10 minutes while they counsel another student who plays the role of a patient with STD.

At the end of the 1-week lecture series, the experimental group will repeat the pretest to assess changes in their knowledge of STD prevention methods. The control group will repeat the pretest after their first 2 weeks of the rotation. Their time will be spent in full-time laboratory work with no patient contact to control for bias that could result from control group exposure to patients. The experimental group will devote half-time to seeing patients and the other half to laboratory work.

'The mounting incidence of STD in Los Angeles County, coupled with deficiencies in teaching communication skills and prevention methods to medical students, justifies the proposal to increase STD education.'

SAMPLE 14.8 continued

Checklist for Course Content, Lectures 1 Through 5

Lecture 1, overview: Time began, time ended, number of students. Syphilis, gonorrhea, AIDS, herpes simplex II, chlamydia, genital warts, trichomoniasis, gardnerella vaginalis, moniliasis, nongonococcal urethritis, viral hepatitis, and pubic lice

Lecture 2, safer-sex practices: Time began, time ended, number of students. Knowing the partner's sex history, reducing the number of sex partners, using low-risk sex practices, avoiding exchange of body fluids, knowing the importance of regular check-ups, inspecting the partners genitals, washing genitals before and after sex, using spermicide, using condoms

Lecture 3, condoms: Time began, time ended, number of students. History, how condoms prevent transmission of sperm, advantages, disadvantages, shared responsibility, film "How to Be a Better Lover," use of model penis, seven steps to encourage condom use (desensitizing the patient, offering a limited amount of specific information, explicitly endorsing the use, rehearsing specific

scenarios, providing behavioral scripts, suggesting gradual practice, and alternatives to unprotected intercourse)

Lecture 4, basic communication variables: Time began, time ended, number of students. Beginning of interview, seating arrangements, body posture, eye contact, interruptions, use of facilitation, maintaining relevance, psychosocial and cultural sensitivity, empathy, use of silence, personal and social issues, verbal and nonverbal cues, warmth, question style, clarity, end of interview, and facts to consider when taking a STD history (baseline level of sexual activity, history of specific STDs, knowledge of specific STDs and how acquired, participation in sexual activities that increase risk, and knowledge about condoms)

Lecture 5, patient–physician relationship: Time began, time ended, number of students. Patient's sexual preferences and behaviors, treatment failure, recidivism, case reporting, confidentiality, dealing with partners, and review

---

A videotaped posttest will be used to assess changes in students' communication skills. The experimental group will be videotaped on the Monday following the last workshop (the beginning of the third week of rotation). The control group will be videotaped on the Monday of their third week.

In order to evaluate the videotapes, two evaluators will be trained to assess each communication variable that was discussed in lecture 4. The evaluator's rating scale is shown. Each variable will be rated on a four point scale that will be tabulated by a biostatistician. Lecture, workshop, and teacher evaluations will be completed by the experimental group after the program to assess their effectiveness and student satisfaction with the program. The results will be tabulated by the

biostatistician and used by the health educators to improve the curriculum.

Based on discussions with professors in the School of Medicine, the researchers expect that 90 percent of the students will attend four out of five lectures and 90 percent will attend four out of five workshops. Because of this, attendance will be taken for each lecture and workshop session.

At the beginning of each lecture, students will be given an outline of the day's activities. The outline will be returned after the lecture to minimize contamination among the second experimental group. To ensure that the lectures and workshops are delivered according to the plan, the checklist will be provided for the health educator for each activity as it is conducted in class. The checklist for

SAMPLE 14.8 continued

the content of the course is shown in the box on the previous page.

A patient evaluation form will be given to each clinic patient before being seen by a student who participated in the intervention. One side of the form is in English and the other in Spanish. Patients will be asked to rate the student's effectiveness in communication and in discussing preventive measures. Six months after the rotation, a follow-up evaluation form will be sent asking the students to assess the usefulness of the methods they learned about preventive and communication skills in their other rotations.

Because almost half of all physicians will encounter STD patients in their practices (5), the proposed intervention is expected to have high relevance for medical students. Selection bias may be a factor in the program because it will involve students who want to work with STD patients and in a clinic setting, rather than those who do not. Medical students' time restraints may be a limiting factor as the program would be an addition to the curriculum. This may be partly offset by offering only components of the lecture and workshop series.

The demanding schedules of medical students may lower the response rate for the questionnaires, but follow-up telephone calls will be made to nonrespondents. Based on a discussion with the medical school registrar's office, we expect to reach 90 percent of all program participants by mail or telephone using an updated medical student directory.

Performance on the posttest video would be rated using the interaction scale described by Verbi and coworkers (13), which measures communication skills. Each component to be measured, such as body posture, will be scored between 1 and 4, with a total of 64 possible points. Student performance on the posttest questionnaire that assesses knowledge of prevention methods will be measured by the number of correct responses on each questionnaire, with one point for each correct answer.

The mounting incidence of STD in Los Angeles County, coupled with deficiencies in teaching communication skills and prevention methods to medical students, justifies the proposal to strengthen STD education. An innovative program that combines these two important but undervalued skills does not exist elsewhere in the UCLA School of Medicine. Since training in communication skills and preventive techniques is valuable to physicians in all fields of medicine, the program could lead to formal integration into the curriculum. The benefits of such a program could help reduce the incidence of STD, enabling many people to improve the quality of their lives and be more productive members of society.

References ....

1. Public Health Service: Promoting health/preventing disease: objectives for the nation. Office of the Assistant Secretary for Health. U.S. Government Printing Office, Washington, DC, 1980, pp. 25–29.
2. Public Health Service: Healthy people 2000: national health promotion and disease prevention objectives. Conference edition. Office of the Assistant Secretary for Health. U.S. Government Printing Office, Washington, DC, 1990, pp. 493–508.
3. Scott, J.: Young, poor suffer silent epidemic. Los Angeles Times, Sept. 4, 1989, p. 1.
4. County of Los Angeles (CA): STD fact sheet. Department of Health Services, 1988.
5. Blount, J.: Compilation from CDC statistics, 1987. Centers for Disease Control, Center for Prevention Services, STD. Atlanta, GA, 1989.
6. Margolis, S.: Initiation of the sexually transmitted diseases prevention training program. Sex Transm Dis 8: 87–88 (1981).
7. Centers for Disease Control: Project to increase STD education in schools of medicine. Associated schools of public health cooperative agreement internship proposal to Center

SAMPLE 14.8 continued

for Prevention Services, Division of Sexually Transmitted Diseases, 1988.

8. Stamm, W. E., Kaetz, S. K., and Holmes, K. K.: Clinical training in venereology in the United States and Canada. JAMA 248: 2020–2024, Oct. 22/29, 1982.

9. Evans, B. J., and Stanley, R. O.: Lectures and skills workshops as teaching formats in a history-taking skills course for medical students. Med Educ 23: 364–370 (1989).

10. Roter, D. C.: Participation in the patient/provider interaction: effects of patient question-asking on the quality of interaction, satis-faction, and compliance. Health Educ Monogr: 281–315 (1977).

11. Andrist, L. C.: Taking a sex history and educating clients about safe sex. Nurs Clin North Am 23: 959–973 (1988).

12. Garcia, K. J.: Experts see disaster in family planning cutbacks. Los Angeles Times, Nov. 29, 1989, p. 1-B.

13. Verby, J. E., Holden, P., and Davis, R. H.: Peer review of consultations in primary care: the use of audiovisual recordings. BMJ 1: 1686–1688 (1979).

SAMPLE 14.8 continued

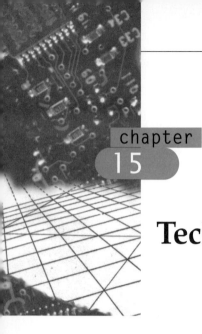

Technical Manuals

CHAPTER OUTLINE
▼▼▼▼▼▼▼▼▼▼▼▼▼▼▼▼▼▼▼▼▼▼▼▼▼▼▼▼

Step One: The Audience-Action Analysis
Step Two: Writing a Task Analysis
Step Three: Developing Graphics for the
    Manual
Step Four: Planning and Presenting Oral
    Instructions
Step Five: Developing the Technical Manual
Reviewing Manuals

CHAPTER OBJECTIVES
▼▼▼▼▼▼▼▼▼▼▼▼▼▼▼▼▼▼▼▼▼▼▼▼▼▼▼▼

After you have worked through this chapter,
you should be able to do the following:

- Recognize the purposes and typical
  audiences for technical instructions
- Write a task analysis that will serve as the
  core document for a technical manual
- Following the CORE method, develop the
  task analysis into a full technical manual

Like the proposal, the manual—or how-to book—is a standard feature of democratic and bureaucratic cultures. In a mobile society in which people move from place to place, from job to job, and from skill to skill, new ground is constantly broken. People must learn to use new products, perform new tasks, fill new roles, and implement new policies. In a literate society, written instructions often supplement or replace mentors, guides, and bosses in the process of initiating novices.

The technical manual is therefore a means of instruction and an instrument of change. It prepares people to perform efficiently in new areas of action and to take on new responsibilities. The manual that comes with a personal computer, for example, helps the user handle tasks that once could be performed only in an office environment with elaborate technological and human support. The aim is to make the home user an independent worker. Likewise, a manual on company policies and procedures helps new employees to perform more independently. They can use the written guide rather than relying on supervisors and senior workers.

Manuals are like textbooks in their instructional focus. But textbooks are usually used in a classroom and supplemented by a teacher's instruction. By contrast, manuals often have to do their work alone. This means that they must provide complete information in accessible prose, packaged in an easy-to-use format.

As Chapter 1 suggested, the manual writer in the context of production is always in danger of losing sight of the context of use and, as a result, producing a worthless document. This chapter is designed to help you avoid that pitfall. Applying the CORE method, we will discover ways of bringing the user into the earliest stages of the process of developing clear instructions. Starting with the production of a core document—a brief *task analysis,* or "recipe"—you can engage readers (either actual users or trial audiences) in user tests and document reviews. Then you can build your complete document a little at a time, following the five basic steps of the CORE method:

1. *Do an audience-action analysis,* making a set of notes or developing a profile to guide the writing process.
2. *Write a task analysis,* getting feedback from trial users and revising accordingly.
3. *Develop graphics for the manual,* putting as much information as possible into graphic form.
4. *Deliver an oral overview or a training session* based on your task analysis and graphics.
5. *Write the full manual,* amplifying the materials you have developed in your task analysis, graphics plan, and oral presentation. (See Figure 15.1.)

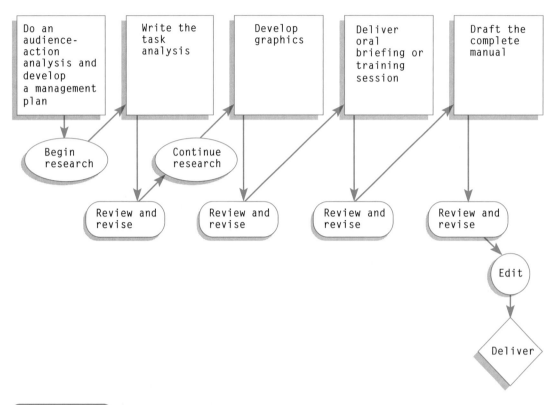

FIGURE 15.1

Applying the CORE method to the development of a technical manual.

Step One: The Audience-Action Analysis

The purpose of a manual is to provide a *guide* that leads a *user* (usually a novice) through a technical *procedure* for the following purposes:

- using *products* (such as computers, telephone answering machines, or garage-door openers)
- carrying out *policies* (in such areas as office communications, nondiscrimination, income tax, and so on)
- taking advantage of *services* (such as health programs, automated banking, and library use)

For the document to work as a *guide*, the expert author must effectively project a tone of authority, helpfulness, and efficiency:

- The guide is *authoritative* if it adequately covers the procedure.
- The guide is *helpful* if it demonstrates understanding of and respect for the user's needs.
- The guide is *efficient* if it provides the shortest and clearest possible path for the user to accomplish his or her goals.

Analyzing the Audience

As you begin defining the specific audience for your manual, keep in mind that all manual users share some fundamental characteristics. By definition, the user of a manual is a technician—a person who employs a technique to solve a problem or develop a product. The user may not be a technician all the time, but that is the role he or she assumes when using the manual. Even a manager is a technician when sitting at a computer console to produce the annual report. A home user is a technician when carrying out the monthly accounting chores.

The user needs to get a job done, is usually in a hurry, and is likely to be impatient, easily frustrated, or even hostile. A novice who turns to a manual is at least one step—and most often several steps—removed from a desired action.

For example, if you have to read a manual before you can use the new database software that the boss has forced on everyone, and if you have to know about the database before you can log your weekly activities, and if you have to log your activities before you can return to doing what the company supposedly hired you for—designing bridges or carrying out polymer research— then you are four steps away from your desired action. Reading the manual seems to be just another hindrance along the way. You just want to know how to make the machine work to your benefit as quickly as you can.

When you make planning notes to write your own manuals, then, remember that your readers don't want long explanations or descriptions. They want explicit *procedures* for effective action. And they want the procedure to follow a chronological sequence that allows them to translate reading into action as fast as possible. For this purpose, imperative, easy-to-read prose that says, in the simplest terms possible, "Do this; then do that" is most effective. If the book is designed so that users can lay it open beside their workstations and read each step as they do it, all the better.

Analyzing Purpose in Two Types of Manuals

All of these general audience characteristics are true for users of the two basic types of manuals: the *tutorial* and the *reference* manual. Tutorials lead readers through a step-by-step lesson on how to perform a single unified procedure, such as how to write a paper using a word processor or how to perform CPR on a drowning victim. Reference manuals have a more complex overall structure. They cover a number of procedures associated with a single product (a computer software package, for example) or with a particular job (repairing copy machines, for example) or working environment (a manufacturing plant with safety regulations, for example).

The tutorial is meant to be used only once by a single user (unless the user needs several trips through the procedure to learn the process), whereas the reference manual provides a resource to which the user can return again and again. This difference in purpose is reflected in the organizational structures of the two types of manuals.

The tutorial is organized chronologically, taking the reader one step at a time toward the mastery of basic procedures for solving a problem. The pace is fairly slow; the instructions are detailed; the tone is friendly and helpful. A general outline for a tutorial looks something like this:

1. Introduction

    1.1. Statement of Purpose and Audience
    1.2. Statement of Objectives
    1.3. Overview of Procedure

2. Task 1

    2.1. Step 1
    2.2. Step 2
    2.3. Step 3, etc.

3. Task 2, etc.

Ideally, when all the tasks are completed, the user's problem will be solved.

The reference manual shows how to solve several problems, which are usually covered alphabetically. Within each section the reader still learns how to perform a procedure, but the pace is somewhat faster, and the instructions are a little more general. In writing instructions for particular procedures, you may assume that the reader has a general familiarity with related processes. (By contrast, a tutorial must constantly review even the simplest points of usage.) But just as tutorial users are often in deep water for the first time, reference manual users usually pick up the book in a moment of frustration, seeking solutions to problems that caused their work to stall out. Your tone, therefore, should remain friendly and helpful.

Essentially, a good reference manual contains a mini-tutorial within each of its sections. The user can go to the manual for a set of refresher courses. So a general outline for a reference manual looks something like this:

1. Introduction

    1.1. Statement of Purpose and Audience

    1.2. Statement of Objectives

    1.3. Overview of the Manual's Organization

2. Problem 1

    2.1. Explanation of the Problem (Policy)

    2.2. Solving the Problem (Procedure)

        2.2.1. Task 1

            2.2.1.1. Step 1
            2.2.1.2. Step 2
            2.2.1.3. Step 3, etc.

        2.2.2. Task 2, etc.

3. Problem 2

    3.1. Explanation of the Problem (Policy)

    3.2. Solving the Problem (Procedure), etc.

4. Problem 3, etc.

Despite their differences, however, tutorials and reference manuals do share a similar style and format. Both types of manuals use the writing style and page design that are standard in technical communication. Overall, the style is user-directed, task-oriented, and graphically varied. Specifically, they both do the following:

- address the reader directly as "you"
- make frequent use of action verbs (*use, set, choose, signal, adjust,* etc.)
- organize the presentation around tasks (adjusting sound and volume), which are divided into a sequence of steps
- give steps in the imperative (do this, do that, etc.)
- Integrate the step-by-step instructions with simplified graphics that show at a glance the relationship between what is on the page and what is happening in real life at the appropriate time of action (as, for example, computer manuals show what should be on the screen the user is actually seeing when the computer is in use)

Because tutorials and reference manuals share these key features, we have chosen to concentrate on the shorter of the two: the tutorial. If, after assessing your audience's need, you decide that you should write a reference manual, you can follow the same steps, adjusting for slight differences in organization and the pace at which you present the information.

Step Two: Writing a Task Analysis

The core document for the manual is a *task analysis*. You can think of it as a "recipe" for the successful completion of a technical problem-solving procedure.

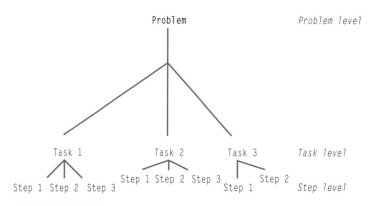

The task-analysis model divides a problem into tasks and chronological steps that the user must complete to solve a problem.

Remember the distinction made in Chapter 14 between an ill-formed problem and a well-formed problem? In preparing to write instructions, you should first present the problem that you intend to solve as a well-formed problem. That is, you need to divide it into a series of chronological tasks and then divide each task into a series of chronological steps. (See Figure 15.2.) When the user has completed all steps and tasks, the problem should be solved.

To develop your own task analysis, take the following steps:

1. Type your title, audience, and purpose at the top of a new file in a word-processing program.
2. Then type a one-line command for each task. Use the imperative form for each sentence: "do this; do that," etc.
3. Under each task, type a series of numbered steps (also imperative commands) for carrying out the procedure.

Keep the number of steps under each task relatively small. No task should contain more than about seven steps. Divide complex tasks into two or more separate tasks.[1]

When you are ready to produce the complete manual, you can use the statement of audience and purpose as the heart of your introduction. The task analysis itself will become the key headings and the main steps to be covered in the

---

1. The so-called 5 ± 2 rule of cognitive science says that the optimum number of items in a series of instructional points usually falls between three and seven. Or, as Howard Gardner puts it, using the mind-computer analogy that is so popular in cognitive psychology: "Short-term memory is viewed as a buffer consisting of about seven slots, each of which can hold a single piece or chunk of information." See *The Mind's New Science: A History of the Cognitive Revolution* (New York: Basic Books, 1987) 122.

GUIDELINES AT A GLANCE 53
Writing and reviewing the task analysis

1. Use the following questions to review your task analysis:
   - Is the concept of the audience narrow enough? Too narrow?
   - Can the intended audience handle the steps required for the procedure?
   - Are the steps covered thoroughly enough to make doing the job seem reasonable?
   - If you were asked to perform the tasks, what objections would you raise just from reading the instructions?
2. If it is practical, perform a usability test on your task analysis, using Guidelines at a Glance 54.
3. After the discussion, revise your paper and submit it to your instructor for additional comments.
4. Use these comments to develop graphics, an oral presentation, and then a full-scale manual, as discussed below.

body of the manual. A procedure for preparing and reviewing task analyses is summarized in Guidelines at a Glance 53.

Samples 15.1, 15.2, and 15.3 are sample task analyses developed by three groups of student authors for audiences of fellow students. Sample 15.1 uses the symbols T and s to make explicit the division into tasks and steps. Sample 15.3 just uses typographical variation to distinguish between the two levels of instruction so that when it's time to produce the manual, the authors do not have to erase all the symbols. Sample 15.1 is a task analysis for a tutorial manual. Sample 15.2 beginning on page 530 is a task analysis for a reference manual. Sample 15.3 on page 533 is another tutorial, but one with alternative tracks for action.

Performing a Simple Usability Test

Once you have prepared your task analysis, you are ready to engage trial users in your composition process. A usability test ensures that the problems, tasks, and steps you have laid out in your task analysis

- are complete, but not excessive,
- can be clearly understood by the reader,
- are placed in the most effective and efficient order, and
- convey a sense of authoritative, yet compassionate helpfulness.

If you have done a through job of rhetorical or market research of your audience, the manual should guide the reader smoothly through the process of performing the specified actions, without giving offense or causing frustration. The usability test is a way of getting feedback from one or more sample manual users about the content and order of your problems, tasks, and steps before you begin to add graphics, develop an oral briefing, or consider layout issues.

<u>Title:</u>   Quickstart for Microsoft Windows™

<u>Audience:</u> First-time student users of Microsoft Windows™

<u>Purpose:</u> 1. To teach first-time users enough about Windows™
               to write a short paper and integrate simple
               graphics.
            2. To introduce users to multiple applications.

<u>Task Analysis:</u>
<u>Purpose 1. Write a short paper with a simple graphic</u>
T1. Get started
    s1. Type WIN at the DOS prompt
    s2. Press ENTER

T2. Start Microsoft Word
    s1. Double-click on the Applications icon
    s2. Double-click on the Microsoft Word icon
    s3. Begin typing text

T3. Start Paintbrush
    s1. Click the down arrow
    s2. Double-click on the Accessories icon
    s3. Double-click on the Paintbrush icon
    s4. Draw a graphic using the tools
    s5. Click on the scissors tool
    s6. Make a box that covers the graphic
    s7. Choose Cut in the Edit menu

T4. Exit Paintbrush
    s1. Choose Exit in the File menu

T5. Insert graphic
    s1. Double-click the Microsoft Word icon
    s2. Place the cursor where you want the graphic to
        appear
    s3. Choose Paste from the Edit menu

T6. Save a document
    s1. Choose Save from the File menu
    s2. Select the correct Drive
    s3. Click OK

T7. Print a document
    s1. Choose Print from the File menu

<u>Purpose 2. Introduce multiple applications</u>
T1. View application choices
    s1. Open Program Manager
    s2. Click on Arrange Icon
    s3. Click on Tile command

SAMPLE 15.1   Student task analysis for a computer manual (tutorial)

T2. Open an application window
    s1. Double-click on the application icon

T3. Run more than one application
    s1. Shrink the current window with the minimize button
    s2. Open another application or accessory by double-
        clicking its icon

T4. Switch between applications/accessories
    s1. Press CTRL and ESC to display the Task List dialogue
        box
    s2. Highlight the application you want
    s3. Click the Switch To button

T5. Move text within an application
    s1. Select the text to be moved by dragging the mouse
        across it
    s2. Choose Cut from the Edit menu
    s3. Place the pointer where you want the chunk of text
        to be
    s4. Click the mouse

T6. Move text between applications
    s1. Follow the first two steps in T5
    s2. Minimize the current application
    s3. Open the new application
    s4. Move the pointer to the new location
    s5. Click the mouse

T7. Copy text
    s1. Follow the procedures outlined in T5 and T6 except
        choose Copy from the Edit menu

T8. View clipboard's contents
    s1. Open Program Manager
    s2. Open the Main Group window
    s3. Double-click on the Clipboard icon

T9. Access your document later
    s1. Open the application that contains your file
    s2. Choose the Open command from the File menu
    s3. Move within the dialogue box to select the file
    s4. Click the OK button

SAMPLE 15.1 continued

TASK ANALYSIS FOR A REFERENCE MANUAL

EVALUATING A KNEE INJURY

Authors: Stephanie Obst, Cristine Smith,
Eddie Palomarez, and Ryan Pillans

Introduction

Problem: Knee injuries are among the most commonly occurring injuries in most types of sports. It is the job of athletic trainers to evaluate these injuries. However, because the tests for these types of injuries are used extensively and considered easy, there is no practical manual describing the method of performing the test.

Purpose: The purpose of this reference manual is to describe for students beginning the athletic training program at Texas A&M University the step-by-step techniques for evaluating knee injuries.

Objective: We want this reference manual to assist these students in deciding what test to perform and how to perform it. We intend to present the information in an accurate and easy-to-follow format.

Topics: There are three main topics addressed in this manual: 1) anatomy of knee, 2) selection of a test to perform when experiencing knee pain, and 3) instructions on how to perform each test.

The anatomy section of this manual includes diagrams of the knee with relevant parts of the knee labeled. There are also diagrams showing the orientation of the knee (medial, lateral, posterior, and anterior).

The second section will cover the proper steps to be taken in deciding what test or tests should be performed for different knee problems. These problems include medial and lateral knee pain and slipping or locking of the knee to the front or back.

The last section will cover the steps involved in performing each test. These tests include the Lachman knee test, Varus stress test, Valgus stress test, knee joint test, pivot-shift test, and drawer test. When performing these tests, it is best to do so bilaterally, on both the injured and uninjured knee, so the differences between the two knees can be recognized.

Body

Problem I: What knee anatomy do I need to know to perform these tests?
    Task I: Review the following diagrams. (See figures)
    Task II: Review various orientations of the knee and terminology used to
        describe it. (See figures)

Problem II: What tests do I perform for each knee problem?
    Task I: Medial knee pain
        Step I: Perform the Valgus stress test.
        Step II: Perform the Lachman knee test.
    Task II: Lateral knee pain
        Step I: Perform the Varus stress test.
        Step II: Perform the Lachman knee test.

SAMPLE 15.2 Student task analysis for a reference manual

Task III: Knee slipping and/or locking to the front
    Step I: Perform the Lachman knee test.
    Step II: Perform the knee drop test.
    Step III: Perform the pivot-shift test.
Task IV: Knee slipping and or locking to the back
    Step I: Perform the drawer test.
    Step II: Perform the knee drop test.
    Step III: Perform the Lachman knee test.

Problem III: How do I perform these tests?
Task I: Performing Varus stress test. This test will indicate injury to the lateral collateral knee ligament (refer to knee anatomy).
    Step I: Lay the patient in the prone position (on his or her back).
    Step II: For right knee (injured knee), place your right hand inside of the knee joint and grab the outside of the ankle with your left hand (see picture).
    Step III: Tell patient to relax.
    Step IV: With the knee at full extension, push the knee outward with your right hand and pull in with your left hand.
    Step V: At 30 degrees of flexion, push the knee outward with your right hand and pull in with your left hand (see picture).
    Step VI: Perform the test on the left knee (uninjured), switching placement of hands, while comparing the laxity of the lateral collateral knee ligament.
Task II: Performing the Lachman knee test. This test will illustrate injury to the anterior and posterior crucial knee ligament (refer to knee anatomy).
    Step I: Lay the patient in the prone position (on his or her back).
    Step II: For the right knee (injured knee), place your left hand on the lateral aspect of the knee, cupped over the femur above the patella, and place your right hand cupped on the tibia below the patella (see picture).
    Step III: Tell the patient to relax.
    Step IV: With the patient's knee in full extension, push down with your left hand and pull up with your right hand.
    Step V: With the patient's knee in full extension, push down with your right hand and pull up with your left hand.
    Step VI: Repeat Steps IV and V until you have given a good test (a licensed trainer can help you know when it is enough).
    Step VII: Repeat Steps IV–VI on the left knee (uninjured) while switching hand placement to compare the injury to the anterior and posterior crucial knee ligament.
Task III: Performing the Valgus stress test. This test will indicate injury to the medial collateral knee ligament (refer to knee anatomy).
    Step I: Lay the patient in a prone position (on his or her back).
    Step II: For the right knee (injured), place your left hand inside of the knee joint and grab the outside of the ankle with your right hand (see picture).
    Step III: Tell the patient to relax.
    Step IV: With the knee at full extension, push the knee outward with your left hand and pull in with your right hand.

SAMPLE 15.2 continued

Step V: Bend the knee to 30 degrees of flexion, push the knee outward with your left hand, and pull in with your right hand.

Step VI: Perform the test on the left knee (uninjured), switching placement of hands, while comparing the laxity of the medial collateral knee ligament.

Task IV: Performing the knee drop test. This test will indicate injury to the posterior crucial knee ligament (refer to knee anatomy).

Step I: Lay the patient in a prone position (on his or her back).

Step II: Tell the patient to relax.

Step III: Grab the big toe on the leg of the injured knee.

Step IV: Pull straight up on the big toe, lifting the leg.

Step V: Watch to see if the tibial tuberosity disappears, indicating injury to the posterior crucial knee ligament.

Task V: Performing the pivot-shift test. This test will indicate injury to the anterior crucial ligament (refer to knee anatomy).

Step I: Lay the patient in a prone position (on his or her back).

Step II: For the right knee (injured), position the knee 90 degree flexion and the hip at 90 degrees.

Step III. Place your left hand on the outside of the bent knee and your right hand on the inside of the ankle (see picture).

Step IV: Tell the patient to relax with the knee naturally positioned slightly outward (see picture).

Step V: Guide the knee inward with your right hand while extending the knee with your left hand.

Step VI: Feel for a "clunk" prior to full extension. This sound of the tibia and femur colliding determines if the anterior crucial ligament of the knee is injured.

Step VII: To compare damage, switch hand placement and perform the test on the left knee (uninjured).

Task VI: Performing the drawer test. This test indicates injury to the anterior crucial ligament or the posterior crucial ligament (refer to anatomy).

Step I: Lay the patient in a prone position (on his or her back).

Step II: Place the injured knee between 90 and 100 degrees of flexion

Step III. Sit on the foot of the injured knee.

Step IV: Take both hands and place them behind the knee joint, with thumbs placed on the front of the knee joint line (see picture).

Step V: Tell the patient to relax.

Step VI: Pull forward on the patient's knee as if opening a drawer and then push back on the patient's knee as if closing the drawer.

Step VII: Observe to see if the tibia bone slips forward, indicating injury to the anterior crucial ligament. If the tibia bone slips backwards, this indicates injury to the posterior crucial ligament.

SAMPLE 15.2 continued

Title: A Guide to Yard Composting

Audience: Home owners and renters with access to yards

Purpose: To describe the composting process, individual advantages,
environmental impact, and a simple 7-step method for beginning
composting

Task Analysis:

Step One: Obtain required materials
1. Single square
   - 12 wooden pallets
   - fencing wire
   - pliers
   - drill
2. Fence cylinder
   - 28 1/2 feet fencing
   - 6 wood posts
   - fencing wire
   - pliers
   - hammer

Step Two: Assemble bin (single square)
1. Drill a hole in the top and bottom of the outside planks.
2. Cut wire into 12-inch lengths.
3. Stand pallets with planks vertical next to each other at a 90-degree angle.
4. Feed wire through the pre-drilled holes, and twist it to tighten the pallets together.
5. Repeat step 4 to create a square.

Step Two: Assemble bin (fence cylinder)
1. For each of three bins, drive the wood posts in the ground about 3 feet apart.
2. Stand the fencing upright in a circular fashion so that the fencing touches the
   posts.
3. Use the fencing wires to connect the fencing to the posts at the top and bottom.
4. Connect the two ends of the fencing with the wire to complete the cylinder.

Step Three: Locate the bin site

Step Four: Identify compostable materials

Step Five: Identify proper mixture

Step Six: Begin composting
1. Put a layer of sticks or branches in the bottom of the bin.
2. Add compostable material to bin.
3. Add water every 3 to 4 days.
4. Turn the pile with a shovel or pitchfork.
5. Once the first bin is full, start the second.

Step 7: Use the compost

SAMPLE 15.3   Task analysis for a manual on composting (tutorial)

GUIDELINES AT A GLANCE 54

Procedure for a simple usability test

1. Select people to fill four roles in the test:
   - *A test user.* This person will try to perform the actions using the task analysis. If you are working in a group to produce the manual, you can select your test user for the initial test from among your coauthors. Choose the person who knows least about the procedure. Use only the task analysis.
   - *A guide.* This person will answer questions and help the user when needed.
   - *A recorder.* This person either takes notes on user questions and the guide's responses or actually tapes the session (using either audio or video equipment).
   - *An editor.* This person takes notes on the manuscript of the task analysis, indicating problem spots and pointing out possibilities for further coverage and revision.
2. Have the test user perform the tasks while the other participants perform their various duties.
3. After the test user has completed the work, question the user about his or her overall impressions.
4. Discuss the problems that became clear during the test, and consider what you will need to add, delete, or change to make the manual workable.

If you are working as part of a production team, Guidelines at a Glance 54 offers a procedure for conducting a usability test of your core document. The procedure requires that three or four people serve specific roles in the test (test user, guide, recorder, and editor). However, if you are developing the manual on your own, you will need to recruit assistance. You should be the guide, but you will still need a test user. Since the test user and guide will actively converse, you will also need an external observer to serve as recorder. Or you can use an audio or video tape recorder to capture the interchange. Later, you can edit your task analysis, incorporating revisions suggested by the usability test.

Step Three: Developing Graphics for the Manual

The third step in the CORE method is to plan and design graphics for your manual project. As you work on your graphics, you will be working simultaneously on Step Four of the CORE method. You can use some of the graphics you produce in combination with your task analysis to do an oral presentation of your instructions.

To develop graphics for a technical manual, focus on the following approaches (covered at greater length in Chapter 6 and summarized in Guidelines at a Glance 55):

- *Use graphics and hybrid displays to preview and review your tasks and steps.* For the oral presentation you can develop your analysis of each task into a trans-

GUIDELINES AT A GLANCE 55
Graphics for technical instructions

To preview and review your main points, use outlines, lists, tables, and flowcharts.

1. Starting up with minimum expense, maximum safety
   • the on button
   • the safety switch
   • the power source
2. Maintenance: An Ongoing Process
   • tools
   • procedures

For a discussion of outlines and other preview devices, see Chapter 6, pages 140–142.

To orient the audience in space and time, use diagrams, line drawings, maps, and flowcharts.

Top
Lobe

For a discussion of line drawings and other orientational diagrams, see Chapter 6, pages 142–147.

To instruct users on the performance of actions, use functional diagrams, photographs, and line drawings.

For a discussion of functional graphics, see Chapter 6, pages 147–149.

To reinforce important points and emphasize warning statements, use typographical variation and low-density graphics.

For a discussion of icons and other highlighting devices, see Chapter 6, pages 159–163.

To motivate the reader, use drawings, photographs, and other high-impact graphics.

For a discussion of photographs and other high-impact graphics, see Chapter 6, pages 163–171.

parency with the numbered chronological steps. For the written manual, use these simple outlines as the skeleton of your instructions, adding more text and graphics where you need them. As with reports and proposals, you can use flowcharts and tables as forecasters and summarizing devices in both oral and written presentations. Lists of objectives (such as the ones at the beginning of each chapter in this textbook) are also useful forecasters.

- *Use graphics to orient the audience in space and time, showing readers where and when actions take place.* For both written manuals and oral instructions the best orientational graphics include maps, flowcharts, and diagrams. Diagrams with labels that help users find their way around machines are especially helpful.
- *Use functional diagrams to show important relationships among parts of machines and to depict correct actions.* Diagrams with arrows and other directional indicators make an effective complement to instructional texts.
- *Use low-density graphics to emphasize important actions and to reinforce warnings and cautions.* Icons, cartoons, drawings, and photographs can capture readers' attention for especially important information, such as notes that keep them from falling into personal danger or frustrating errors.
- *Use motivational graphics to create an atmosphere of "user-friendliness," taking the edge off of frustration and user lethargy.* Photographs, drawings, and cartoons can keep reluctant readers moving forward in the text—and in the procedure.

Step Four: Planning and Presenting Oral Instructions

The fourth step in the CORE method is to plan and present a short oral overview or training session for your project, integrating the material developed in your task analysis and your graphics plan. You should use the guidelines for effective oral presentation in Chapter 8, paying special attention to the section on training sessions.

The task analysis makes a good guide for discovering the right mix of demonstration and participation. For example, the student authors of the task analysis on how to use Microsoft Windows (Sample 15.1) divided their training program into two sessions (each having its own purpose) with each session organized as follows:

```
Session 1.   Write a short paper with a simple graphic.
Tasks to be demonstrated, then practiced:
1. Get started.
2. Start Microsoft Word.
3. Start Paintbrush.
4. Exit Paintbrush.
5. Insert graphic.
6. Save a document.
7. Print a document.
```

```
Session 2.   Introduce multiple applications.
Tasks to be demonstrated, then practiced:
1. View application choices.
2. Open an application window.
3. Run more than one application.
4. Switch between applications/accessories.
5. Move text within an application.
6. Move text between applications.
7. Copy text.
8. View clipboard's contents.
9. Access your document later.
```

For each of the numbered tasks, the trainers first modeled the steps, then gave time for the users to try the procedure out.

Your task analysis, with its effective chunking, also serves as a basic outline for your overhead transparencies. In drafting your visuals, make sure they have the following characteristics:

- They are clear, easy to read, and not overcrowded.
- They use primarily low-density graphics to reinforce major points.
- They provide effective forecasters and summaries.
- They give enough information so that if the audience were not able to hear your oral presentation, they would still get a good sense of what they are supposed to do.

With your visuals completed, you are ready to prepare and rehearse your presentation. Keep in mind the following guidelines:

- Review your task analysis and audience-action notes until you have a strong command of your subject matter and approach.
- Practice looking forward, making eye contact with the audience, maintaining a fluent delivery, and varying your pitch and tone for emphasis.
- Keep your own talking to a minimum; use most of the time for demonstration and user practice.
- Time your report carefully, staying within your assigned limits.

As you develop your presentation, remember that a *helpful persona* is even more important in training sessions than in oral reports and proposals. Let your own enthusiasm motivate the user, and avoid suggesting that only experts like you are capable of completing the difficult tasks you describe. Bring the users into the action with encouragement and solid advice.

Step Five: Developing the Technical Manual

After you have completed the task analysis, graphics plan, and perhaps an oral training session, you are ready to develop a full-length instructional document.

This chapter shows you how to write a complete manual by amplifying the simple recipe for action and your accompanying set of graphics.

To create a usable and attractive final manual, you can begin by considering several rhetorical elements that nearly every manual includes: definitions, explanations, warnings, troubleshooting advice, options and alternatives, examples, previews, and reviews.[2]

Definitions

Since many manuals (especially tutorials) present new information to novice users, manual writers often have to define unfamiliar terms or terms used in special ways with respect to the procedure in the manual. The purpose of definitions is to make the unfamiliar familiar, so you should either use well-known terms to define the new term or use words that connect the new term to the user's experience.

The student manual on Microsoft Windows, for example, begins with the heading "What is Windows?" It answers the question this way: "Windows is a graphical environment that allows you to easily start up and work with software applications, run more than one application at a time, and transfer information between applications." Instead of a technical definition, the authors chose to develop a user-oriented definition that includes information about action.

Definitions are necessary, but they tend to slow down the pace of the manual. Avoid large clusters of definitions (especially at the beginning of a manual, where you may be tempted to define everything at once). More than two or three definitions in a row can lead to cognitive overload. The reader might conclude, "The procedure must be hard if I've got to learn all these new terms before doing it."

If you must include a lot of definitions, also include in the manual a *glossary* that lists all key definitions, either at the end of each section or at the end of the manual. If your manual is brief, simply use **bold type for important definitions** so that readers can easily scan the pages to check definitions.

Explanations

In addition to defining key terms, you may need to add sentences that describe the causes or effects of tasks and steps in a process, sentences that tell *why* as well as *how to* accomplish an action. An instruction in a tool manual might say, for example, "Keep the drill bit straight while you make the hole," and then add the explanation: "The straight hole will make the routing work easier."

---

2. These concepts were originally developed in an article by Jimmie Killingsworth, Michael Gilbertson, and Joe Chew, "Amplification in Technical Manuals: Theory and Practice," *Journal of Technical Writing and Communication* 19 (1989): 13–29. For those who are interested, the article draws on classical rhetoric as a theoretical foundation for the amplification of technical manuals.

Explanations help the user to master the process, and they can also be used to slow down the pace. For novice users (or in tutorials) a slow pace can be helpful. However, if you make the going too slow, you will frustrate users who are eager to get the job done. So use explanations sparingly and only where they are really helpful.

Warnings

Warnings are extremely important elements in any manual. They can prevent users from doing harm to themselves or to others, sometimes in matters of life or death. But sometimes they simply prevent the frustration that results from making a "wrong turn" in the process. Use warning statements whenever you need to do the following:

- indicate the consequences of a wrong step (physical hazards, etc.)
- point out the importance or timing of a crucial step
- alert the user to possible errors and suggest ways to avoid or correct them

Your trial tests can help you spot places to insert warnings.

It is the ethical duty of a manual writer to keep any potential harm or property damage to a minimum (or to eliminate it altogether—the ultimate goal). It is also a legal duty. More than one lawsuit has arisen because a user, or the user's lawyer, accused a company of failing to provide sufficient warnings in a manual.

In every manual you write, make sure warning statements are clearly and crisply written. Place them before the command to which they relate, and highlight them with large, bold type or a recognizable icon (such as a stop sign). A typical warning appears in our student manual on CPR:

> **Caution: Be careful not to close the victim's mouth while lifting the chin.**

In this case the warning can make the difference between life and death; the victim would smother if the user did not follow the instruction. Typically, manuals use three labels to issue the different kinds of warning statements:

- **Caution** indicates a possible danger to human life.
- **Warning** indicates a possible danger to human property.
- **Note** indicates the possibility of making a mistake in an operation that will frustrate the user or spoil the results of the procedure without necessarily endangering property or life.

Though it is important to protect the user, take care not to add warnings to too many steps, or they will lose their effect. In documenting a dangerous process, for example, you might be tempted to append a warning to every step because every step is crucial. But a warning is not a device for emphasis; rather, it is a signal that something might go wrong with a process if the user does not carefully follow the procedure at a certain point. You should save warnings for steps to which the reader must pay very special attention to keep the procedure from going off course.

Troubleshooting Advice

If trouble does develop along the way, your reader will need troubleshooting advice. Again, trial testing can help you see where to insert such advice.

In the Windows manual, for example, the student authors decided to add shaded boxes labeled "Troubleshooter" after seeing a number of test users stumble at a particular point in the procedure. In one place, for example, a Troubleshooter advises the user as follows: "If Windows does not start, you may be in the wrong directory on the computer's hard drive. Ask for assistance from a helpdesk person."[3]

Some manuals include a troubleshooting table, like the one in Sample 15.4.

You may be able to make a troubleshooting table even easier to use by adding graphic symbols. See Sample 15.5.

Alternatives

There's usually more than one way to accomplish a task. By presenting optional ways of doing a job, you can accommodate users who have slightly different needs or different levels of expertise. The student composting manual, for example, covers two ways of building a composting bin: the single square method and the fence cylinder. Some users may not have access to the kind of wooden pallets that the single square method requires, so the manual gives another way of building the bin.

Sometimes, however, alternatives can create organizational havoc and confuse the user. The composting manual, for example, covers "locating the bin site" as the third task (Step 3) after the task of building the bin. But since the second option for building the bin—the fence cylinder—requires that the user drive posts into the ground, for this option the job of locating the site must come first. The organization of the manual is obviously driven by the first method—the single square—which produces a relatively mobile bin. The authors try to compensate by adding the phrase "after determining where to put the composting bin" to the first step of the fence cylinder method. A better solution would probably have been to switch the second and third tasks (Steps 2 and 3).

---

3. The students developed this strategy under the influence of John Carroll's fascinating work on the "minimalist manual." See his book *The Nurnberg Funnel* (Cambridge, Mass.: MIT Press, 1990). Another good treatment of troubleshooting, which covers the use of troubleshooting tables, appears in Donald H. Cunningham and Gerald Cohen, *Creating Technical Manuals* (New York: McGraw-Hill, 1986).

Problem	What's Wrong	How to Fix It
You can't reopen your file.	You forgot to initialize the open function.	Press control-CLEAR to start over; then use the O command to start up.
The cursor has disappeared from the screen.	You have used up available memory on your hard disk.	Press the ESC key to get out of the file. Then clear some space on the drive you are using.
More text disappears than you intended to delete during a CUT procedure.	You accidentally invoked the CLEAROUT command.	Select REPLAY from the FILE menu, retrieve the text you cut, and redo your editing.
Your screen goes black.	Something is wrong with the power source.	Quit working, unplug the computer, and call a certified repair specialist.

SAMPLE 15.4   Troubleshooting table

When this symbol appears in an error message ...	... it means ...
	stop current activity, save your work, and check error status.
	stop current activity and check the power source.
	proceed slowly and with caution.
	the time you have programmed is insufficient.

SAMPLE 15.5   Troubleshooting table with graphic elements

In general, before you offer too many paths toward completing a single procedure, you should carefully weigh the advantages of including alternatives against the possible difficulties they may cause for you as a writer *and* for typical users. Experienced users are typically more interested in options than novice users are, so reference manuals, which cater to experienced users, tend to include more options for handling a problem than do tutorials, which instruct novices.

Examples

Examples demonstrate how to accomplish an action by providing a real or hypothetical context, either with short presentations of a few sentences or with extended descriptions or case studies. Examples are very useful for making abstract instructions concrete and easy to imagine.

Tutorials for computer manuals, for example, frequently offer sample business letters or brochures that show how to apply the principles of word processing or desktop publishing. Terms such as "headers," "sections," and "fonts" seem much less confusing when the user can see them in a real letter.

Examples may be either *demonstrative* or *interactive.* A demonstrative example simply shows the reader how something looks when it is finished or at a certain stage in the process. An interactive example invites the user to manipulate the example in order to learn the process. The student manual on Microsoft Windows, for instance, invites the user to write a letter to David Letterman as a way to learn the program's features. (See Sample 15.6 at the end of the chapter.)

Previews and Reviews

Like all other technical documents, manuals benefit from forecasting statements, or previews, which put the reader into the right "frame of mind" to follow the structure of the task or document. Reviews, or summaries, are also useful; they serve as memory aids once the user leaves the manual for the field of action. The student CPR manual, for example, uses an excellent mnemonic device that doubles as a preview and a review: the ABCs of CPR (Airway, Breathing, Circulation). In an emergency this simple structuring device could help to jog the user's memory and save a life.

Another commonly used preview device is a list of *objectives* at the beginning of each section or chapter. Notice that we use a *performative* model for the objectives listed at the beginning of each chapter in this textbook. We try to say what the reader should be able to *do* after reading each chapter. Since manuals are action-oriented, their objectives should also be performative.[4]

The *checklist* is a common review device. It simply lists the main steps that have been covered in a section. You can include a little box beside each item so that users can actually put a mark beside each action they have correctly com-

---

4. For a further treatment of objectives, see Cunningham and Cohen, *Creating Technical Manuals.*

pleted. The student CPR manual closes with a checklist (though it omits the little boxes):

CPR Review

A. Check for unresponsiveness
B. Shout for help
C. Position the victim
D. Open the airway
E. Check for breathing
F. Give 4 full breaths
G. Check for pulse
H. Locate correct hand position
I. Begin chest compressions
J. Check for return of breathing and pulse

Such a list could even be printed on a small card for the user to carry and use in case of an emergency.[5]

Style Pointers

In drafting a full-length manual, keep in mind a few points of style to save yourself some editing time later:

- *Avoid the passive voice altogether.* Although passive constructions have their uses in scientific and technical prose, they have no place in instructions and can lead to dangerous confusion. If you say, for example, "The data are now entered," do you mean that the user should enter them, that the user has already entered them, or that they are automatically entered by the machine? Keep actors and actions straight by using active sentences.
- In the same spirit, *use the simple imperative* whenever you can. Just say, "Enter the data now."
- *Use definite and indefinite articles—"the," "a," and "an"—in their proper positions.* Instructional writers in the past (especially in the military) created an unnatural kind of pidgin English that has come to be associated with technical manuals. They omitted the articles, presumably in an effort to save words, in sentences like "Insert screw into cylinder." The result is nonconversational and even intimidating prose that obstructs understanding and slows reading rather than speeding it up. Just say, "Insert the screw into the cylinder."[6]

---

5. For a more extensive treatment (and demonstration) of the use of checklists, see Jonathan Price, *How to Write a Computer Manual* (Menlo Park, Calif.: Benjamin/Cummings, 1984).

6. Prose that omits articles always reminds us of the speech of Boris Badanov, the villain in the classic cartoon of the Cold War era, "The Rocky and Bullwinkle Show": "Get secret formula from Moose and Squirrel!" Boris's language is a weak imitation of Russian speakers of English. The Russian language lacks our complicated system of definite and indefinite articles.

- *Use graphics to display; use words to narrate, direct, and explain.* You can't realistically expect graphics to do all the work in instructional documents, but do try to keep reading time to a minimum, especially for procedures that require extensive muscular involvement (such as performing CPR or making a compost bin).

Ethical Issues

As in reports and proposals, instructions in manuals should adhere to ethical practices whenever they make promises or claims about knowledge, objectivity, or judgments. In addition, since manuals require the user to perform definite actions on the spot, the manual writer has further ethical responsibilities:

- The instructions must be clear enough to keep users from injuring themselves or others.
- The instructions must be complete enough to satisfy the reasonable expectations of the user; that is, the document must deliver what it promises in statements of objectives and what it claims as advantages for the product or service represented. (This is a variation of the principle of "truth in advertising.")
- The instructions should make clear the consequences of the actions they recommend.

case 15.1 ETHICS IN MANUAL DESIGN

In the ethics column from a 1995 issue of the Society for Technical Communication's monthly magazine, John Bryan offers an interesting hypothetical case.[7] Here it is in a nutshell. To gain a niche in the highly competitive personal computer market, a company develops an ad campaign to appeal to "the klutz in each of us." The claim is that the company's computer is so simple that any home user can open the plastic case and add features such as internal modems or expansion cards. The ads offer testimonies from technophobes and children saying how easy it is to use the new computer and even work on it. The manual that goes with the computer is a central feature of the marketing strategy. Designed like a comic book, it portrays the adventures of SuperKlutz, who succeeds heroically in every task from installation to adding a memory board.

The campaign goes well until a user is electrocuted after ignoring the manual's warnings and poking a screwdriver into the computer's power source. The lawyer of the victim's family has decided to file a suit focusing on the documentation, claiming that the SuperKlutz manual is "irresponsibly cute," "criminally negligent," and "misleading about the dangers of novices working with electrical devices."

7. John G. Bryan, "Ethics Case: Even Klutzes Can Do It." *Intercom* (March 1995): 4–5.

The author of the case asks some provocative questions. Regardless of legal liability, is the company ethically at fault for the user's failure to use the documentation in the manner intended? Does the company incur more responsibility than usual by marketing to "klutzes"?

What the case demonstrates above all is that the tradeoffs that we accept in designing our information products can have ethical consequences. Every designer faces the choice between a tone of seriousness and one of friendly fun, for example. In the case of the SuperKlutz manual the technical writers and their marketing advisors decided to come down hard on the side of friendly fun. After all, even a deadly serious manual cannot save a careless user. But the comic book format—which, as we saw in Chapter 6, has been used successfully in the U.S. Army—runs the risk of distracting users from warnings and cautions by placing them on pages where many visually stimulating elements compete for attention.

The case also suggests that technical writing, while cooperating with marketing, cannot do its work properly when the marketing issue of user motivation is an overriding concern. Moreover, although clarity is certainly important, its close relative—"user-friendliness"—must always take a back seat to efficiency, effectiveness, and safety. •

Reviewing Manuals

These ethical issues offer a good place to start the review process, whether you are considering revisions to your own manual or performing peer reviews for other authors or for an authoring team. For other issues to consider in the review process, consult Guidelines at a Glance 56.

Doing the audience-action analysis.   Select a topic for an instructional manual for using a product, taking advantage of a service, or following policies and procedures. As clearly as you can, write down a statement of your purpose and audience in a sentence based on the following model:

EXERCISE **15.1**

"This manual shows [a particular audience] how to [do something]."

Example: "This manual shows college students unfamiliar with the Windows™ software how to use the program to complete a short writing assignment."

Once you have specifically narrowed your purpose and named your audience in this way, make a set of notes that answers the following questions:

- What kinds of information does my user need to know to complete the desired actions?
- What kinds of information about my topic can my user do without?

GUIDELINES AT A GLANCE 56
Review questions for technical manuals

- Does the information provided in the manual seem adequate? What else might a user need to know to carry out the procedures effectively?
- Do any of the instructions overload the user with information? What could be eliminated?
- Does the organization lead the user clearly through the process?
- Are alternatives and options (if they appear) carefully worked into the organization?
- Are there sufficient warnings and cautions?
- Are the warnings and cautions effectively placed and clearly written?
- Can the warnings and cautions be read quickly?
- Are keywords effectively defined?
- Does the manual need a glossary? If a glossary is included, is it easy to use?
- How could the graphics be improved?
- What about the style? Does the manual make appropriate use of
  —active voice?
  —imperative instructions?
  —definite and indefinite articles (*the, a,* and *an*)?
- Is the manual designed for easy use, with the page as the basic design unit? Can the reader move easily from instructions to action and back to the manual without missing crucial steps and warnings?

- What are the activities the user will need to do to achieve the overall action goal of the manual?
- What kinds of frustrations is the user likely to bring to the task, and how can I keep from making these worse (and avoid others)?
- What kinds of strengths will my typical user bring to the task, and how can I build on these in delivering the instructions?

When you have completed your notes, share them with your discussion group. Use the feedback you receive to refine your notes. Then write a full profile (as described in Part One).

EXERCISE 15.2

Part 1. Practicing task analysis and usability testing. In a small group of your classmates, select a simple task that a user could perform easily in class in about ten minutes, such as making a special kind of sandwich, polishing shoes a certain way, or setting up a simple lab experiment. Write a task analysis for the procedure, and bring it to the next class along with the materials needed to carry out the tasks. Using the steps in Guidelines at a Glance 54, do an initial usability test within the group itself. Revise as needed, then exchange your task analysis with another group and do the usability test with subjects from that group. Discuss the results with the class as a whole.

Part 2. Writing and testing the task analysis you will expand into a manual.   You should now be ready to write a task analysis for the topic you analyzed in Exercise 15.1. Follow the procedure outlined in Guidelines at a Glance 53.

Part 1.   In your discussion group, return to the simple task analyses that you created with your group in Part 1 of Exercise 15.2. Make a list of the graphics that could have been helpful in the instructions. Which of these would be most appropriate for an oral presentation and which would be best for a written manual? Compare your list with those of other groups.

Part 2.   Now consider your own task analysis (developed in Part 2 of Exercise 15.2). Make a list of graphics you will use first in your oral presentation and then in your full written manual. Share the list with your discussion group and get further ideas. Complete the list, and submit it to your instructor for further advice.

EXERCISE
15.3

Preparing and delivering an oral training session.   Take the following steps:

EXERCISE
15.4

1. Use the task analysis and graphics plan that you developed in Exercises 15.2 and 15.3 to develop a plan for a short oral training session. Draft a set of overheads for the presentation. (See Chapter 8.)
2. Show your tentative transparencies to your discussion group. For every group member's transparencies, ask the following review questions:
   - How is the intended audience likely to respond to each visual?
   - Do the visuals clearly convey the steps the user must take to complete each task?
   - Would the audience be able to perform the actions just from reading the visuals?
3. After the discussion, revise your visuals and rehearse the presentation until you are ready to present to the class or to a small group from the class.
4. Take turns delivering your presentations. In an informal group discussion, critique each presenter, using the guidelines in Chapter 8 and the special considerations discussed above.

Part 1. Evaluating technical manuals.   At the end of this chapter are the three complete student manuals we have been discussing, Samples 15.6, 15.7, and 15.8. Though far from perfect, they provide a reasonable starting place for novice manual writers by applying many of the principles outlined in this chap-

EXERCISE
15.5

ter. Read them with this question in mind: "How could I have done this better?" Then, in your discussion group, undertake a more systematic review of one of the three manuals. In your review, try to answer all of the questions in Guidelines at a Glance 56 as completely as you can. Develop a set of notes for how to improve the manual, then compare your notes with those of other groups.

Part 2. Developing your own manual.   Take the following steps:

1. Using the techniques covered under Step Five of the CORE method, develop your task analysis and graphics plan into a draft for a full manual.
2. Share your draft with a small group of classmates. After everyone in the group has read all the papers, spend some time discussing each one. Answer the review questions in Guidelines at a Glance 56.
3. With help from members of your discussion group or your authoring team, conduct a usability test on the completed draft. Follow Guidelines at a Glance 54.
4. Discuss the results of the test with your group members, then revise your manual and submit the final draft to your instructor for evaluation.

Recommendations for Further Reading

Barker, Thomas T. *Writing Software Documentation: A Task-Oriented Approach.* Boston: Allyn and Bacon, 1998.

Cunningham, Donald H., and Gerald Cohen. *Creating Technical Manuals.* New York: McGraw-Hill, 1984.

Nielson, Jacob. *Usability Engineering.* New York: Academic Press, 1993.

Price, Jonathan. *How to Write a Computer Manual.* Menlo Park, CA: Benjamin/Cummings, 1984.

Redish, Janice C., and David A. Schell. "Writing and Testing Instructions for Usability." *Technical Writing: Theory and Practice.* Ed. Bertie E. Fearing and W. Keats Sparrow. New York: Modern Language Association of America, 1989: 63–71.

Skees, William D. *Writing Handbook for Computer Professionals.* Belmont, CA: Lifetime Learning Publications, 1982.

Weiss, Edmond H. *How to Write a Usable User Manual.* Philadelphia: ISI Press, 1985.

Additional Reading for Advanced Research

Barker, Thomas T., ed. *Perspectives on Software Documentation: Inquiries and Innovations.* Amityville, NY: Baywood, 1991.

Carroll, John M. *The Nurnberg Funnel: Designing Minimalist Instruction for Practical Computer Skill.* Cambridge, MA: MIT Press, 1990.

Doheny-Farina, Stephen, ed. *Effective Documentation: What We Have Learned from Research.* Cambridge, MA: MIT Press, 1988.

Documents for Review

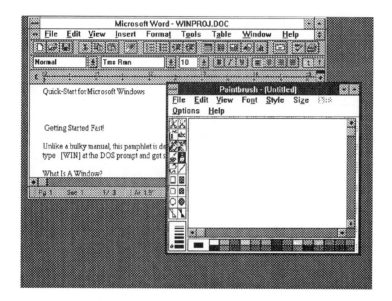

Quick-Start for Microsoft Windows™

Quick-Start for Microsoft Windows

Getting Started Fast!

Unlike a bulky manual, this pamphlet is de
type [WIN] at the DOS prompt and get s

What Is A Window?

What is Windows?

Windows™ is a graphical environment that allows you to easily start up and work with software applications, run more than one application at a time, and transfer information between applications.

In Windows™ your computer screen is referred to as a *desktop*. The desktop displays all your work in rectangular areas called *windows*. You work with your applications and documents in these windows.

You arrange windows on the desktop just as you move work items around on your actual desk. For example, to review a spreadsheet and a report from two separate

Each page of the Windows manual takes up half of a regular 8 ½ by 11-inch sheet of paper. The smaller sizing of the pages creates a "handbook" that can be carried easily from station to station. Using standard-sized paper keeps costs down.

SAMPLE 15.6   Quick-Start for Microsoft Windows™ (a tutorial manual)
Source: Screen images are reprinted with permission from Microsoft Corporation.

applications, you change the size of their windows so that they can fit side by side on the desktop.

If you want to leave an application for a while, shrink its window to an application icon and the application will keep running. When you shrink an application, Windows™ places its application icon at the lower edge of the desktop until you restore it to window size again.

The first two graphics are designed to get the reader used to seeing screenviews in the computer program.

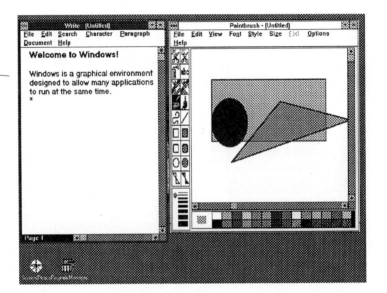

With Windows™, you can run several powerful applications at once and switch quickly among them. For example, you can switch from a word-processing application to a drawing application, and then to a data base application with only a few simple moves and without waiting.

The examples and illustrations used here are from Microsoft's *Windows 3.0 User's Guide.*

SAMPLE 15.6 continued

Get Started Fast!

Unlike a bulky manual, this section will help to acquire the skills necessary for your project.

To get started:

1. Type [WIN] at the DOS prompt.

D:\WIN> **win_**

2. Press [Enter↵].

Troubleshooter: If *Windows* does not start, you may be in the wrong directory on the computer's hard drive. Ask for assistance from a helpdesk person or another person with PC experience to put you in the WIN directory.

The Program Manager window opens on your desktop. If not, double-click on the icon labeled [Program Manager] with the left mouse button.

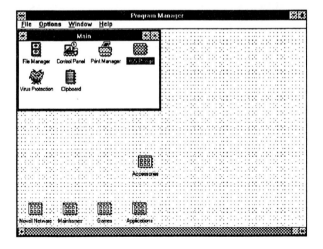

Notice what was added to the original task analysis. The amplifications in Steps 1 and 2 on this page mainly deal with outcomes. They provide feedback for the user, saying what should appear on the screen and what to do if it does not.

SAMPLE 15.6 continued

To learn the word-processing program (Microsoft Word), you can write a short fan letter to David Letterman or you can begin typing your own project.

To start Microsoft Word:

1. If the Microsoft Word icon is not visible, double-click on the [Applications] icon.
2. Double-click on the [Microsoft Word] icon.

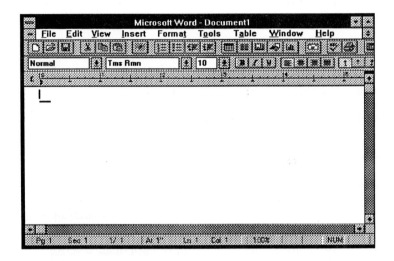

3. Type.

Dear Dave,
   Did your hear the joke about the Aggie English Professor? Well, one Aggie asks another Aggie:
   "How do you know when an Aggie English professor has been using your word processor?"
   Second Aggie: "I don't know."
   First Aggie: "There is white-out all over your screen."
   HA!

                                                    Sincerely,
                                                    Your Late Night Aggie

SAMPLE 15.6 continued

Let us add a graphic to Dave's letter or your project.

To start Paintbrush:

1. Click the arrow down button located at the top right hand corner of your screen.

Don't worry, the Microsoft Word program is still running, it's just been reduced from an application to an icon.

2. Double-click on the [Accessories] icon .

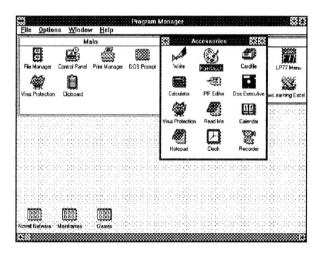

3. Double-click on the [Paintbrush] icon.
4. Draw a graphic using the tools on the left side of the screen.
5. Click on the Scissors tool.
6. Holding the mouse button down, make a box that covers the graphic.

SAMPLE 15.6 continued

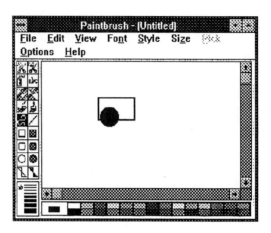

7. Open the [Edit] menu, located at the top left corner.
8. Click on [Cut].

You have just removed the graphic from the [Paintbrush] window, and placed it on Windows™ [Clipboard].

The "notes" in the shaded boxes provide expla- nations and hints that keep the user from going astray.

> **Note:** The Windows Clipboard is a temporary storage location that is always available when you want to transfer information during a Windows session. You can cut or copy information from an application onto the Clipboard and then transfer that information from the Clipboard into other applications.

To exit Paintbrush:

1. Click on the [File] menu located at the top left of the window.
2. Click on [Exit].

> **Note:** Exiting an application does not shrink it to an icon—it quits the program.

SAMPLE 15.6 continued

Going back to our letter to David Letterman . . .

To insert your graphic:

1. Double-click on the [Microsoft Word] icon at the bottom of the screen.

> **Troubleshooter:** Troubleshooter: If you don't see the Microsoft Word icon, it is
> hidden behind the Program Manager window.
>
> 1. Click on the double arrow 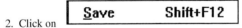 located at the top of the screen.
>   (This resizes the window to a smaller work space.)
> 2. Double-click on the [Microsoft Word] icon.

2. Place the cursor where you want the graphic to be.
3. Click on [Edit] located on the menu bar at the upper left corner.

4. Click on

Paste	Ctrl+V

Before we print our letter, let's save it.

To save a document:

1. Open [File] menu located on the menu bar at the upper left corner.

2. Click on

Save	Shift+F12

The [Save As] menu should be similar to the following:

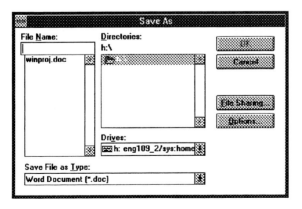

SAMPLE 15.6 continued

3. Beneath the word Drives, click on the ⬇ icon.

4. Using the scroll arrows, find the drive that your disk is in.

5. Click the drive.

6. Click [**OK**]

Now, let's print it out.

To print a document:

1. Click on [File]

2. Click on

Dave will love your letter.

Unfortunately for Mr. Letterman, we can't continue with fan mail; your project awaits. This concludes your introduction into Microsoft Windows.

Problems/Questions???

Most of the Microsoft Windows manual is stored in the Windows' Help menu. It is an excellent source to answer most of your questions.

Remember: "When facing a difficult task, act as though it is impossible to fail. If you're going after Moby Dick, take along the tartar sauce."

H. Jackson Brown, Jr.

The quotation is one instance of motivational material included in this manual. Can you find others?

SAMPLE 15.6 continued

The following section contains additional information to help you with your Windows™ project. Use it as a reference/help guide for unfamiliar procedures.

Working with Applications

Windows offers many software or application choices such as Write, Paintbrush, and Calendar.

To view application choices:
1. Open [Program Manager].
2. Click on the [Arrange Icon] command from the Windows pull down menu.
3. Click on the [Tile] command of the same menu.

Note: New users are often intimidated by dialogue boxes that ask them to make a decision or execute a command. Click the [Cancel] or [Undo] buttons to get out of a sticky situation. Most options, like Write, and Calendar, are easy to figure out. Others, like PIF Editor, may require advanced knowledge and outside assistance.

Once you decide which application or accessory you want, you must open its window.

To open an application window:
1. Double-click on the application icon.

Multiple Applications and Switching Among Applications

To run more than one application:
1. Shrink the current window with the minimize button.
2. Open another application or accessory by double-clicking its icon. (Your old application is still running, but hidden.)

The two-page spread is the basic unit of design in this manual (see Chapter 7). We come to a good stopping place before turning any page. No task is broken by page turning.

SAMPLE 15.6 continued

Note: You may need to use the [Arrange] and [Tile] commands again to view all available application icons.

To switch between applications/accessories:

1. Press [Ctrl] and [Esc] to display the [Task List] dialogue box.
2. Highlight the application you want.
3. Click the [Switch To] button.

Moving Information with the Clipboard

To move text within an application:

1. Select the text to be moved by dragging the mouse across it.
2. Choose the [Cut] command from the [Edit] menu.
3. Move the pointer to the new location.
4. Click the mouse.

To move text between applications:

1. Follow the first two steps above.
2. Minimize the current application.
3. Open the new application.
4. Move the pointer to the new location.
5. Click the mouse.

To copy text:

1. Follow the procedures outlined above except choose the [Copy] command from the Edit menu.

Note: Anything cut or copied remains on the Clipboard until you clear it, copy more information onto it, or exit Windows.

To view the Clipboard's contents:

1. Open Program Manager if it is not open.
2. Open the Main Group window if it is not open.
3. Double-click on the Clipboard icon.

SAMPLE 15.6 continued

Working with Documents

By now you have created and saved your first document.

To access your document later:
1. Open the application which contains your file.
2. Click the [Open] command from the [File] menu.
3. Move within the dialogue box to select the file.
4. Click the [OK] button.

This guide was developed at Texas A&M University by:

A+
Publishers

Norman Woody
Eve Rickenbacker
Heather Parsons

SAMPLE 15.6 continued

Basic
CPR

Presented by:

Kathy Huebner
Angela Arterberry
Timothy Smith
Scott Untrecht

SAMPLE 15.7   Basic CPR (a tutorial manual)

BASIC CPR

If you are reading this manual, then you have probably already made the commitment to learn how to perform cardiopulmonary resuscitation (CPR). Perhaps someone you love is suffering from heart disease, or maybe your job requires that you be prepared to handle medical emergencies. Or you may just feel that knowing the skills of CPR makes you a more useful member of your community. For whatever reason you chose, it is important to remember that *CPR can help save lives.*

Cardiopulmonary resuscitation is a process by which you can keep vital tissue alive until the Emergency Medical System (EMS) arrives. The process involves providing air to the lungs and a heartbeat to someone who has lost both heart and lung functions. With simple modifications, this procedure can be performed on anyone.

This manual will go through each step of the CPR process. After reading this manual and obtaining the proper "hands-on" training, you should be able to perform CPR. The chances for survival are increased by 40% when early intervention by someone with CPR skills is coupled with an efficient EMS. For this reason, you should learn CPR and keep your certification current to remain skilled. You can receive your certification from either the American Red Cross or the American Heart Association.

The introduction to the manual states the purpose and scope and hints at the qualities needed in the ideal audience (commitment, etc.). It also provides motivation.

Instructions

The type is fairly large, so that a user could conceivably read the manual with it lying on the floor beside a practice dummy during a training session. Again, the single page is the basic unit of design.

1. *Check for Unresponsiveness.* This can be done by tapping or gently shaking the person. Then shout "Are you okay?" (Be careful when tapping or shaking because you can cause severe damage to the victim if he/she has suffered a head or neck injury.)

2. *Shout for Help.* If the victim does not respond, then call out for help. If someone else is around you, send them to activate the emergency medical system. This is typically done by calling 911.

3. *Position the Victim.* You can provide effective CPR only with the victim flat on their back. If the victim is not lying on their back, you must roll them over. When doing so, you need to support the head and neck while turning the body over as a unit. The head *cannot* be above the level of the heart.

After determining unresponsiveness and properly positioning the victim, you must begin the ABC's of CPR. The ABC's of CPR are:

It is important for the user of this manual to remember the procedures, so mnemonic devices such as "the ABC's of CPR" are helpful.

SAMPLE 15.7 *continued*

AIRWAY, BREATHING, AND CIRCULATION.

AIRWAY

You must immediately open the victim's airway using the head-tilt/chin-lift method, as shown in Figure 1.

The graphics complement the text by showing things that are hard to describe in words.

Figure 1. Head Tilt/Chin-Lift Method

Head-tilt/Chin-lift method:

A. Place your hand on the victim's forehead and apply firm backward pressure to tilt the head back.

B. Place the fingers of your other hand on the bony part of the lower jaw near the chin and lift forward until teeth are almost together.*

Breathing

After you have opened the airway, you will need to check for breathing. To do this, you will use the Look/Listen/Feel method that is demonstrated in Figure 2.

The engagement of many of the user's senses (look, listen, feel) helps the transition from reading to action and also aids memory.

Look/Listen/Feel method:

A. Place your ear near the victim's mouth with your face looking toward the victim's chest.

* Caution: Be careful not to close the mouth while lifting the chin.

SAMPLE 15.7 continued

Figure 2. Look/Listen/Feel

The figures in this manual come from the American Red Cross. Notice their functional simplicity.

B. *Look* at the chest to see if it is moving.
C. *Listen* for breathing noises.
D. *Feel* for air movement from the victim's mouth.

Each step contains only one action.

If you determine that the victim is not breathing, you will need to provide four full breaths using the mouth-to-mouth technique shown in Figure 3.

Figure 3. Mouth-to-Mouth Technique

SAMPLE 15.7 continued

<u>Mouth-to-Mouth technique:</u>

A. Gently pinch the victim's nose with the hand holding the forehead. (Continue to hold the head back with the fingers on the victim's chin.)
B. Seal your lips tightly around the victim's mouth and give four full breaths.
C. Watch for the victim's chest to rise.
D. Pause long enough between breaths to completely refill your lungs.

After providing the victim with four full breaths, you will need to determine if the victim has a pulse. You will find the pulse by locating the carotid artery as shown in Figure 4 and described below.

Note the use of the arrow in the orientational diagram.

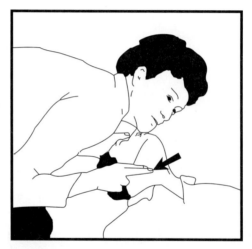

Figure 4. Checking the Pulse

<u>Taking a Pulse:</u>

A. Place 2 or 3 fingers on the Adam's apple (voice box) just below the victim's chin. (Do not use your thumb, because you could confuse your own pulse in your thumb with that of the victim.)
B. Slide your fingers into the groove between the Adam's apple and the neck muscle on the side nearest you.
C. Maintain the head-tilt with your other hand.
D. Feel for the carotid pulse for 5 to 10 seconds.
E. If the victim has a pulse but no breathing is evident, then you must give the victim one breath every five seconds.

SAMPLE 15.7 continued

If the victim does not have a pulse, then you must activate the EMS and begin compressions to stimulate circulation of the blood.

Circulation

A. Kneel at the victim's side near the chest.
B. With the middle and index fingers of the hand nearest the legs, locate the notch where the bottom rims of the two halves of the rib cage meet in the middle of the chest.
C. Place the heel of one hand on the sternum next to the fingers that located the notch. See Figure 5.

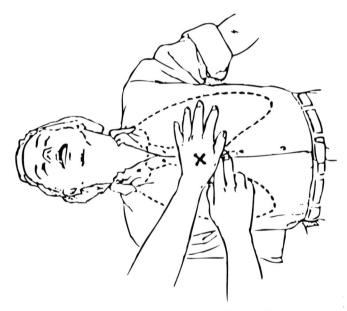

Figure 5. Locating Hand Position for Compressions

D. Place the other hand on top of the one that is in position on the sternum. (Keep your fingers up and off of the chest; interlocking your fingers will help you do this.)
E. Bring your shoulders directly over the victim's sternum as you compress downward, remembering to keep your arms straight. Refer to Figure 6.
F. Depress the sternum about 1½ to 2 inches.
G. Relax pressure on sternum completely without removing your hands from the sternum.

SAMPLE 15.7 continued

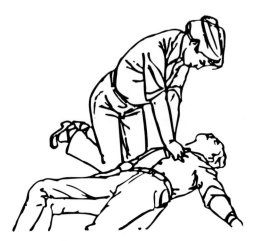

Figure 6. Correct Position for Chest Compressions

H. Maintain a ratio of 15 chest compressions to 2 breaths. (You must compress at the rate of 80 to 100 times per minute.)
I. After completing 4 sets of 15 compressions and 2 breaths, check again for pulse and breath using the methods described previously.

Continue CPR Until:
A. Qualified relief arrives
B. Physician arrives and assumes responsibility
C. You become totally fatigued
D. The victim responds to CPR breathing and a pulse returns

The checklists function both as reviews for first-time users and reminders for experienced users.

CPR Review
A. Check for unresponsiveness
B. Shout for help
C. Position the victim
D. Open the airway
E. Check for breathing
F. Give 4 full breaths
G. Check for pulse
H. Locate correct hand position
I. Begin chest compressions
J. Check for return of breathing and pulse

SAMPLE 15.7 continued

Conclusion

Cardiopulmonary resuscitation is a very important skill to acquire. When CPR is started within 4 minutes, the victim's chances of leaving the hospital alive are 4 times greater than if the victim did not receive CPR until after 4 minutes. **This manual does not constitute a CPR course.** You must practice on CPR mannequins, with certified instructors to guide you, in order to acquire certification of your CPR skills. Also, **never** rehearse or practice the skills in this manual on another person.

Should the bold-face words in the conclusion be presented as a warning or a caution?

SAMPLE 15.7 continued

A Guide to Yard Composting
In Bryan–College Station

Why put all this information on the cover page?

This manual is published in hope of educating you about composting and its advantages. It describes the composting process, individual advantages, environmental impact, and a simple seven-step method by which you can start composting in your own backyard.

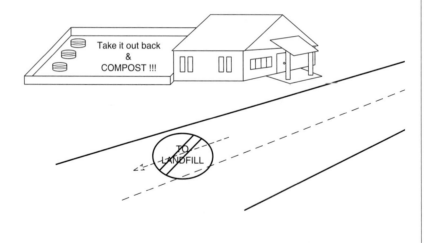

SAMPLE 15.8   A guide to yard composting (a tutorial manual)

What is composting?

Composting is a process whereby biodegradable waste, such as leaves or vegetable matter from your kitchen, is collected and piled together to decompose. This process creates a nutrient-rich material which can be used for fertilizing soil.

Why should you compost?

Approximately 26% of all waste generated by the public each year is food and yard waste. This waste is almost always sent to landfills. Due to stricter regulation of land-fills by the Environmental Protection Agency in 1991, garbage disposal costs will in-crease, costing you more money.

By composting you can:

- reduce household waste by approximately 40%
- reduce yard waste by nearly 100%
- save money
  - eliminate expensive fertilizers
  - eliminate plastic bags
- make a positive impact on the environment
  - eliminate plastic bag waste
  - replenish soil
  - save landfill space

Is all this motiva-tional material necessary in a manual like this? Why or why not?

How do you start composting?

There are 7 easy steps to follow:

| STEP 1 | Obtain Required Materials

First you must obtain the materials essential to make 3 composting bins. There are many different types but only two types are described below.

Could some prob-lems be avoided if more informa-tion about the differences in the two types of bins were included here?

TYPE 1   *Single Square*
Materials:
- 12 wooden pallets (given away free in many stores)
- fencing wire (8 coat hangers work well)
Tools needed:
- pliers
- drill

2

SAMPLE 15.8 continued

TYPE 2   ***Fence Cylinder***

Materials:

- 28½ feet fencing (ask salesman for 14-gauge if available)
- 6 wood posts, 2 inch × 2 inch, 5 feet long
- fencing wire (coat hangers work well)

Tools needed:

- pliers
- hammer

STEP 2   Bin Assembly Instructions

Single Square

Assembly: *Refer to FIGURE 1.*

(1) Drill a single hole at the top and bottom of the outside planks as shown in FIGURE 1.

(2) Cut wire into 12-inch lengths.

(3) Stand 2 pallets with planks standing vertical next to each other at a 90 degree angle.

(4) Feed wire through pre-drilled holes at top and bottom as shown and twist wire with pliers which should tighten the pallets together.

(5) Repeat step (4) to create a square.

Fence Cylinder

Assembly: *Refer to FIGURE 2.*

(1) After determining where to put the composting bin, using a hammer, drive the two wooden posts into the ground approximately 3 feet apart as shown in FIGURE 2.

(2) Stand the fencing upright in a circular fashion between the two posts so that the fencing touches the posts.

(3) Use the fencing wire to connect the fencing to the posts at the top and bottom as shown.

(4) Connect the two ends of the fencing with the wire to complete the cylinder at the top and bottom as shown.

(NOTE: When you have completed STEP 2 you should have 3 bins.)

STEP 3   Locating the Bin Site

Once you have chosen the type of bins you will use, determine the best location for the composting bins. Take into consideration the following factors:

3

Are the graphics and the accompanying instructions well integrated? How could the integration be improved?

SAMPLE 15.8 continued

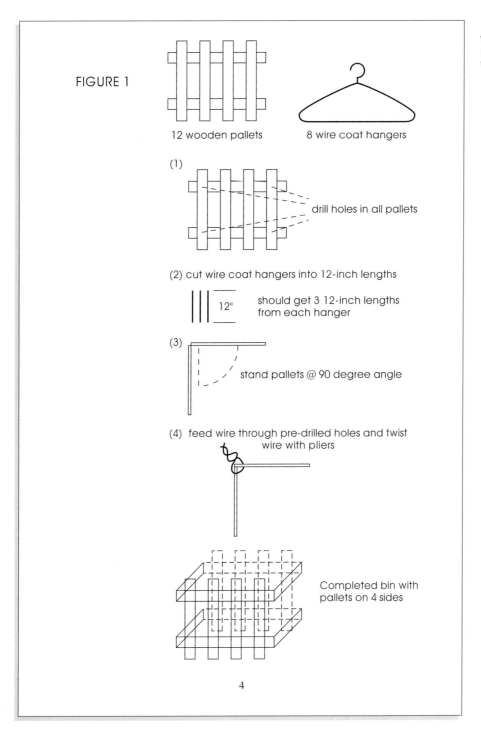

FIGURE 1

12 wooden pallets

8 wire coat hangers

(1) drill holes in all pallets

(2) cut wire coat hangers into 12-inch lengths

12" should get 3 12-inch lengths from each hanger

(3) stand pallets @ 90 degree angle

(4) feed wire through pre-drilled holes and twist wire with pliers

Completed bin with pallets on 4 sides

4

What is the basic unit of design in this manual?

SAMPLE 15.8 continued

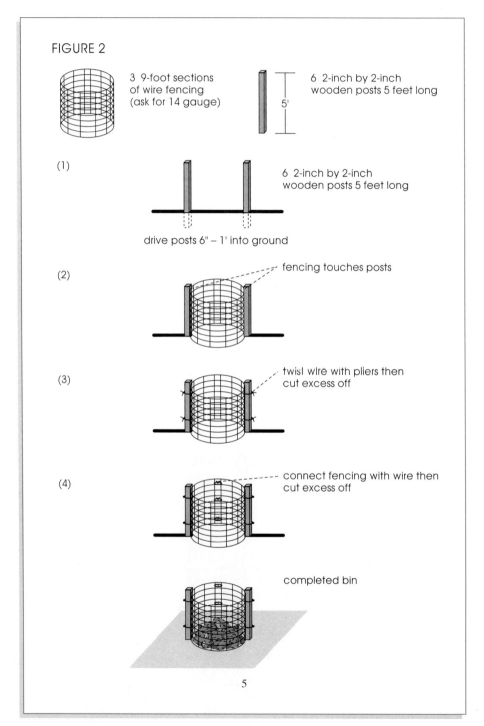

FIGURE 2

3 9-foot sections of wire fencing (ask for 14 gauge)

6 2-inch by 2-inch wooden posts 5 feet long

5'

(1)

6 2-inch by 2-inch wooden posts 5 feet long

drive posts 6" – 1' into ground

(2)

fencing touches posts

(3)

twisl wlre with pliers then cut excess off

(4)

connect fencing with wire then cut excess off

completed bin

5

SAMPLE 15.8 continued

♦ odors generated by the compost
♦ possible insects or rodents
♦ site flooding (i.e. nearby stream)
♦ water supply
♦ existing structures

Some Do's and Don'ts

Don't
- put the bins too close to the house because of the possibility of foul odors, insects, and rodents.
- locate the bin in an area which floods because the excess moisture will slow the composting process.
- place the bins against an existing structure because it may begin to rot or may be damaged by the heat and moisture which the compost generates.

Do
- locate the bin near a hose so dry compost can be moistened.
- put the bins in an area easily accessed.

STEP 4   Identify Compostable Materials

The following is a list of common materials that are readily compostable:

grass clippings	pine needles	flower clippings	vegetables	fruits
leaves	soils	shredded paper	bread	most leftovers
small branches	garden stalks	sawdust	table scraps	

While most organic materials are compostable, there are several which should be avoided:

♦ *Meat scraps & bones*—This matter may contain organisms or parasites that can spread disease through your garden.

♦ *Vegetation sprayed with toxic pesticides*—This matter may contaminate the compost and can adversely affect vegetables grown in the compost later.

STEP 5   Identify Proper Mixture

The decomposition of a compost pile depends on the mixture of nutrients present. **Green material,** such as kitchen scraps, is nitrogen-rich. Too much green material will decompose quickly but will not produce good compost. The list below identifies some green material:

wet grass clippings	eggshells	vegetable peelings
lettuce	coffee grounds	moist bread

6

SAMPLE 15.8 continued

Brown material, such as dead leaves, is nitrogen-poor and decomposes very slowly. While the green material provides nitrogen, the brown material provides carbon. The list below identifies brown material:

straw	dead leaves	sawdust
branches	dry grass clippings	pine needles

The Ideal Mixture

The ideal mixture ratio is approximately 30 parts carbon to 1 part nitrogen. Since most households generate more yard waste than kitchen scraps, it is easy to stay near the proper mixture.

| STEP 6 | Begin Composting
|---|

- Add a small layer of sticks or branches at the bottom of the bin to allow oxygen to reach compost from below.

- Add compostable materials to the bin.

- Add water to the pile to keep it slightly moist, every 3 to 4 days.

- Turn or mix the pile with a shovel or pitchfork every week or two to keep oxygen flowing to it.

- Once the first bin is full, start filling the second. The first one should be kept moist and mixed as stated above until the compost is ready to use. (Approximately 1–2 months.)

- When the second bin is full, start filling the third. The second bin should be maintained as usual. By this time the first bin is ready for application and can be emptied, so that when the third bin is full the process may start over.

Once you have gotten well into the composting process, you will have three bins in varying stages:

      (1) a fresh, hot working compost
      (2) a cool maturing compost
      (3) a brown completed compost ready for application

How does the "NOTE" function in the manual?

NOTE: The pile will get hot but should not smolder. Exposed food will attract flies, so a light cover of leaves or grass clippings can be placed over the food to solve this problem. Proper maintenance of the composting bin will keep insects, rodents, and odors away.

| STEP 7 | Uses for Compost
|---|

Within 10 days to two weeks, the pile will shrink to half its original size and within a month it will decrease to a fourth. It takes approximately 1 to 2 months to get a brown completed compost suitable for application. FIGURE 3 shows the final compost.

7

SAMPLE 15.8 continued

FIGURE 3. Compost Product

Now the compost is ready for application. You can use the compost for:

food garden application   flower garden application
planting soil   lawn application
potting soil   topsoil use

How much will this cost?

Starting an individual backyard composting site is very inexpensive. The amount of money invested is negligible compared to the amount saved on plastic bags and lawn/garden fertilizer. This will be very evident when the city finally changes to volume-based garbage pick-up. TABLE 1 shows what is spent on waste disposal and fertilizer without composting and TABLE 2 shows what is spent on starting an individual yard composting site.

8

TABLE 1

Approximate Costs Without Composting
Expenditures Per Year

Garbage bags for compostable material—50 bags per year @ $0.17/bag	$8.50
Fertilizer costs for lawn & garden application per year per 2500 sq feet lot	$36.00
Total Cost	**$44.50**

TABLE 2

Approximate Costs With Composting
One Time Investment
(Fence Cylinder)

Garbage bags for compostable material—0 bags per year	$0.00
Fertilizer costs for lawn & garden application per year	$0.00
Cost of wire fencing	$30.60
Cost of wire/use of wire coat hangers	$0.00
Cost of six 2 × 2 inch wood posts 5-feet long	$9.00
Total Cost for 3 composting bins	**$39.60**

The composting bin materials shown in TABLE 2 are one-time costs, while the costs shown in TABLE 1 are paid yearly.

By composting your food and yard wastes, you will save approximately $4 the first year and over $43 the following years. Additionally, you will have a positive impact on our environment so your children and grandchildren may enjoy it also.

9

SAMPLE 15.8 continued

Sources of Information

Hartman, R. A. 1989. *Good Riddance.* College Station: privately printed.

United States Environmental Protection Agency. 1989. *Decision Makers Guide to Solid Waste Management.* U.S. Government Printing Office: 1990.

10

SAMPLE 15.8 continued

Index